THE QUALITY OF

URBAN LIFE

Volume 3, URBAN AFFAIRS ANNUAL REVIEWS

THE QUALITY OF

URBAN LIFE

Edited by

HENRY J. SCHMANDT and
WARNER BLOOMBERG, Jr.

Volume 3   URBAN AFFAIRS ANNUAL REVIEWS

SAGE PUBLICATIONS, INC. / BEVERLY HILLS, CALIFORNIA

For information address:

SAGE PUBLICATIONS, INC.
275 South Beverly Drive
Beverly Hills, California 90212

Standard Book Number 8039-0007-4

Library of Congress Catalog Card No. 69-18752

FIRST PRINTING

*This series is distributed,
outside the Western Hemisphere, by:*
PERGAMON PRESS LTD.
Headington Hill Hall
Oxford, England

Contents

Part IV   URBAN ORDER
Present and Potential

INTRODUCTION

1

Introduction

The Problem of
Contemplating the
Quality of Life

WARNER BLOOMBERG, JR.
HENRY J. SCHMANDT

□ "IS THERE A life before death?" With that ironic twist on a traditional theological inquiry, some of the spokesmen of the hippy movement have asserted that the existence of most urban Americans, in spite of their affluence, is actually so alienated as to be almost lifeless, almost totally lacking in those kinds of meaning and passion which make one intensely aware of and committed to being alive. Convinced that the quality of life depends mainly upon one's inner state and intimate interpersonal relationships, they link up intellectually and emotionally with those other critics of the city, pragmatic and prophetic, for whom an earthly paradise would have a countryside location. Some have actually founded rural communes, which are likely to suffer the same fate as the other utopian villages and settlements of the past century. Others, while preferring not to move out of the impersonal metropolis in body, have taken to "tripping out" with drugs. Even those who abjure travel by hallucination have clearly "dropped out" of the alienated middle-class civilization of the modern city. Whatever else the hippy movement may or may not be, it is unequivocally an exodus from urbanism as we today experience it in the United States and in most of the other Western industrialized nations.

But the hippies are only the most extreme and perhaps bizarre of the critics of our contemporary way of life. It is doubtful if there has ever been a time in our history when this nation was involved in as much self-examination and self-criticism as today. The discontent, moreover, is not peculiar to intellectuals, or to a particular class or age strata, or to a religious or ethnic category. Such indices as the consumption of pep pills, barbiturates, tranquilizers, and alcohol, the incidence of neurotic and psychotic behavior, and the occurrence of divorce and of illegal acts all suggest a society in which discontent is chronic and epidemic. Such a society is hardly what most of us have in mind when we speak of "the good life."

But what are the qualities of "the good life" for contemporary Americans? It turns out that very little investigative time and energy have been devoted to this question. Pollsters seldom explore it and most social scientists avoid it, as if it would contaminate their objectivity and possibly gag their computers. Even present-day social philosophers, with some exceptions (Haworth, Goodman, Mumford, for example), have done little to explore intensively the question of what the quality of urban life ought to be in an age of mass societies, computers, atomic energy, and emerging genetic control. We are immersed in a vast flood of literature describing and analyzing the multitude of problems which now beset us while a new rivulet of publications—one likely to become a minor torrent—is delineating what will probably bemuse and bedevil us as we cross the millenial calendar bench mark into the year 2000. But is this purported likelihood of the future where we want to go, or ought to go? If we now have knowledge of at least how to attempt to domesticate the evolution of human societies, is not the utility of this knowledge dependent upon arriving at some judgments about the kinds of institutions and social systems we would like to create?

Of course, as Kenneth Kenniston has pointed out, we have all but lost the capacity for utopian thought. Most of our hopes for the future are often little more than extrapolations of the present: we imagine an increase of those things we like and a decrease of those we don't. But when we attempt to conjure up images of a truly different social order, our dreams too often turn into nightmares and we end up with the "futopias" (ideals of futility) of *Brave New World* or *1984*.

Several years ago one of us had occasion to lead for a few days a seminar of college and university faculty and their spouses in the consideration of what would constitute an ideal community.

Composed mainly of younger, junior-level professors but representing much of the whole spectrum of academic specialties, this was an educationally elite group not tied by age or status to things past. Yet the opportunity to dream aloud individually and collaboratively was, as could be predicted, initially paralyzing even for these persons who were habituated to operating with ideas and engaging in verbal exchanges. Nevertheless, with a good bit of coaxing the members of this seminar began to reveal at least some parts of such wishful and wistful visions as they allowed themselves. Here is a summary of their thoughts:

Homes in a good community would be of our own design, self-cleaning, and provide controls over all immediate environmental nuisances as each family defined them—bugs, noises, odors, and the like. Every single component of such a dwelling place would embody good design in the least of its details, and good maintenance service would always be available. These criteria for the house of utopia not only reflect the distaff side of the seminar, but also provide convincing evidence for the thesis that visions of the perfect future are likely to be the presumably pleasurable opposites of present pains and irritations.

Some thought perfection included a country side location with a few acres of land, while others, no longer burdened by practical restrictions, foresaw both a town house and a country house for all who wished to enjoy more than one life style. In any case, it would be easy to get to work; indeed, some technological visionaries suggested a means for instant transportation to almost any elsewhere. Others wanted a community scale small enough to eliminate most of the need for cars or their future functional equivalents. All agreed that there should be easy access to an abundance of educational, recreational, and (no doubt in case you played too hard) medical facilities; and the physical environment with its cityscape full of trees and parks, would also be a safe place in which to live and thus to stay alive.

The residents of our utopian community would all be relatively affluent, would come from "everywhere" and yet be neighborly people, and would help to provide for one another the stimulation that results from variety and complexity. Freed also from the necessity of being consistent or resolving paradoxes, the group opted for a life in which people were less compartmentalized than in the present daily round, had their needs for privacy secured, enjoyed equal exposure to and wide freedom and opportunity to choose from a broad range of alternatives in all areas of life with a minimum of coercion and conformity. A few wanted access to full information about the systems that sustained existing conditions and how they operated.

In such a social order there obviously would be a life for parents apart from their children, which is a special case (strongly pushed by some of the mothers present) of the broader opportunity to become, from time to time, "someone else." Indeed, some participants proposed that there should be two types of dwelling place available to each person: communities of identification in which one enjoyed such valued experiences as neighbors and "belonging," and communities of experimentation in which one could live while trying out other identities for a continuously developing self.

In their collective utopia one would find at last "worthwhile" television as part of a generally constructive leisure; work would be creative but no longer viewed in the puritan tradition as purifying. There would be challenges to insure human growth, but they would be comprehensible challenges. There would be tolerance of all kinds of "otherness," and thus no age cults or age exclusions. In particular, women would enjoy a lively relationship with men based on differentiation without becoming compartmentalized as women. Indeed, there would be more opportunity, within the context of fundamental male-female differences, for individuals to partake of both modes of response; but with this there would also be more and more authentic sensuality and thus sexuality than we have today. Love would be the predominant value in human relationships, but also stressed would be reverence and respect for the wonder of life. And yet there would also be a sense of relativity, and this would be a continuing source of further change [Bloomberg, "The Emerging Urban Society: What Options for Democratic Man?" *Illinois State Journal* 3 (April 1968): 3-5].

What is especially striking about this compendium of not too unusual wishes and visions is the virtual absence of business and politics. There is no mention of the giant economic and governmental corporations which surely would have to exist to provision and maintain this appealing environment with its multitude of opportunities for personal development and family fulfilment. At the same time there is no specification that the massive institutional systems of production, distribution, and social control which now do so much to shape the character and quality of our urban places would somehow have withered away. This latter view might have been expressed by some of the young radicals of the latter 1960's with their psychedelic and simplistic versions of a neo-Marxist perspective; but this was a few years earlier and the daydreamers were already of the older generation, albeit the younger of the older. They simply turned their backs on those huge bureaucratic organizations which the New Left so often identifies and berates as the source of the lifeless existence of modern urban man; indeed, they apparently assumed a most benign management for their city of satisfactions.

Nor is their vision seriously at variance with what appear to be some of the great long-run tendencies that have been underway in the urbanizing, industrialized societies for many generations and are likely to persist for many more: rapidly increasing productivity; greater equalitarianism and more spatial and social mobility; declining puritanism and increasing sensuality, accompanied by the diminishing centrality of work and the extension of leisure; the predominance of secular orientations, and the further aggrandizement of science and science-based technology in order to increase the effectiveness of control systems in all spheres—physical, biological, and social; the continuing development of continental and subcontinental "super-state" systems of societal organization within which "megalopolitan" urban regions evolve.

One could expect a number of "side effects" to accompany these persisting trends. In a more affluent and equalitarian society, power disparities between minorities and majority would diminish, promoting a reduction of internal conflict. More mobility and increased and increasingly accessible national and international communications could further accelerate the decline of differences based on religion, race, and ethnicity as direct and vicarious "culture contact" became the norm for more and more people. Within this context, and with the support of an essentially benign and well-managed welfare bureaucracy, leisure life-styles, personal development and interpersonal relationships could gain even more salience as the valued forms of variety in an otherwise largely conformist population. There would still be minorities and deviant personalities, but the already evident trend toward greater formal freedom and informal toleration for such individuals and groups would make them politically irrelevant. Indeed, along with the maintenance and extension of legal political rights, elections and referendums have become increasingly ritualistic, and more and more policy is likely to be made through the massive corporate organizations, both public and private (although the distinction is already blurred and may disappear) which would manage and maintain this service-welfare oriented "city of satisfactions."

A small minority of politically inclined activists imbued with an ideology of public responsibility would manage these corporate structures and would thereby increasingly manage the social order itself—not by authoritarian or dictatorial control of each individual through a police-state apparatus, but by controlling the context within which a highly conformist majority could make only small adjustments in the several available patterns of work and leisure. Since part of the managerial function would be to react to the determined and felt needs of the majority, general satisfaction would be high, with changes constantly measured through computer-analyzed surveys of the population. Thus, the computer can be used to help sustain a "contented cow" citizenry; indeed, it is already being employed in this way. In such a future metropolis most persons would feel that the opportunity to select from the available alternative patterns and make small adjustments in them constituted a high degree of freedom and individuality—the kind of individuality that people display in their homes today, consisting mainly of a multiplicity of variations on several main themes of interior decoration and leisure life-style.

The managers of the great corporate structures would be accorded much respect instead of being denounced, and perhaps would even be viewed with real affection as servants of the people. Those whose personalities lead them to enjoy having power over others would compete with one another for the managerial positions, often using the rituals of elections to legitimize their success and determine final choices. Critics and dissenters inevitably would appear, but not in any large numbers. Not only would the rights of the latter be protected, but some of them would actually be subsidized by various corporate organizations, including foundations, in response to the conforming majority's desire for intellectual titillation and a surface feeling of freedom.

Thus, the academicians' alleged "utopia" in large part assumes only that present trends be brought to completion. And as this takes place, some blurring of the difference between the advanced world and the "third world" of modernizing societies might well occur. In such an event, there would not be the same "three worlds" as today because the society just described is not only very similar to the society we seem likely to become; it is also very much like that which Russia, along with the other socialist and communist societies in the West, appears to be moving toward. The more advanced developing countries of the present would then likely constitute a new "second world," while the less developed nations would remain as a "third." Certainly the differences between "capitalist democracies" and "people's democracies" become with each passing decade more difficult to ascertain and less easy to assert except in terms of certain kinds of ideological rhetoric.

But there clearly is a second possiblity. It reflects a greater emphasis on means of controlling people: genetic control is on the horizon; informational control, including the spy devices and computer tracking systems, is already with us; psychological control reaches its most obvious and ugly form in what has been called "brainwashing" but takes place in more subtle and effective ways which we are inclined to call "therapeutic." Such forms of control would become more and more available in the society along with highly developed variations of such devices as are increasingly being used by the military and police forces—mace, nerve gases, chemical conditioning, and the like. Such a society, then, would not be quite as "nice" as the one just described. It would be much more authoritarian and centralized in its control and would provide fewer options even for those limited choices that people now make on behalf of their individuality. There would always be the likelihood of serious and disruptive problems and of resistance movements going beyond the tolerance limits of the managerial society and of the international institutional control systems. These developments, in turn, could lead toward an increasingly controlled, programmed social order—a kind of combination of the hedonism of Huxley's *Brave New World* and the authoritarianism of Orwell's *1984*. At least this is an evolutionary alternative that is technically and sociologically available as we try to visualize the transformation we are now experiencing projected only a few generations into the future.

A third possibility is even uglier. Recurring and ever more serious internal disruption arising from the conflict between dictatorial managers and increasingly militant minorities within the technologically advanced and highly urbanized societies would have the effect of shifting policy and mood from welfare and manipulative control to more coercive and punitive patterns. In such a context it seems unlikely that development programs for the still impoverished nations would increase or that foreign policies would become less militaristic. But both the new "second world" of still developing nations and the remaining "third world" societies would eventually acquire nuclear capabilities. Then both internal and external pressures would promote armed conflict, making the atomic holocaust all too likely. The results of this possibility are, of course, not predictable; we might review our histories of the emergence of what we have called the Dark Ages to help us imagine what might follow.

A fourth possibility depends on the success of the reformist minorities and their advance guard, the radical cadres in today's society. This would be a society in which the new technology is used by benign but nonetheless powerful managers to extend participation by people in general rather than to increase control over them. The new possibilities for socialization would be employed to promote individual autonomy and what some psychologists now call "self-actualization." Productivity would be utilized to expedite equality of opportunity and of power both within and between societies. Such a possibility combines the "city of personal satisfaction," imagined by the seminar of professors and their spouses, with the vision of a social order whose institutions are committed to decentralization, variety, and participation. Again, this is technologically, sociologically, and psychologically possible as we seek to peer ahead a few generations. Probabilities, however, are quite another matter. The contributors to this book are most often concerned with what they sense to be the probabilities, although attention also is paid to some of the less probable possibilities.

We have suggested such a large frame within which to view the contributions to this volume because it serves so well to remind us that the character of urban life (at least in the highly urbanized and industrialized society which is our main focus) depends relatively little upon what is done or not done at any one time by "the power structure" within any particular city. It is clear that major changes now take place through regional and national (and even international) systems of action. The individual urban system per se can only react, whether it resists, enhances, acquiesces in, or develops some variation on the mainline of more general tendencies. Thus, when we talk about the "quality of life" in our cities, we are necessarily considering the character of the larger social order of which the local communities are but a part. This frame of reference also forces us to keep in mind the contemporary interpenetration of institutional systems which the conventional wisdom of social science so often tempts us to consider separately (always "for the sake of analysis").

Yet one cannot help wishing for some way to move from these necessarily abstract representations of long-run historical tendencies to the more immediate level of tangible human beings of flesh and blood—the ephemeral little seminar, the people one knows through daily encounter, the face in the mirror. This transfer may not make sense methodologically and may be impossible theoretically, at least for the present; but nothing in our culture has prepared us to contemplate these large-scale concerns without asking the question: how will they affect *us,* what will happen to *me*? Some of the authors do try to keep individuals in mind while contemplating institutional and social systems; that is perhaps the most difficult task of all.

In any case, contributions to this volume, not uniformly but with remarkably few exceptions, represent responses to the demands of such a broad frame of reference—even though the editors did not provide it in advance to those who have written the essays which follow. Moreover, the analysts who decided to join with us in seizing hold of this intellectual nettle, this concern for the quality of urban life, had to face additional dilemmas. One could simply expound on what he believes to be proper criteria for judging quality and then deductively explicate from them a set of idealizations about what urban life ought to be; but we've all been there before—such purely theoretical exercises avail us little in our efforts to domesticate the evolution of human societies. Another alternative would be to describe prevailing values and discuss how they could better be implemented; but far too many of us are already convinced that we can never achieve the prevailing values we most prize unless we can change others (which, unfortunately, some of our fellow citizens probably hold very dear). Finally, one could focus on how evaluative sets relevant to the question of quality evolve and change or resist change; but such an approach, if left unqualified by an examination and affirmation of alternative possibilities, tends to produce the negative counterpart of pure idealism: an assertion of pure inevitabilities.

Most of the essays in this volume to some degree connect with all of these concerns. Appropriately in an age marked by disillusionment, relativity of value orientations, and awareness that the evaluative sets people actually use are only in part subject to deliberate development and indoctrination (even in dictatorships), the formulation of idealisms per se is largely neglected. Much greater concern is shown for the problems and possibilities inherent in contemporary urban value systems given what we know (or at least think we know) about cause and consequence in human affairs. Recurring throughout the volume are two questions: how much of what we think we want can we actually get, and by what devices?

How did we arrive at where we are, given what we thought—or at least asserted—we wanted? The three papers which deal with education, taken together, illustrate the full range of these concerns: one delineates the immediate situation in the United States, focusing on our preoccupation with the education of the disadvantaged; a second deals specifically with how actual priorities get established, at least in public educational systems, and with how the operating criteria of schools are in fact likely to continue to be determined; and a third, unconstrained by either the presently predominant values of the schoolmen or the prevailing politics of educational institutions, looks outward toward a more distant future with much more concern for what ought to be and possibly might be.

The present volume covers a wide range of subject matter—a reflection of the complexity of urban life itself—from the mass media to the performing arts; from the nature of community leadership to the problem of conceptualizing urban management; from the impact of technology to the role of the private foundations; from the quality of neighborhood life to the larger question of governing the metropolis; from the desiderata for planning the physical city to the mental health of its inhabitants. If any reader cares to indict this collection as an assemblage of pieces only partially or loosely related to one another rather than a tightly knit, cohesive group of contributions unified around a central theme, the editors will readily admit to the charge but will feel no sense of guilt in doing so. Some subjects yield to editorial pressure for thematic cohesion even when they are dealt with by assembling component parts or chapters from many different authors. The problem of the quality of urban life, however, is not amenable to such an approach. Its collective treatment can be given no greater coherence than that of the culture and institutional systems of our urban civilization. Only if the editors were arbitrarily to impose some one ideological version of this complex reality on the authors could a superficial semblance of unity be achieved, for life in the modern city is itself an assemblage of disparate pieces, often poorly integrated.

A major purpose of this volume is to emphasize and explore the diversity of concerns which must be dealt with if we are to stop evading an assessment of what we have wrought, both intentionally and inadvertently, in building a society of cities. In saying this, the editors by no means wish to imply that the challenge to "make sense" of our multiplicity of dilemmas should be avoided. Indeed, the concluding section of the volume contains several efforts to suggest some sensible ways in which we might begin to reconstruct the urban order (or disorder) to enhance the quality of urban life. But here, as in the preceding parts, we can present only pieces of a whole which is neither unified in fact nor satisfactorily comprehended by one abstract theoretical or philosophical system. Those who are seeking some overall formula for reform will be as frustrated by this concluding component of the book as those who desire a unified, all-encompassing analysis will have been by the preceding sections.

What follows, then, is an effort to establish a set of intellectual outposts in a territory which is surprisingly poorly explored given our almost desperate need for the resources it must contain. For we are increasingly certain that what we mean by "progress," by "improvement" in the human condition, is not guaranteed by any inevitable social evolution. The future of what we understand to be a more humane and democratic social order consists of alternative probabilities and possibilities, some of them hardly imagined prior to the era of the fully urbanized society. Our capacity deliberately to affect the probabilities and the options actually exercised is obviously far less than that which the early prophets of industrially based (though often de-urbanized) utopias once imagined. But it is far from negligible. Yet all our potential for determining more rather than less of our own destiny, and for further domesticating the processes of social evolution, depends for its manifestation as historical events upon the cultivation of more powerful conceptions of what we wish to become. This task cannot be delegated to an endless sequence of presidential commissions. More and more of us must be stimulated and educated to become involved in the argument about the quality of urban life as we have come to know it—however frustrating and incomplete our efforts inevitably must be.

This book is intended to make a contribution to that stimulation and involvement.

Part I

PERSPECTIVES ON THE URBAN MILIEU

Introduction

TO some of today's disenchanted youth, the past is not simply dead, it is also irrelevant. History is meaningless and the experience it reflects useless if not deceptive. The communicated culture of centuries, the record of civilization-building, the chronicles of human deeds and events stand merely as mute testimonials of man's failure to come to terms with life itself. For those so disposed, intellectually or emotionally, this volume will have little meaning, set as it is within the larger frame of history.

The three essays which open this volume deal with the question of urban quality from an essentially historical perspective. Although they are written from the vantage points of different disciplines and with different objectives in mind, they are, in the broader meaning of the term, urban history writ large. Each in its own way looks at the factors that have conditioned the quality of urban life through time and at their linkages to the present and their import for the future. Professor Dyos directs his attention to these factors with particular reference to the Western world of the last century and a half. Jean Gottmann writes and speaks about the long-term and emerging trends that urban man has been experiencing and where—in a kind of prospective history—we seem to be headed. And Scott Greer sweeps across a wide panorama of social and cultural history to link conceptually the urbanizing forces and developments of the past to the America of the present.

Chapter 2 provides a broad setting for much of what follows in the succeeding contributions. Defining the quality of urban life as the collective attributes of social and eonomic behavior in the cities and towns, it pays particular attention to three influences which serve to explain in large measure the principal features of this quality: growth and distribution of the population; technological innovations; and the zeal and competence of government. As the author indicates, the first was not a significant factor until after 1750 when the expanding number of urban dwellers began to matter—as a strain on urban resources, an acceleration of civic ambitions, a determinant of the scale of public amenities and cultural conditions, an element of social intercourse, and the basis of mass communication.

With population growth, urban life became conspicuously subject to the impact of the second factor—technological change—affecting the household and community. Most of these changes conferred social benefits but some imposed social costs and none of them affected the different classes of city dwellers equally. The social implications of technological developments had wide ramifications, some of them indirect, like the retreat of the countryside, the pollution of the environment, and the separation of home and work. Others were more directly involved in the life-style of urban residents, such as the durability and obsolescence of housing, the standards of comfort and sanitation, the means of transportation and communication, the public utilities, and even the street life itself. Finally, the quality of life enjoyed by city-dwellers depended in varying degrees on the quality of their governors, those who had charge of the administration and direction of the community, and on the political realities of an imaginative response to the newly recognized problems of urban life.

In his essay and in the discussion with the editors which accompanies it, Professor Gottmann looks at the modern metropolis and the forces shaping it from his wide-ranging international experience. His is the perspective of a planner, humanistically oriented and sensitive to the relationship between human needs and aspirations and the physical environment which provides the locus for man's activities. Examining the major historical trends which find their expression in the huge population aggregations of contemporary times, he shows how they portend significant changes in the ways of urban life. As is evident from his careful analysis, the efforts of the planners and the conceptual tools with which they work have been incapable of dealing with this rapidly growing and increasingly intricate system of metropolises and megalopolises. Most of the solutions to cope with contemporary needs are, in fact, aimed at the problems of a society which has been vanishing in the last half century. The metropolitan structure cannot rest on the planner's concept of a system consisting merely of housing, green spaces, less pollution, and decent transportation from home to work; the requirements are far greater. Environment should express rather than determine the ways of urban life, but the reverse is too often true.

In the dialogue with the editors, Gottmann emphasizes that the quality of urban life is less influenced by the factors of scale and density than by the values and organization of the society itself. The world is expanding for each of us and our contacts are multiplying across space. We are exposed more and more to people with different values, different life styles, and different cultures or subcultures. Regardless of our desires or dispositions, we are compelled to live in a spatially shrinking society that is filled with "exotic" elements—a situation always disturbing to the many who find propinquity to others who are "different" uncomfortable and threatening. Our ability to find ways of adjusting to these changes is critical to the urban future. Finally, it is becoming ever more apparent that man is still very limited in his capacity to determine and control his own historical evolution even for relatively short periods of time. In such circumstances, the rational course is to direct our urban planning at "domesticating" the evolutionary process rather than trying to control it in total.

Scott Greer, in Chapter 4, dissects urban history to lay bare the key developments that have brought us to where we are today. In incisive fashion, he takes the reader on an analytical tour de force beginning with a clarification of the meaning of urbanization and ending with a discussion of the new social character emerging in the United States. Urbanization, as he points out, stands not merely for the concentration of people in cities and the processes which lead them there but also for a certain kind and quality of living we call "urbanity." The city-dweller of the past was more sophisticated, more exposed to variety, and more wont to see the world in larger perspective than his country kin. Today all men, urbanite and rural villager alike, are exposed to the communication flows of the larger society. No longer is the city the prime shaper of the dominant culture; it is only a part of the more inclusive national system which sets the prevalent tone of the society and dictates its conventional mores.

Tracing the parallel evolution of social change and social character (defined as typical ways of acting in important social situations), the author shows how the latter now fits neither the situations of the individual actor nor the demands of the large-scale system. What has happened is that the social character appropriate to the rational producer, the laborer, the small businessman, and the peasant has been applied to the rapidly urbanizing society of the twentieth century with its technology, its huge bureaucracies, and its fabulous affluence. The result is a lack of congruency between the reality of the present and the norms of a bygone age. Greer, however, strikes an optimistic note in posing the possibility of a new American social character—one organized around the armature of intellectual control and rational consensus—emerging in response to the recurrent problems of everyday life. But for the immediate present, we can expect a complete mix of social characters: old styles, new styles, and the tolerant middle, the changelings arbitrating between the extremes and providing the cushion "which prevents the war of norms from utterly hamstringing the society." It is in this milieu that the question of urban quality must be resolved.

—W. B. Jr. and H. J. S.

2

Some Historical Reflections on the Quality of Urban Life

☐ I WRITE THIS in the English Midlands, in a temporary building circa 1915 which was erected as quarters for nurses alongside the former lunatic asylum then doing duty as a military hospital. Around me stretch the nobler architectural edifices of the latest tenant of this site, the University of Leicester, and beyond it the ancient city of that name, now containing nearly 300,000 people set in a conurbation half as numerous again. The making of this complex of buildings and of this fourteenth city of the kingdom is not difficult to explain, even to the narrow point at which the towering memorial arch in the park, originally dedicated to the dead of World War I, came to exercise its peculiar hold over the height and character of the university buildings which have crept towards its base. But how far can one go in discerning the quality, not simply of these artifacts, but of the life of the community that created and used them?

A diverse economy of small and moderate-sized firms, unruffled industrial relations, a high level of female employment, an escape from serious damage from bombing, have meant for the city and its inhabitants little acquaintance with poverty, no overgrown fortunes, a hundred years of prosperity, and a hard core of Victorian buildings.

Local people still refer unconsciously to a League of Nations report of the 1930's that put Leicester second in riches in Europe after Lille. Be that as it may, the local community has both the satisfaction of thinking it is so and some of the attributes that come from really believing it. The city's weekly bank clearings are indeed about double those of its larger neighbor, Nottingham, and building societies flock here to scoop up the savings.

The city itself embodies the qualities of bourgeois success. It has a rather puffy pride without a real sense of community, a desire to lead without any real taste for adventure, some suspicions of the intellectual life without any declared antagonism towards it, a complete spectrum of religious sects without much feeling of piety. The city has demolished all three of its theaters since the war and given their sites to shops and a building society. It now gives its tentative support to a tiny repertory company, temporarily housed on the edge of a bus terminus but with imaginative energies unconfined, and substantially more to bingo halls and workingmen's clubs, which flower here supremely well. It has destroyed practically every one of its pre-Victorian buildings and, the town hall excepted, put virtually nothing of architectural merit in their place. It barely manages to stay in the first division of the football league. Yet there is a core of thrift, a tenacious spirit of independence, a feeling for natural beauty, some compassion for the handicapped, an earnest of pleasantness in most things, a kind of urbanity. To some it is clean; to others dull. Its motto pledges allegiance to both. *Semper eadem.*

There is much more to be said about Leicester than this, but significantly less could be said with confidence by anyone who had not lived in the place long enough to use it well, make friends, and come to belong to it. Londoners like myself need longer because they are usually full of metropolitan conceits, especially a disbelief in provincial variety and excellence—the notion that to be provincial is to be deprived. They have also to combat local chauvinism, which in Leicester is leveled against Nottingham and London in that order. These things are an illustration of something important in the discussion of the quality of urban life in its historical perspectives. They point to the fact that there is a limit to the range of places that can be known sufficiently for the purpose, and they contain a warning that, like index numbers, their interpretation cannot be taken very far at second hand.

For these reasons, the horizons of this chapter are not very distant ones in time or space. They do not stretch much beyond 150 years from the present. The examples given come mostly from Britain and a very few from the United States. Little even of this can be traversed descriptively for reasons that we are about to discuss. Almost the nearest approach for the historian must be through a consideration of some of the more important factors conditioning the quality of urban life—not so much those elusive qualities themselves. However, an incidental advantage of this is that the analysis is capable of being extended, *mutatis mutandis,* to include a wider area and conceivably a longer period. In what follows the discussion will center on three main themes that are interwoven with each other: size and distribution of urban populations; technological influences bearing on the urban household and the community; and public administration and amenities.

THE HISTORIAN AND URBAN QUALITY

There are things about urban life that historians can never tell. It is one thing to discover and explain why people have collected together in one place rather than another, why they have manufactured these things rather than those, how they have added to their numbers with unnatural speed or watched them slowly ebb away, how such communities have organized their social space and how they have occupied it. These matters left their marks, as often as not, either in documents or on the ground—not always very legibly perhaps but in some discernible palimpsest which told of what once went on. They assumed, that is, a verifiable public aspect of urban life. Such materials have sometimes permitted historians to write vivid, dispassionate urban biographies which come near to activating the urban past with the same flickering locomotion and newfound excitement of the early films. We see the likeness to ourselves and are satisfied with its authenticity. Sometimes the images speak and we hear what they have to say. They tell us how the various urban containers were made and a little of what went into them, and we get in consequence a more solid sense of form, if not of substance.

Seldom can the historians tell us all the things that mattered to the objects of our curiosity, especially how good or bad it *felt* to be alive in the conditions of the time. What did these urban contents think of their urban containers, without our hindsight and within their own preoccupations and scales of values? Who can tell? Rarely, if ever, has the historian been able to light up those private worlds of individual thought and feeling in a way which enables us to tell what consciousness of beauty or ugliness or sheer actuality was there, what sense of justice or oppression, what sense of community or deprivation, what satisfactions or hurts. These things have remained in historical darkness since cities began and they probably always will. The evidence for them is hardly ever tangible enough to be looked at in true perspective. Indeed, it is rarely possible to look directly at them at all and the circumstantial evidence for the simplest bread-and-butter questions bearing on them is invariably so partial that they remain in the limbo of controversy.

It has to be recognized that the quality of life can probably be interpreted in all its pregnant meaning only by the creative artist, especially by the novelist, who is above all imaginatively concerned with the contemporary scene. The best of them penetrate the very structure of feeling and make everyday situations bear a kind of responsibility for conveying those qualitative things that no ordinary historical document can record. In Mrs. Gaskell's first novel, *Mary Barton* (1848), for example, there are points where we no longer feel observers—as we do in her second novel, *North and South* (1855)—but real participants in working-class family life at the full onset of the new industrialism. The quality of that experience, the real meaning of community among industrial workers lately sucked in from rural villages, and something of the start of a tradition of working-class solidarity have an undeniable objectivity. Even among the other industrial novels of that period this objectivity seems comparatively rare, and for the most part we discover instead fictionalized commentaries on social values or imaginative propaganda for social reform. The original possibilities of going beyond rhetoric and giving sober meaning to that actual quality of life that belongs to what Carlyle had named industrialism are realized more vividly still in the early D. H. Lawrence:

> The real tragedy of England, as I see it, is the tragedy of ugliness. The country is so lovely: the man-made England is so vile. . . . It was ugliness which betrayed the spirit of man, in the nineteenth century. The great crime which the moneyed classes and promoters of industry committed in the palmy Victorian days was the condemning of the workers to ugliness, ugliness, ugliness: meanness and formless and ugly surroundings, ugly ideals, ugly religion, ugly hope, ugly love, ugly clothes, ugly furniture, ugly houses, ugly relationship between workers and employers.

Here, from the grain of one imaginative son of a Midlands mining area, is an analysis of social values and the quality of the life they framed in terms that are honest but unlikely to have been true for the mass of the people who were subject to these things. It is only human to find ways of fending off harsh experience from which there is no real escape, and familiarity with squalor must have bred indifference or blindness or self-rewarding deceits at least as often as contempt or protest. However, such avenues do not remain open forever, and we must hope to see one day that the revulsion from ugliness had some kind of positive historical dimension too.

There are some clues to this already in the way in which urban life invites ambivalence. The city appears in ancient and modern writing as the genius of degradation as well as civilization, the place of destruction as well as salvation, the object of aspiration and the source of despair. Almost any reflections on city life bear this out. Take a quite inconspicuous example—the unsigned comments on the Post Office London Directory which appeared in 1849 in an improving little magazine named *Hoggs's Weekly Instructor:*

> London! word of wonderful import—symbol of a mighty apparent unity—of one mass of human beings, vast and world-like, gravitating together by a system of mutual dependency—yet word, motive with ideas, diverse as the poles and opposite as positive and negative in thought—"of life's extremes, the grandeur and the gloom," the boldest and most striking phenomena are seen in thee. London! the word describes, as with the graver of a Prometheus, palpable and distinct, a circle replete with unequals and antagonisms. . . . Here wealth, groaning with satiety, sighs for some new world of pleasure, and poverty, maddened with hunger, seeks from day to day any world, it cares not which, save this.

Rusticity was never, so runs the underlying theme, like this. It was therefore easy for the city to assume more often the literary role of symbolizing the downard rather than the upward tendencies in human nature. That bias, accentuated by the horrors—both real and imaginary—of the industrial cities of modern times, has provided one of the persistent themes of recent literary history. The city has had few friends among modern novelists, fewer still among poets. It has had, to use the modern jargon, a poor image, and it is difficult to make due compensation for this in assessing its qualities. Our own imaginations trip us up. Did Victorian towns really stink as unfailingly of sulphur and open sewers as Port Sunlight did of soap? Was there never any innocent fun in the slums nor dark crime in the suburbs? Can the cultural slope always have run, like the trains, *up* to

London? There is a strong spirit of anti-urbanism in both British and American thinking which is rooted, no doubt, in more fundamental parts of the national psyche, just as its outcome leads in all kinds of unexpected directions.

In Britain this bias helps to explain, for example, why the English (and still more markedly the Welsh) should have been such reluctant city-dwellers, accepting their lot with a devious determination to bring as many rustic features as possible onto the urban scene. The English social historian, David Eversley, has put it this way: "The example of the Eisteddfod is to me a very mild example of our pretentious rurality; we bring up our children on a diet of the Bible, Shakespeare and the pastoral poets; we never have conceded that we have industrialized. What is the semi-detached house except the desperate nineteenth-century attempt to try and maintain an outlook on your own plot of ground?" It is the kind of romantic yearning that can move whole generations and never be expressed in words. The English hardly knew they had such a preference until drawing-board architects tried to stand streets on end and lift public housing toward twenty stories in the last few years. The tenants then noticed that no birds sang so high in the sky and neighbors dropped in less. Feet on the ground, a bit of green, curves and slopes, chance encounters, the children in sight, a place of one's own—these might almost have been written into the genetic code of the ordinary people whose great-grandparents had in so many cases left the country to join the town. It is no accident that the English seem to prefer to speak of town-and-country planning. The garden city movement, whose antecedents go much earlier than Ebenezer Howard, very properly became the chief contribution which the English had to make to the art of living close together. Perhaps this fact helps to account for the relatively graceless failure of British town planners, both professional and amateur, in using pure urban forms. The long debate over rebuilding the bombed environs of St. Paul's in London in recent years has unconsciously plumbed depths like these.

One more comment in this connection will help to broaden the historical perspective a little. When he was painting his brilliant miniature of Manchester in *Encounter* some years ago (March 1957), the English historian A. J. P. Taylor could not help turning for its true epitome to a stretch of green:

> The most revealing spot in Manchester is not the historic centre, not even the Royal Exchange, but Victoria Park. This is still a private estate, with

toll-gates and keepers in uniform. Gothic palaces jostle each other; gardeners dust the soot from the leaves of the trees; and the ghosts of merchant princes walk in the twilight. These were the men who gave Manchester its historical character. We think of them in retrospect as Radicals; and so they were in lack of respect for traditional authority or in their ruthless destruction of whatever stood in their way. But they were far from a belief in economic equality or even in democracy, if we mean by that putting the needs of the majority first. They had succeeded by their own energy; and they supposed that the duty of society was discharged if it gave others the chance to do the same. It did not worry them that, while the rich man was in his mansion, the poor man at the gates of Victoria Park lived in a slum. The road to success lay open for those who wished to take it. Like the men of the Renaissance they exalted the individual. They lacked one Renaissance characteristic. Of all dominant classes, they were the least equipped with aesthetic taste. Perhaps Money is less beautiful than Intrigue or Wickedness—the Renaissance routes to power; or perhaps it is so beautiful in itself as to destroy the need for beauty elsewhere. At any rate, the result is the same: Manchester is irredeemably ugly. There is no spot to which you could lead a blindfold stranger and say happily: "Now open your eyes." Norman Douglas had a theory that English people walked with their eyes on the ground so as to avoid the excrement of dogs on the pavement. The explanation in Manchester is simpler: they avert their eyes from the ugliness of their surroundings.

The municipal park, however regimented its borders and quaint its taboos, has been the saving grace of a good many of the ugly bits of England. It was a Victorian transmutation of the English landscaped park of the eighteenth century and of the botanic garden which developed with such force in the late Georgian and Regency period. Before long, it also formed a focus for ideas on gardens from all over Europe and collected its flora from all over the world—perhaps the most beautiful residue of the English Empire. It provided a blend of the exotic and the indigenous in a way which no other English art could excel and contributed more subtly than we tend to think to the quality of urban life both then and since. The park open to the public was an unexceptionable way of making public gestures. Sometimes parks were produced simply by public demand, sometimes by a patron's deed of gift, sometimes simply by opening the gates of a private or royal estate. The Victoria parks of England were above all expressions of good manners. Commons were different. They were survivals in the suburbs of an agrarian culture which had been overrun, like ancient fairs reduced to roundabouts and palmistry. Their scrubby acres tended to be less inviolate than parks barricaded with ornamental railings, at least until preservation societies rose late in the nineteenth century to defend what was left of them against railways, brick-makers, and allotment holders. Recreation grounds came later still and were, like public baths, municipal all through. Collectively, the parks, commons, walks, burial grounds, made a decent difference to the places that had them. They gave a place style. What happened when the railings were removed for scrap metal during World War II was another matter. They no longer closed at dusk.

These few reflections throw the historian's ordinary qualifications for making judgments of value about the quality of urban life into sharper relief. Rarely does the evidence which he normally handles help very much. His main struggle to discover what really happened, or how attitudes actually changed and why, may require him now and then to navigate as best he can the little unfathomed inlets of intangible things that impinge on his theme; but he cannot cast himself adrift on such a sea, however benign. The truth is that there can be no reliable *historical* chart to the quality of urban life without a new discipline for connecting the historical and literary traditions of scholarship, and it would be idle in the course of these few remarks to try and create one. An essence like the quality of urban life, secreted in every joint and carried on by every nerve of the urban organism, is bound to be more fugitive than its own minutiae. The state of the streets, the program of concerts in the civic hall, the efficiency of the postal service, the proximity of friends, all help to make life more or less agreeable and add to or subtract from its quality, but they cannot be aggregated or synthesized into a neat overall result. The historian cannot do what Joyce has done in *Dubliners* or *Ulysses* or what T. S. Eliot did with such economy in "The Waste Land"—however much he admires their grasp on what eludes him in the self-same London or Dublin. For the historian it is one thing to recognize the factors germane to a sense of the quality of life; it is another to take readings of the satisfactions they gave.

THE PROBLEM OF QUALITY

Quality is, after all, a slippery concept. In one sense it can mean degrees of excellence in almost anything; in another simply the sum of its attributes; but in ordinary usage it has gathered vague overtones that have given it an historical perspective of its own. The only way of describing this is to say that the whole notion has been democratized. Trollope described in one place in *Barchester Towers* (1857) how "the quality, as the upper classes in rural districts are designated by the lower with so much true discrimination, were to eat a breakfast, and the non-quality were to eat a dinner." As a substantive, used without sarcasm, to indicate those in the higher ranks of society, the term had become a colloquialism and was now beginning to go out. It had not, apparently, become an urban term, perhaps because "the quality" remained essentially a landed aristocracy. The pursuit of quality as the hallmark of approval, the only true sign of breeding, as distinct from its natural, unconscious expression, arose with the middle classes and when it was duly pinned down became bourgeois. In the Victorian period "high class," "select," "family," "quality" became interchangeable tradesmen's terms for their middle-class clientele and the stuff they bought. It is difficult to find writers of that time who were free of such constraints and who were ready to describe the things the working classes—or more accurately the poor—bought or the way they lived in terms of the quality of the goods or of their lives. To have spoken of quality when writing about their *condition* would have been a serious misuse of terms. This is no longer true. Outward appearances matter less, except to the undistributed middle, and the quality of life, be it urban or rural, is now the way it is lived at large.

There is also a sociological difficulty. What meaning are we to attribute to urban life itself? Clearly, it is impossible to go into the whole ambience of urbanism, though it does have an obvious bearing on the quality of life that occurred. The question is directed rather at the generic differences between urban and rural life. In what important respects has the experience of living in towns differed, if at all, from life in the country, and are such differences those of quality rather than degree? Is there some urban continuum along which villages, market towns, small industrial towns, industrial cities, metropolises, are strung out, and in which some discernible differences in the quality of life may be found? Any such budding idiosyncracies tended to be blurred in the beginning, of course, by the interpenetration of rural and urban economies: urban life could not begin at any level before cultivators produced more food and raw materials than they themselves consumed; it could not go on unless urban products or services found outlets beyond the communities that provided them. Moreover, the beginnings of modern industry occurred for the most part outside the towns and penetrated them only when advancing technology and entrepreneurial demands made the towns themselves gigantic machines for both living and working in. The historical patterns are in fact quite complex. Wherever it occurred, the transition from country-dominated town to town-dominated country was neither smooth nor sudden, and rural and urban features have persistently occurred out of context. Nor has this been confined to economic life.

Although a large literature has already grown up on the subject, the distinctive attributes of urban and rural life, the resemblances and differences between life in the village and in the city as well as between towns of different sizes and functions cannot be discerned with real clarity even for those places still under our own gaze. And as for towns used by generations now dead, generic distinctions have simply not yet been identified by sustained historical research. We do not know enough by way of observation how to isolate distinctively *urban* behavior from rural behavior, or know how far to go in rejecting such distinctions altogether. There are several hypotheses. We may tentatively accept the proposition that urban life is essentially impersonal, secondary, contractual, but we can easily find a whole range of mixed examples in which it is not so. Though common sense tells us that we cannot fail to recognize a city when we see one, the inhabitants of certain city neighborhoods—locations often better known by reference to individual streets than by more general place-names like Highgate or Greenwich Village—know that they form a community which for all the ordinary purposes of social life is indistinguishable from that of the authentic village enclosed by fields. There is that same knowledgeable intimacy between people, the same sense of forming a little universe, the same tacit conviction of durability that is questioned only when another world seems about to collide or a familiar landmark or villager is being taken away.

It is even arguable in the situation created by the flight of the middle classes to the suburbs and beyond, and the social isolation of the Negro and the poor in the core of so many American cities, that the traditional territorial distinctions between village and metropolis are in some sense being turned inside out. The impersonal transience of metropolitan life is being transferred to the outer rings of the cities, where cosmopolitanism assumes new forms and flourishes vicariously through television or some other means, and its social networks begin to transcend suburbia altogether. The close face-to-face relationships of the traditional village are often now more palpable in the self-contained and self-containing slums at the center than in the distant villages submerged by their suburban newcomers, dead in the week and alive at the weekend.

THE CONCEPT OF COMMUNITY

There is here one long perspective of importance. The cities, which in some places have been losing their geographical tidiness for two or three generations, are now perhaps in the process of losing their separate identities altogether. The technologies of our own day are having a much more profound impact on the social ecology of cities and the quality of life within them than earlier advances of a more primitive kind once had. The structure of advanced economies with their increasing emphasis on services rather than products, communications systems that extend drastically the potential scope of many of the old business markets, the nationwide supply of power through electrical grid systems, the indefinite extension of the means of personal mobility—these are exercising one kind of influence, most noticeably of all in the United States. The urban repellent of acute congestion, incipient public violence, pollution of air and water, importunate social demands by the poor and deprived, increasing sensitivity to environment in whatever form or for whatever purpose—these are exercising influences of another kind. Between these positive and negative influences the physical artifact of the city is beginning to disintegrate. There are parallels here between physical form and social substance. The wider social networks that are bound to develop in a society making full use of its technological possibilities are now beginning to form on the regional, continental, and intercontinental scale. For the members of such a society, personal relations have fewer and fewer territorial limits.

The sense of community, which has an almost tactile quality in village society, has not been eroded everywhere with equal force. Among the mining communities in the northeast of England, in places that have been purged of all other rusticity, the fierce loyalties that come from common disasters and unjust treatment by remote governments hold them together still. In other places, uncharacteristically urban events, like the August bank holiday walk by thousands together out of Leeds onto the moors, go on under their own momentum or a desire for a common identity. There are even signs among the urban poor of that kind of secret understanding between strangers that Tocqueville recalled in his *Souvenir* when he wrote of "a kind of morality peculiar to times of disorder, and a special code for days of rebellion." It would be a mistake to regard the sense of community as having been rendered entirely archaic even among the thoroughly blasé and urbane, much less among the urban villagers of the slums, where the disorientation of social life from more bourgeois styles has only been carried through by a complex organization peculiar to their own (sometimes "antisocial") needs.

It is clear all the same that *community* is a term which has been changing its meaning for a very long time, and in some contexts it now means very little beyond the old crowd at the pub or the regulars in the self-service laundries or the little adolescent groups that collect in the light of back-street fish-and-chip shops. The concentrative forces that produced the cities as we now see them have in some places already come to an end. The city, once the Renaissance state of its own, aloof from its surroundings except for purposes of provisioning, has ceased even to be a single component in the nation-state, a center for administration of a wider whole, the locus of industry and trade. It has become instead its controlling mechanism and motive force, and its changing structure is perhaps already beginning to provide an entirely new social framework for entire nations, as it is certainly now putting new items onto the agenda of central and local governments. Urban problems, which once looked small-scale, fragmented, almost inert, now seem to have congealed—and not in the United States alone—into the "problem of the cities," forbidding, intractable, volatile. The city is now history's looking glass and what it reflects are the conflicts of class and race and ideology that are the products in our time of historical processes far transcending it.

THE ECOLOGY OF THE CITY

The chief dimension of the urban community is social distance. This is being felt more acutely now than it was during the formation of the first industrial cities, if only because basic conditions have improved somewhat. In Britain, for example, the housing problem had barely been perceived by the end of the nineteenth century, and the eight million people who were, according to Sidney Webb, being housed, washed, and watered in them worse than horses, had scarcely begun to measure the distance that had to be closed. The middle classes had given that distance a physical dimension by their retreat to the suburbs. The physical residue they left behind helped to fix it more concretely, for the houses they vacated were normally ill suited to multiple occupation and therefore deteriorated all the more readily when pressed into use by the lower classes. This not only meant that decaying neighborhoods lacked the stability in local affairs which would have come had the middle classes remained, nor simply that their social problems tended to develop unseen, but that they were cursed by the poverty of their own rateable values. Schemes for equalizing local rates, which tended to be regarded as one of those empty gestures that blessed neither givers nor receivers, did not develop anywhere to any marked degree until the very end of the nineteenth century.

It must be remembered, of course, that the mere juxtaposition of different living styles has never noticeably reduced their differences, not even when, as in ancient Athens, physical contact between them was almost inescapable and the living accommodation deemed fit for persons of high and low rank was barely distinguishable. What clearly did happen later was an accentuation of such differences as the growth of private property took place and was reinforced by the award of public honors. These developments proclaimed social differences among citizens in quite unequivocal ways. The range of common experience among town-dwellers seems to have diminished as the scope for private consumption opened up. A glance at the irresistibly widening gulf in modern Rio de Janeiro between the *favelas* of the squatter poor and the luxury flats of the rich, sometimes within range of the common fly below them, illustrates the point clearly enough. What has tended in relatively recent times to unbridle the frustrations of the poor in so many places has been a sustained awareness of differences in social conditions, in expectations of life and of advancement in the world, in degrees of hope—in a consciousness, one could almost say, of the removable differences in the quality of human life. Sooner or later cities have always enabled such things to be seen. Television cameras now ensure that they are demonstrated for the whole world to witness.

The historical process of dividing the urban ground plan into socially segregated compartments has not occurred in the same institutional setting everywhere nor produced identical results with equal speed. There is, for example, the obvious difference between the ecological patterns of cities north and south of the Rio Grande: in the former, the rich have tended to take to the suburbs; in the latter, they have kept their hold on the center while the working classes and the poor suffer the inconveniences and miseries of life on the perimeter. The latter pattern also existed up to the eighteenth century in cities of the Old World large enough to have suburbs; then the merchants in particular, looking for the best of both worlds, urban and rural, within daily coaching distance of their London counting houses, began to leapfrog the suburban communities already sprawling beyond the ancient city limits. A number of familiar geometrical variations of the zones created by such tendencies can also be clearly seen in the ground plan. It is a geometry which simply exhibits the economic and cultural diversities to be found between cities which are no less varied than their age profiles and occupational structures.

The ground plan of a city is in truth a kind of effigy of the social system from which it has been struck. The benefits as well as the costs of pursuing particular ends, not to mention the social and legal disciplines that society accepts in distributing its rewards and penalities, show up vividly in the detail. Britain's pursuit of industrial power and overseas expansion in the nineteenth century, for example, was unimpeded by the constraints of investing heavily in the housing and welfare of the labor force at home. Little time was spent in maturing industrial workers and virtually none in superannuating them. One of the social costs accepted by the industrial pioneers was, therefore, poor housing for the ordinary people. This was a condition that might have been different had more capital been freed from more directly productive use—but perhaps not so radically different as might seem at first sight. Another way of putting this situation is that the quality of the lives of those still using these houses and neighborhoods two or three generations later has been impaired in some degree by what was initially done. It is not until very much later that the historical implications of the building contracts which were first entered into, of the terms on which the capital used to meet them was raised, of the low standards of building tolerated, or of the imposts laid on building materials by the government, can be fully understood.

Wherever one looks, the forces of change have operated with peculiar subtlety and have produced constantly varying patterns on the ground. Neighborhoods go up and come down in social status in ways that have seldom been subject to close historical examination.

But it is clear that these mutations have always had real meaning for those involved. The signs were money lost or made by those on the move, neighborliness increased or diminished among those left behind, local amenities enhanced or destroyed, and those indefinable nothings that might have gone on drawing people to the spot thrown unthinkingly away. Here, in fact, are some of the most mercurial qualitative changes of all. They suggest many more minutely significant historical perspectives to the quality of urban life than this brief chapter can possibly reveal. They are overlaid by personal patterns of life that might change, or not, in phase with the neighborhood. These aspects of the matter are a reminder that the quintessence of the quality of urban life is to be found, not in the general circumstances with which a paper like this must deal but in their particularity. The true dimensions are matters of detail. The nearest approach to them lies through their local history, especially in that nearly contemporary field, where historians are so ill at ease, in which the verbal testimony of old and young have to be attended to with even more care than the public-conscious written records of the administrators.

URBANIZATION AND QUALITY

One of the implications of the discussion so far is that the most fundamental influences on the quality of urban life were often matters of quantity. The numbers of people in a given space, the speed at which they increased, and the extent to which this development occurred in isolation or in a system of cities combined to make the growth of cities beyond a certain size a different kind of human experience and to provide for a new sort of urban culture. However it be defined by the leading authorities, the urbanization of the world's population has achieved a different order or magnitude over the last 150 years or so from that prevailing before. During the more than six thousand years since cities began, the isolated unstable pockets of human concentration that had formed around local potentates or regional markets were completely overwhelmed in number by people working the land. This was true even where rapid urban developments were about to occur, as in England in the early eighteenth century. By the beginning of the nineteenth century there were only seventeen towns of 20,000 inhabitants or more and only one of more than 100,000—London—which was about to become the first city anywhere to contain a million people. Throughout the world there were probably no more than twenty places at that time with populations exceeding 100,000. By the middle of the next century the number was approaching 900 of which 50 held more than a million people. According to Kingsley Davis, the increase in the proportion of the human population living in cities of 100,000 or more is now accelerating, and in the decade 1950-1960 the rate of increase became twice that of the preceding half-century. Davis calculates that over a third of the world's inhabitants already live in urban places of some size and that the distribution of this population will, by the last decade of this century, be roughly comparable to that prevailing in the most highly urbanized country, Great Britain, in 1851, when about half the population lived in towns. The process of redistributing these human aggregates will then be within reach of its end, for the proportion of the population living in cities and their suburbs in the most advanced nations of the world already appears to have reached a ceiling at around 75 percent of the total.

There is no need to amplify this familiar outline with local detail in order to make the point that the quality of human life in general is not on the verge of assuming an urban condition everywhere. The statistics are in some respects ahead of the qualitative changes involved in the process of transposing and modifying agrarian institutions to an urban setting. This is true not only for countries in which the process has been accompanied by industrialization but in those still struggling to achieve it. In the latter, the speed of urbanization is really a function of simple population growth, and the cities are now having to absorb the resulting numbers much faster than ever was the case among developing countries in the past. On the other hand, the more pervasive massing of human populations has already had two kinds of effect, the one demographic and the other aesthetic. It may also have had a number of psychotic effects of the kind observed in overcrowed populations of other primates. Edwin Chadwick, the social reformer of the nineteenth century, once came very near to saying that the violent tendencies of the lower classes of his day were related to their overcrowding, and told a story about one owner of a court of dwellings who had found it necessary to give it two entrances "so that people in feud could avoid meeting each other."

PHYSICAL FACTORS

The most direct aesthetic effects of concentrated numbers, noise, and smell have no historical dimension whatever. It is a matter of common observation that noise levels have almost certainly been rising everywhere within living memory despite the muting of road drills, the banning of motor horns after hours, or vague legislation about the number of decibels of sound to be emitted at any point. Even now there is hardly any scientific evidence about actual levels of noise and their impact on human life. Around the closes of selected cathedrals in England today, the simulated sounds of sonic booms are being let off, not to see whether anyone objects (which they will learn not to do) but to ensure that stained glass and eroded pinnacles are not actually brought down. Advertisements for double-glazing reinforce their appeal to customers wanting extra warmth with the bonus of quietness. It is a commodity now worth marketing, but it is doubtful whether the quality of urban life has been seriously reduced until fairly recently by the lack of it. It is the so-called beauty spot within reach of a million cars that is now more threatened by transistorized "pop" and the deafening discothèque that offers for pleasure more noise than can be contained in any street.

What is less certain is whether the combined assault on the senses of a profusion of sights, smells, noises, crowds, moving objects, changing levels, constraints, commands, is an important source of disturbance to people. How widespread is that characteristically urban malady, agoraphobia—unidentified before 1873—in its mild or its extreme forms? Some writers have argued that the urban scene has more sources of disturbance than is commonly thought. Kevin Lynch, for example, has reason for disquiet over what he calls the lack of visible identity that comes from overplanned spaces or excessively standardized components. Can one be certain whether he is leaving Blackpool or arriving in Chipping Sodbury? Rational answers to pressure of numbers and competing claims on spaces have tended to produce their own irrationalities and perplexities. The baffling tangles of street signs and other public paraphernalia, the concrete arteries that curtail human movement, the dissonances of color and form which go with commercial eagerness (especially now that advertisers know we must be approached "in depth"), the planing away of the coarse natural grain of individual places and the veneering of the surfaces in a plain monochrome—these have combined to create a habitat that differs markedly from the public centers of things of barely a generation ago. The most obvious historical factor here, of course, is the motor-car, whose demands on social space have risen on an exponential curve that shows little sign of flattening out. The general failure to recognize the high social costs incurred in meeting this inordinate demand has to be explained, however, in terms of social ethics and political realities or even plain hedonism rather than transport economics. It seems fair to say that the automobile has been winning its claims for space because most people want it to win (for reasons that are far more inscrutable) and there are political penalties for denying the fulfillment of these desires. What the automobile has actually meant in terms of the quality of life can be assessed in part by reference to the figures for road casualties, but what it has meant aesthetically or emotionally is beyond historical reckoning. The historian, like anyone else, can but look at the multiple freeways converging on Los Angeles or at the arterial roads that meet in the new heart of Birmingham in England and wonder. The effect appears to have been disastrous in terms of human values since the circulation of traffic has become more important than the circulation of people.

The aesthetic aspects of the quality of urban life have something to do with the way in which the parts are put together. Among the chief historical changes that have occurred in this respect in modern times has been the introduction of discontinuities. It would be unduly romantic to insist that the fitting together of the old small towns of England—or of New England too— was always a comfort to contemporary observers, yet in many of their forms and settings, they have an undeniable seemliness to modern eyes. This sense of aptness springs from the basic affinities in the purposes of the buildings on the main streets, the dimensions to which they were built, and the character of the local materials. Within these overall conditions there usually developed a variety of buildings held together by a kind of rhythm of their own making which corresponded to the human stride. The size of the openings in the walls for doors or windows or between the buildings for squares or funnelled-out places for markets, the height of the structures, the pursuit of the natural contours by the streets, all played a part. It was almost as if the whole scene could have been composed by reference to cubits and spans.

In the course of the eighteenth century, the unity, though not the form of such developments, was disciplined and refined in Bath and Edinburgh, in the spas and watering places, and in the West End of London, to a degree that is cause for astonishment still, given the advance at the same time of ideas about the perfection of things in a state of natural growth. The unbroken stride of the Georgian terrace and the classical idiom of its Palladian counterpart, the suburban villa, were transmuted by means of builders' pattern books into innumerable doll house replicas of what were thought to be their essential features until they were modified almost out of existence. The archaeological evidence for this is recognizable still throughout the inner suburbs of London and in the older parts of Baltimore. The principal streets of Victorian towns were slow to put on any respectable uniform. The street improvements which were beginning to take place, more generally as a means of disinfecting a slum or arching over an offensive sewer than as an artery for traffic, did provide new opportunities for regular street fronts in some places, even though taste itself—in John Betjeman's phrase—was thickening so fast that the opportunity tended to be missed. No one could now pretend that the central parts of these towns, where the prevailing values and business interests of the community were both on show, were values of aesthetic harmony. The chatter of advertisement in every form, including giant-sized autographs and trade symbols across commercial buildings (seldom as a kind of architectural headgear, as in other parts of the English-speaking world), the greatly varying scale and modules of buildings that came with the wider substitution of iron for timber and the greater variety in warehousing and office needs, the use of building materials brought in by railway, and the growing volume of horse traffic on the streets, were none of them composing influences.

Yet composure of this kind is really an afterthought. It occurs to a generation that took over still grimier buildings and more tortuous streets in relation to their needs and looks back now across the waste land of the interwar years, when mass unemployment in the older industrial towns meant a standstill for everything. What really mattered to those pre-Keynesian generations was work. To the workingman, the quality of life in the towns depended entirely on a job; and artisans on the tramp or agricultural laborers turning their backs on the land were responding to the call of higher wages or the hope of them. It was their first lesson in the arithmetic of the city. In his brilliant exposure of the hidden forces of city life, Georg Simmel wrote of the urban economy as being imbued with a calculating quality: "Only money economy has filled the days of so many people in weighing, calculating, with numerical determinations, with a reduction of qualitative values to quantitative ones." The force of that inexorable process, alas, has not yet been spent. City life has always had a legendary intensiveness, a sense of things twice their size, an enticement, a reward for everyone somewhere between overstuffed charities and underused capital. The French sociologist, Durkheim, taught that the division of labor in urban society which created a greater variety of jobs was related not only to the massing of people but to the social attractiveness of it. He saw an inseparable connection between what he called moral density and material density, and which he stated in terms of a "law of gravitation in the social world"—the social mass exercising an attraction by its own density.

There is no room here to test the truth of Durkheim's statement even were there the means. The simple picture that one must have of the nineteenth century in this regard is of movements by the million, most conspicuously transatlantic ones, virtually all of them wheeling on the cities, most of them ending there but others filtering through. The voracious appetites of the towns of Britain, gulping people in from the country or from other towns and disgorging them, or some of them, in successive bouts of prosperity and depression, was repeated within the entirely different orbits of the cities on the eastern seaboard of the United States and of the moving frontier to the west. The cities were bound to be marked by this huge drama, if only because they were nowhere ready to take in the numbers that came or to give breathing space to those that remained. The sharpest demographic effect of this was expressed in terms of mortality.

The high loss of life in the towns of the eighteenth century had been one of the most effective hindrances to their more rapid growth. London's reputation as a devourer of souls was borne out by its Bills of Mortality which told plainly of the great excess of deaths over births. "London will not feel any want of recruits," wrote Corbyn Morris in 1751, "till there are no people in the country." By the middle of the nineteenth century there were, in a contemporary phrase, other "burial pits of the human species." Liverpool, Leeds, and Manchester each grew by 50 percent in the 1820's, Bradford and West Bromwich at around 70 percent, and Glasgow at 30 percent or more for each decade between 1801 and 1841. During the 1830's, Glasgow's population was increasing twice as fast as its houses while the ratio of typhus cases admitted to the Royal Infirmary (which had been under 10 percent of total admissions from 1800 to 1815) reached 50 percent. Typhus meant dirt and destitution, and its pronounced appearance at a time when the industrial towns were being packed so tight with people is significant. So also is the fact that the incidence of tuberculosis remained almost entirely urban throughout the nineteenth century. Cholera, which killed only in thousands while frightening tens of thousands more between the 1830's and 1860's, occurred in brief epidemics and did not correspond as closely with conditions of overcrowding, though it was yet another demonstration of the capacity of towns to waste human life. It is impossible to correlate statistically the density of these urban populations with the diseases they suffered because mortality depended on diet, occupation, and age structure, as well as domestic conditions. There was, nonetheless, an impressive difference in the figures that were collected on the death expectations of the rich and the poor. One local doctor in Leeds, for example, calculated around 1840 that the mean age at death for the upper classes was forty-four, for the middle classes twenty-seven, and for the lower classes nineteen.

The tendency of urban death rates to drive up national averages did not abate in Britain before the twentieth century, but urban birth rates had long been moving the other way. As was the case more generally in the Western world, the age at which both sexes married was significantly lower in the towns than in the countryside and the proportion of the urban population marrying was greater. Yet both the size of completed families and the birth rate were lower in the towns than in the countryside, a difference in fertility which appears to have stemmed from the deliberate limitation of families by the urban middle classes and, a generation or more later, by the urban working classes. The English sociologist, J. A. Banks, has argued that this democratization of birth control became possible because of the opportunities which urban life gave to the working classes to make comparisons of living standards with their well-to-do neighbors. What certainly seems the case is that the position of women in society and of children in relation to the family have both undergone changes which have risen out of urban conditions. The impact of these changes on other social relations has also had a feedback effect on the quality of urban life itself—though we may only now be witnessing the culmination of this process in the changing role of adolescents in society.

The Victorian middle classes, as Asa Briggs has pointed out, saw the growth of their cities, and in some senses the quality of what went on in them, primarily in terms of numbers. They also recognized profoundly that both depended on technological achievements. Even their visions of utopia were based on these developments. They took their technologies as far as their own imaginations could reach or thought back to living conditions which dispensed with them in an unimaginable way. The city naturally occupied a central position in these speculations because the possibilities for controlling the environment had never drawn so close. In a purely technical sense, the short historical span that has occurred since then has made every one of these possibilities capable of being realized. There is little now that single-minded effort on a sufficient scale cannot achieve. What is difficult is for our imaginations to reach as far as our own technologies will allow.

The cities have become the special repositories of these ironies. The basic equipment of the home has come to include machinery and power on a scale sufficient at one time for a small factory, and the electrical consumption of a metropolis of ten million in a rich country has come to exceed that used in a poor nation of four hundred million. This material density of city life has been increasing sharply, particularly since the 1930's. The invention of the domestic version of industrial equipment of all kinds and the domestication of the means of entertainment have tended to give social life a new, more private focus while reducing the means of privacy in action as well as in the imagination. The space required for social life on every level has become more uniform, and social distinctions have come to depend less and less on the scale of domestic arrangements than on their style. Domestic servants, who were already giving the Victorian middle classes some trouble, have been disappearing at high speed, though a cheap substitute has been found in *au pair* girls from the Continent who are prepared to do most things in return for the opportunity of visiting the country for a prolonged period. Do-it-yourself appears to have become less of a slogan than a new kind of democracy. Yet appearances in these matters are deceptive. Despite all the technical possibilities, everyone knows that the enjoyments and miseries of city life still often depend on all kinds of trivial and unsophisticated circumstances. It has become something of a political joke to be living in Britain in the white heat of the technological revolution of the 1960's. What we need to do now, therefore, is to examine quite briefly some of the technological and administrative factors that have come to influence the quality of urban life at so many points.

TECHNOLOGY AND URBAN QUALITY

What is immediately striking about so many of the basic inventions of the city is the recent date at which they were brought into widespread use. The water closet, for instance, the most urbane installation of all, though invented in the sixteenth century was not developed commercially before the end of the eighteenth. And substitutes for it—like the Moule patent earth closet, invented in 1860 and regarded as a great advance on the public middens of many of the manufacturing towns in the north—continued in use for generations yet. At the middle of the nineteenth century, London was still perforated with cesspits and had not completed its scheme for sewage disposal in the Thames estuary rather than its own tidal reach. Similarly, provincial towns had as great a variety of means of coping with their sewage as they continue to have in collecting household refuse. Bringing the water closet indoors has an even more recent history.

Drinking water tended to remain a scarce and rather complex fluid in British towns till the very end of the nineteenth century. Microbiology was in its infancy until the 1800's and water analysis was bound to remain rather unperceptive. The inclination in many towns to collect water from the local river grew annually more hazardous as waterborne sewage was carried as often as not to nearly identical points, and shallow surface wells received the percolations from cesspits, graveyards, and industrial premises of all kinds. As late as 1872 a depot for processing dried dung still operated next door to the filter beds of a London water company. The supply of water was generally intermittent and the air space in the cisterns in which water was domestically stored often filled with noxious gases drawn in from nearby sewers and water closets. A continuous supply of water was unobtainable anywhere, even in London, much before 1900. According to the Milner-Holland report on the condition of housing in Greater London in the mid-1960's, only about a quarter of privately rented households had a fixed bath, a wash basin, a hot water supply, and a water closet in or adjoining the house.

Most of the other improvements in domestic facilities that began to confer tangible advantages on city-dwellers have, in fact, occurred within living memory. Despite eye-catching displays at international exhibitions, gas lighting did not begin to supplant oil lamps in homes of all social classes in Britain until the penny-in-the-slot meter was invented in 1892; and kitchen ranges for cooking which had not come into general use until about 1870, were not replaced with gas cookers in any number until the Edwardian period. The triumph of electricity for either purpose followed much later still and belongs to the 1930's. Even in the United States as the beginning of that decade less than three percent of families cooked by electricity. In Britain, where back-boilers to open grates in living rooms were first being installed in any numbers in the 1890's, the widespread adoption of domestic central heating, mainly fired by gas boiler within the home, is only just beginning. The interwar period, it might also be noticed, was also the threshold for much of the domestic equipment now regarded as a necessity of one kind or another—the automobile, vacuum cleaner, refrigerator, washing machine, telephone. These helped to release time, and the gramophone, radio and cinema, which were then having their effective beginning, helped to fill it.

All these technological developments marked some kind of change in the condition of domestic life and presumably led to some enhancement of its quality. When considering the relationship between technology and the quality of the life of the urban community at large, it is difficult to be certain which is cause and which effect. The introduction of steam railways, for example, interconnected cities, enlarged their hinterlands, imposed a new sense of space and time on their economies and their social life, gave new tasks to their municipalities, and determined the whole chronology of their urban growth. To Mathew Arnold they banished feudalism forever. The demands of the urban communities in turn—for enlarged access to the suburbs and places of recreation, for their own provisioning, and for a return on their investments—made a difference in the quality of the services that could be provided. Their ultimate contributions to the efficiency of household management, the welfare of suburban families, the emancipation of women from an all-absorbing commitment to the home, were given first to the middle classes and did not begin to affect the working classes to any marked extent before the very end of the nineteenth century. For the latter, the initial impact of the railway was more often a blow than a blessing. In Britain, the demand of the railways for satisfactory termini was unequivocal. To reach these points they often had to destroy houses and other premises on a tremendous scale. By dislodging their occupants this action tended to start a chain reaction which subsided only when the last ripples of migration set off in this way were absorbed in the suburbs. In the larger cities these demolitions were augmented by others for street-widening, office-building, public works, and by the persistent tendency of high land values at the center to cause the population to disperse for cheaper living room. From the 1860's the populations of these central districts were beginning to decline almost everywhere. The "haussmanization" of Paris and the building of the boulevards performed a similar operation during the Second Empire. The urban freeway program in the United States at the present day provides another echo.

GOVERNMENT AND QUALITY

These upheavals are a reminder that some of the most pervasive influences on the quality of urban life have been acts of government. Some have had direct consequences, for good or ill, that could scarcely have flowed from any other source, such as the whole program of social welfare legislation which has slowly been unravelling itself in Britain since 1906 without ever quite ensuring for everyone a minimum acceptable standard of living. No doubt pensions, sickness pay, unemployment relief, even the workhouse, sometimes made grim realities endurable a little longer or took the edge off more desperate acts by unhappy people. The quality of urban housing, however, has never been ensured by legislation, nor sanitary codes kept by regulations. Indeed, from what we know of housing reform in England before 1914, we are bound to ask whether it had not made things worse in many places by forcing up tiny islands of decent standards here and there at the expense of the quality of housing round about—not to say disturbing the social ecology of the area in such a way as to intensify its worst tendencies. Condemning unsanitary dwellings and turning people out of doors usually meant aggravating the overcrowding nearby, and rehousing them has been difficult to do until comparatively recently, in part because the new rents either could not be afforded by those displaced or were felt too excessive to be met. In times of inflation the quality of working class housing tends to deteriorate if only because working class concepts of a fair rent drag behind their actual movements and families make do with less space than they should have. In many of the cities of Latin America and India the attempt at urban improvement is regularly being frustrated in another way: by squatters taking over cleared ground before any housing can be erected, creating thereby a political presence as well as a human demand.

These are, of course, no more than faint hints of governmental influence as well as impotence in relation to the quality of urban life on its lowest levels. Similarly, they are no more than the barest suggestions of how acts of amelioration have not always had their expected results. The historical record of governmental arbitrariness is actually far more extensive and indirect. It has been conducted by means of budgetary and fiscal measures that have been assuming more comprehensive forms virtually everywhere. The ultimate controls on the advance in the quality of urban life are being determined within fairly wide limits by governments, not only explicitly by discrete grants to various cultural organizations or massive allocations of funds to public systems of education and the like, but implicitly by every other way in which public money is spent. The latter includes global strategies of defense and trade and the extent to which resources are being made available to help poor countries whose needs are so much more acute.

In a more domestic sense, the very performance of government helps to distribute the cultural prizes unevenly. In Britain, the great divide is between north and south, between the kingdom of the government in London, with its crowded southeastern hinterland of rolling suburbia and coastal resorts, the sunniest, most prosperous, most expensive, and most densely settled region of the country to which industry is now gravitating more naturally than ever, and the old center of the industrial revolution on both sides of the Pennines in the north, a more frequently depressed, more intricately textured, more independent country with its own distinct cultures. These sprang from great diversities of social structure and sheer native wit that produced across the country a rich variety of thrusting or conservative or purely supine elected councils that either put money into things or did not, whether it was a town hall, a sewage farm, or a processional avenue that would give mayor-making more style. In the course of the nineteenth century the metropolitan influence of London gradually enthralled all of them, and they are no longer so clearly distinguishable. Yet what has not happened is any devolution of metropolitan institutions. Whitehall has stayed intact. London first became a center of conspicuous consumption during the seventeenth century because it held the crown, the court, the law, the government, parliament, and the largest single share of the country's industry and trade. All this it has retained, and to it has now been added the nation's richest hoard of cultural treasures and treats and the fattest cornucopia of pleasures in the world. In one way much of this has come about by the sheer force of being a world city but the foundations for it were being laid on an imperial basis late in Victoria's reign. No government since has acted with any resolution to reverse the provincial flow. When, by a purposeless stroke of government, performed in 1961 in the teeth of every outraged body of artistic opinion in the country, the very gateway to the northern provinces—Hardwick's classic Doric arch at Euston—was destroyed, it was as if the process had all the unstoppability of a ritual murder. Let the provinces stand back.

A CLOSING REFLECTION

Looked at from this provincial corner in which I write, the quality of urban life is beginning to be influenced by more subtle metropolitan influences. We have been witnessing during the last few years a marked growth in immigration from the British Commonwealth, a kind of tidal movement flowing back from territories taken long ago and lately released from colonial dependence. This migration has produced in a number of cities in the Midlands and the north, as well as London, a new kind of social mass that is beginning to infuse into local affairs a new kind of problem—race relations. The historical perspective here is a long one, coordinate in many ways to that by which Britain's industrial cities must be seen. Both stem from the nineteenth century when commercial expansion overseas began to yield its first rich harvest for the industrial pioneer; and it is now that the full implications of this enrichment and the unconscious collection of an empire under the British crown can be realized. What this has meant in terms of qualitative change cannot yet be properly grasped, for every large and conspicuous inflow of people from abroad in modern times has produced in the country a certain amount of emotional noise while it was taking place.

For the cities that have received substantial numbers of migrants, the impact on their cultural life has profound meaning, a meaning which only begins to become clearer as these communities endeavor to work out in quite pragmatic terms how to make urban society in Britain a composedly multiracial one. Leicester is among their company. In a different direction altogether, the approach to a solution of another crucial issue facing this city—as it does others—might have been clarified a little since I began to write. The future quality of urban life is much embroiled with the automobile, and now Leicester's planning officer has published his vision of the city at the end of the century. It is based on the first scientifically devised traffic plan, for any city, that says "no" to the automobile. We now must wait to see whether the motor car will obey.

REFERENCES

ANDERSON, STANLEY (1962) Britain in the Sixties: Housing. London: Penguin Books.

American Academy of Arts and Sciences (1968) "The conscience of the city." Daedalus (Fall), issued as Vol. 97 of Proceedings of the American Academy of Arts and Sciences.

ASHWORTH, WILLIAM (1954) The Genesis of Modern British Town Planning. London: Routledge and Kegan Paul.

BANKS, J. A. (1968) "Population change and the Victorian city." Victorian Studies 11 (March).

———(1954) Prosperity and Parenthood: A Study of Family Planning among the Victorian Middle Classes. London: Routledge and Kegan Paul.

BRIGGS, ASA (1968a) "The Victorian city: quantity and quality." Victorian Studies 11 (Summer Supplement).

———(1968b) Victorian Cities. London: Pelican Books.

CHADWICK, GEORGE F. (1966) The Park and the Town. Public Landscape in the 19th and 20th Centuries. London: Architectural Press.

CHAPMAN, S. D. (1963) "Working-class housing in Nottingham during the Industrial Revolution." Transactions of the Thoroton Society 67.

CLINARD, MARSHALL B. (1966) Slums and Community Development. London: Collier-Macmillan.

COHEN-PORTHEIM, P. (1930) England, The Unknown Isle (Alan Harris, translator). London: Duckworth.

DURKHEIM, EMILE (1933) The Division of Labour in Society (George Simpson, translator). New York: Macmillan.

DYOS, H. J. (1968a) "The slum attacked." New Society 280 (February 8).

———(1968b) "The slum observed." New Society 279 (February 1).

———[ed.] (1968c) The Study of Urban History. London: Edward Arnold.

———(1961) Victorian Suburb: A Study of the Growth of Camberwell. Leicester: Leicester Univ. Press.

FLINN, M. W. (1965) Report on the Sanitary Condition of the Labouring Population of Great Britain by Edward Chadwick, 1842. Edinburgh: Edinburgh Univ. Press.

FRIEDMANN, JOHN (1961-62) "Cities in social transformation." Comparative Studies in Society and History 4.

GLAAB, CHARLES N. and THEODORE A. BROWN (1967) A History of Urban America. New York: Macmillan.

GLASS, DAVID (1935) The Town and a Changing Civilization. London: John Lane, Bodley Head.

GLOAG, JOHN (1949) The Englishman's Castle (Second ed.). London: Eyre and Spottiswoode.

GUTKIND, E. A. (1962) The Twilight of Cities. New York: Free Press.

HAIG, R. M. (1926) "Toward an understanding of the metropolis." Quarterly Journal of Economics 11.

HANDLIN, OSCAR and JOHN BURCHARD [eds.] (1963) The Historian and the City. Cambridge, Mass.: MIT Press and Harvard Univ. Press.

HIRSCH, WERNER Z. [ed.] (1963) Urban Life and Form. New York: Holt, Rinehart & Winston.

JEPHSON, HENRY (1907) The Sanitary Evolution of London. London: Fisher Unwin.

JOHNS, EWART (1965) British Townscapes. London: Edward Arnold.

LAMPARD, ERIC E. (1968) "The evolving system of cities in the United States: urbanization and economic development." Pp. 80-141 in Harvey S. Perloff and Lowdon Wingo, Jr. (eds.) Issues in Urban Economics. Baltimore: Johns Hopkins Press.

LYNCH, KEVIN (1960) The Image of the City. Cambridge, Mass.: MIT Press and Harvard Univ. Press.

MANN, PETER H. (1965) An Approach to Urban Sociology. London: Routledge and Kegan Paul.

MARSHALL, LEON S. (1940) "The emergence of the first industrial city: Manchester, 1780-1850." In Caroline F. Ware (ed.) The Cultural Approach to History. New York: Columbia Univ. Press.

MITCHELL, R. J. and M. D. R. LEYS (1958) A History of London Life. London: Penguin Books.

MORRIS, R. N. (1968) Urban Sociology. London: Allen and Unwin.

MUMFORD, LEWIS (1961) The City in History. Its Origins, Its Transformations and Its Prospects. London: Secker and Warburg.

PIMLOTT, J. A. R. (1947) The Englishman's Holiday. London: Faber.

REISSMANN, LEONARD (1964) The Urban Process: Cities in Industrial Societies. New York: Free Press.

Report of the Committee on Housing in Greater London (1965) [Milner-Holland Report, Cmd. 2605]. London: H. M. Stationery Office.

RICHARDS, J. M. (1946) The Castles on the Ground. London: Architectural Press.

RIMMER, W. G. (1961) "Working men's cottages in Leeds, 1770-1840." Thoresby Society Publications 46.

ROBERTS, DAVID (1960) Victorian Origins of the British Welfare State. New Haven: Yale Univ. Press.

SCHNORE, LEO F. (1958) "Social morphology and human ecology." American Journal of Sociology 63: 620-634.

Scientific American (1965) Cities. New York: Alfred A. Knopf.

SHARP, THOMAS (1968) Town and Townscape. London: John Murray.

SIMMEL, GEORG (1957) "The metropolis and mental life." Pp. 635-647 in P. K. Hatt and A. J. Reiss (eds.) Cities and Society (Second ed.). Glencoe: Free Press.

SJOBERG, GIDEON (1960) The Preindustrial City. Glencoe: Free Press.

SMIGIELSKI, W. K. (1968) Leicester Today and Tomorrow. London: Pyramid Press.

STEWART, CECIL (1952) A Prospect of Cities, being Studies towards a History of Town Planning. London: Longsman, Green.

STRAUSS, ANSELM L. [ed.] (1968) The American City. A Sourcebook of Urban Imagery. London: Penguin Press.

TAYLOR, A. J. P. (1957) "The world's cities (1): Manchester." Encounter 8 (March).

THOLFSEN, T. R. (1956) "The artisan and the culture of early Victorian Birmingham." Univ. of Birmingham Historical Journal 4.

TOCQUEVILLE, ALEXIS de (1948) The Recollections of Alexis de Tocqueville (J. P. Mayer, ed.). London: Harvill Press.

TUNNARD, CHRISTOPHER (1953) The City of Man. New York: Scribner.

WARNER, SAM BASS, JR. (1962) Streetcar Suburbs. The Process of Growth in Boston, 1870-1900. Cambridge, Mass.: MIT Press and Harvard Univ. Press.

WEBER, ADNA FERRIN (1963) The Growth of Cities in the Nineteenth Century. A Study of Statistics. Ithaca: Cornell Univ. Press.

WENTWORTH, ELDREDGE H. [ed.] (1967) Taming Megalopolis. New York: Frederick A. Praeger.

WHITE, MORTON and LUCIA (1962) The Intellectual Versus the City. Cambridge, Mass.: MIT Press and Harvard Univ. Press.

WILKES, L. and G. DODDS (1964) Tyneside Classical: The Newcastle of Grainger, Dobson and Clayton. London: John Murray.

WILLIAMS, RAYMOND (1950) Culture and Society, 1780-1950. London: Chatto and Windus.

WRIGHT, LAWRENCE (1964) Home Fires Burning: The History of Domestic Heating and Cooking. London: Routledge and Kegan Paul.

———(1960) Clean and Decent: The Fascinating History of the Bathroom and the Water Closet and of Sundry Habits, Fashions and Accessories of the Toilet, principally in Great Britain, France and America. London: Routledge and Kegan Paul.

3

Environment and
Ways of Life in
the Modern Metropolis

JEAN GOTTMANN

☐ THE MODERN METROPOLIS is today the object of much
solicitude. An abundant and multifaceted literature hints at
metropolitan ways of life as manifest destiny for the majority of
mankind. Already the majority of the population in most of the
well-developed countries live in large cities or their environs. The
trends of population migration indicate growing concentration in a
few major metropolitan areas for most industrialized nations. The
presidential message to the United States Congress on the cities, in
March 1965, forecast in its preamble that most of the forthcoming
growth of the American population would gather in and around the
country's major cities [U.S. House of Representatives, 1965].
Recent British studies and official White Papers debate the possibility
and desirability of channeling or deflecting the present flow of

EDITORS' NOTE: *This article is reprinted, with permission, from J. W. House
(ed.),* Northern Geographical Essays in Honour of G. H. J. Daysh *(Newcastle:
University of Newcastle on Tyne Press, 1966). It is used as the basis for the
discussion which follows between Professor Gottmann and the editors.*

people towards the south-east of England, around London [Ministry of Housing and Local Government, 1964]. Similar concerns and government planning are developing in France about the irresistible growth of the region of Paris [Premier Ministre, 1963]. Governments are faced with similar trends and problems for Amsterdam and the Randstad Holland in the Netherlands, Zurich in Switzerland, Vienna in Austria, Toronto in Ontario, Montreal in Quebec, Mexico City in Mexico, Buenos Aires in Argentina, and so forth.

DEFINING THE "MODERN METROPOLIS"

It is rapidly becoming a general feature of our time that small patches of a national territory (not more than 5 per cent of its land area) hold 15 to 35 per cent of the national population. Some countries have only one such region around the main metropolis; other countries may have two or even three such regions, already in existence or rapidly forming. This means that the population is thinned out over vast land spaces and that it gathers in thick densities over small but essential patches of the land. New aspects of both conflict and complementarity develop between the emptying and the congested areas within a country. But these trends also mean great selectivity in the present process of urbanization and metropolitan growth. Not all the conurbations or large metropolitan centres grow in similar fashion and rate. In fact, a distinction must be drawn today between what may be classified as a "central city" and what is indeed a "modern metropolis". The concept of "metropolis" implies for the "mother city" a certain role of dominance, of protection, and, of course, of relative independence. It is not merely a matter of size but rather of function, and, in some way, of authority.

In the United States a school of thought has arisen, comprising leading architects, urban planners, and sociologists, that looks forward to a gradual dissolution of the tightly agglomerated city of the past. The American metropolitan region grows chiefly through its suburban sections, and the central city declines in importance and sometimes even decays. To this school of thought (of which Frank Lloyd Wright was perhaps the most famous and extreme exponent), the outward movement of residences, stores, industrial and warehousing establishments, from the central city towards the metropolitan periphery and beyond, appears an essential sign of the times; it is the product of modern technological progress in transportation, communications, and mass production of consumer goods, including housing; it looks like the ultimate liberation of the individual from the servitudes of crowding and social hierarchy [Martindale, 1958; Elias et al., 1964: 90-117].

Such thinking is largely fostered by the recent suburban sprawl around North American cities, now coming to many other metropolitan districts throughout the world; it entails decentralization from, and anxiety in, the cores of many old central cities. However, all these trends are deeply rooted in the predominant philosophy of urban planning, which stems partly from the 19th century utopian concepts condemning the larger cities and calling for smaller communities set in greener environments. This philosophy in the 19th century permeates the work of such different thinkers as Fourier, Dickens, Karl Marx and Friedrich Engels, and, finally, Ebenezer Howard, whose garden cities may be interpreted as a liberalized, improved, and enlarged version of the socialist "phalansteres."

INDUSTRIALIZATION AND URBANIZATION: SHIFTING TRENDS

The 19th century was one of rapid urbanization, producing too often the unpleasant environment of the "black country" and of the blighted quarters in which the masses lived and laboured who were gathered in the cities by the expanding employment in mining and manufacturing. Still, in 1915, Sir Patrick Geddes, in his memorable volume on *The Evolution of Cities*, foresaw urban growth developing chiefly on top of coalfields. This was written and believed a few years after Winston Churchill had started the shift of the Royal Navy to petroleum fuels and after H. G. Wells had forecast, in the *The World Set Free* (1914), something like the first atomic bomb.

Nowadays the coal-mining district has become a painful sight and a problem for modern urban planning. Mining is still expanding production while curtailing the number of men who are employed. Energy is becoming cheaper and more readily available in a greater variety of forms over most of the world. In the more advanced countries, the number of people employed in manufacturing production is slowly levelling off, while the quantity and variety of goods manufactured increase. Through mechanization, automation, and rationalization in the methods of production, the same process which the industrial revolution brought to bear, first on agricultural production, then on mining, has now come to industrial manufacturing. In the United States the total number of production workers in manufacturing remained approximately stable (between 12 million and 13 million workers) from 1950 to 1964. In the same period the volume of manufacturing production grew by about 65 per cent. A similar trend was observed in the statistics of the labour force in manufacturing in the United Kingdom and Canada. In the West European countries on the Continent the number of manufacturing production workers may have still been slowly increasing, but the point of stabilization appeared near, and employment in the services, particularly in white collar jobs, increased much faster. The American labour force added 13 million jobs in non-agricultural establishments from 1950 to 1964, two million of them non-production workers in manufacturing industries and over 11 million of them in services other than transportation and public utilities.

There can be little doubt that this general trend of employment away from the farm, the mine, and the manufacturing plant is not reversible any more. As the cities grow, the process of urbanization has to be separated in our thinking from the process of industrialization, at least in those countries which have already experienced the fruit of the industrial revolution for a century or more. The trends may be different in the less developed countries which are catching up and are now trying to go through the stages of the industrial revolution and its related development. As we look ahead, however, towards the needs and resources of the modern metropolis, we must expect the economies of the North Atlantic realm to show the way for the immediate future of the large urban regions presently forming.

PORTENTS OF CHANGING LIFE-STYLES

The indicated shift in the kinds of employment provided in the modern metropolis portends a considerable change in urban ways of life. The 19th century brought about a heavy concentration of blue collar workers and in many places the "black country" cityscape. Now the metropolis can afford to be mainly and increasingly a concentration of white collar workers and of rather clean commercial and industrial operations. Massive industrial production has a trend to move out of the large congested urban area, either towards its periphery or towards less congested territory, a smaller town, or even a crossroads in rural surroundings. The competitive nature of our civilization still fosters constant efforts to attract industrial plants and large warehouses to older urban nuclei, threatened with industrial outmigration and in some cases with actual decrease in population. Perhaps the metropolis, remaining a pole of attraction for the white collar employment and "quaternary" activities that are the faster-growing sectors of the labour force, could afford to let those industrial and warehousing plants that prefer outmigration move out to locations of greater efficiency under the new circumstances. Even if most of that migration of industry were to go towards the centres of declining population, which it will not necessarily do, this would not solve the problem of employment in the older centres. The people would steadily keep on moving towards the areas where the new jobs are, and particularly the better jobs, or the jobs amid better living conditions.

THE URBAN ENVIRONMENT

FORM AND FUNCTION

The modern metropolis is chiefly studied and regulated in terms of its economic functions and its forms of habitat. Form and function are, of course, as deeply related as ever. But the modern metropolis is becoming such a huge and intricate system that the old methods of analysis may no longer be adequate. Form and function are both part of the environment of the city. This environment is supposed to express rather than to determine the ways of life in the urban districts. Those who plan, analyse, and design our cities are essentially concerned with the environment in its most material aspects. They would not think, at least in a country that rejects the philosophy of totalitarian government, of determining by planning the ways of life of the inhabitants. For instance, planners hope to provide the people with some choice of their residences, of their work, of the places where they may shop or take their recreation. The speed of urbanization, however, does not permit the responsible authorities to provide all the equipment and the services for the citizens with the variety of means and surplus margins that would allow for free choice by all members of the community. Some choice is still left for a minority of more privileged people, usually the wealthier. But the scarcity of means strictly limits the choices for a majority of the urban folk (Gottmann, 1966a). As these try to escape insofar as possible from the unpleasant aspects of crowding and scarcity of means (such as traffic congestion, slums and blight, air and water pollution, inadequate schools, etc.), an increasingly chaotic condition develops in the large metropolis and around it. Writing in Paris during Haussmann's redevelopment of the city, the poet Baudelaire remarked on "the chaos of the living city", and in another poem, that "the form of a city changes faster, alas, than a mortal's heart."

The mortal's heart is not independent of his mind. But most of our contemporaries, even in the field of urban planning, seldom realize the rapid changes in the city's ways of life. Obviously, a metropolis which chiefly houses a white collar population, with ten per cent or more of its adult inhabitants having had some higher education, cannot be satisfied with an environment that is improved to meet the needs of an industrial blue collar citizenry. In other words, our ideals of environmental design are still attempting to solve the problems of the 1890's or perhaps of the 1910's, and today, in the modern metropolis, these problems are ancient history.

In October 1964, this writer attended the sessions of the symposium on "The Future by Design" of New York City, for which he had been given the task and privilege of concluding with "an interpretative summary." The environment of the great city was discussed in its various aspects by distinguished specialists, and it was apparent that they were all thinking of the future of New York as a great centre of financial management, cultural and governmental activities; a few speakers bemoaned the decline in manufacturing employment and blue collar jobs, but none of them expressed more than the hope of maintaining the figures at the level of recent years. However, no planning was considered to face the fact, which this writer could not refrain from emphasizing, that New York was becoming a white collar city, and that perhaps such a future could be envisioned for it and provided for [New York City Planning Commission, 1964: 155-163].

A year earlier, the writer chaired and concluded a debate in Paris on the future of the French capital region. A good deal of this discussion focused on the redevelopment of the area in the centre of Paris which was going to be vacated by the moving of Les Halles, the central market, to a new suburban location. Several distinguished architects suggested bold new architectural designs for the renewal of Les Halles' site, including high-rise towers. But lively resentment was aroused by any solution creating a skyline in the heart of Paris. The Parisians' hearts and minds were not yet ready for the new forms which are, in fact, functional for the ways of life now manifesting themselves and already taking over the metropolis [See Revue Francaise, 1964].

The great expansion of office work which increasingly makes up the bloodstream of the modern metropolis, and the demand for office space attendant on it, is still a source of amazement and misunderstanding among the professional authorities. Paris does not want to encourage new office buildings in its central district. The White Paper of 1963 and the South-East Study discussing the future of the London conurbation recognize that the expansion of offices in London was unforeseen and unprovided for; it was hoped the trend might be stopped and the offices decentralized. The City of New York officially rejects and resents its destiny of being a huge transactional and cultural center. Altogether, urban planners and architects do not seem to realize what kind of new economic activity and related ways of life are heralded in the modern metropolis by the spectacular and spreading rise of skylines.

ENVIRONMENT AND LIFE-STYLE

The high-rise towers, either lean and lofty or massive and rambling, express the need of concentrating, on favoured locations over a small ground area, of certain activities which are all in what we have already described as "quaternary economic activities" and faster-growing sectors of employment. These are such select services as the management of public and private affairs, research, higher education; specialized medical, legal, and other professional advice and servicing; banking, insurance and mass media. The processes of management and decision-making which involve substantial financial or political interests require physical proximity to the hubs of mass media, advertising, research, legal and technical competence on a wide and varied scale. Obviously research and higher education are closely associated functionally and geographically. The ties between the centres and processes of management, government, and information (in the wider sense of the latter) are just as evident.

These are the activities and the jobs, all of a transactional nature, that gather in the hubs of the modern metropolis and along the shafts of the skyscrapers [Gottmann, 1966b; Vernon, 1959].

All these quaternary activities will also attract such services as highly specialized and sprawling medical centres, specialized retail stores, art galleries, and a wide variety of entertainment. Thus, recreational activities must be located in the vicinity of the transactional hubs. As the labour force evolves towards occupations requiring more background and education, and affording more leisure, the facilities for recreation must also play a bigger role in the design of the metropolitan area.

For the newly evolving ways of life are no longer the privilege of a few but are actually the ways of life of many millions, and the metropolitan environment must be designed to provide for their recreation. To some extent this trend is already obvious in the urban landscape: just as the skyline in the central business districts has become a common feature of the large city, groups of impressive buildings usually arise near the skyline with arenas, stadiums, auditoriums, theatres, museums, some of these structures dwarfing Rome's Colosseum. Also, new metropolises are developing in and around choice spots selected a century or less ago for their special recreation by American millionaires, British lords, and Russian princes. The French and Italian Rivieras, the Gold Coast of Florida, sections of the shores and hills of California, the Bermudas, the Bahamas, and Hawaii are coming into the category of substantial urban centres. In fact, while many of the larger urban areas developing in various parts of the world have shaped up around one or several large commercial and industrial centers, and may be termed regions of a "megalopolitan" character, other urban regions, growing even faster and holding great promise, are developing from a group of fancy resorts. As this new category of urban regions, which may be termed regions of a "Riviera" character, expands, usually along narrow ribbons of land, it comes to encompass older large or medium-size cities, which immediately benefit by the association. Thus, the Rivieras mushrooming along the French and Italian shores, between the mouths of the Rhône and the Arno rivers, are adding to the growth and metropolitan character of Marseille and Nice, Genoa, Leghorn, and Pisa. To some extent, the coast of southern England is also the site of a Riviera-like urban development, which adds to the attraction of the South-East as a whole. Sections of the shores of Spain, of the Adriatic, of the Greek islands are experiencing a similar evolution. The Soviet Union also has its Rivieras, in the Crimea around Yalta and on the Caucasian coast around Sotchi. The list is getting longer every year. Not every one of these resort places along warm shores or snowy mountains (as winter sports develop) will lead to the formation of a metropolitan area. But the geography of recreational amenities is becoming an essential element in the distribution of the centres attracting the migration of transactional activities outside the older established large urban regions [Ullman, 1954].

AFFLUENCE AND ENVIRONMENT

Modern urbanization unfolds in a competitive world. To succeed in the competition and attract the people and the economic activities which nowadays make for progress and growth, any given area must offer, in addition to the basic facilities, an array of physical and cultural amenities, for the rising generations of today are asking for both more opportunity and a better environment at the same spot. The means of technology and of the affluent economy of the better developed countries authorize such demands. Unless the next generation fails to take advantage of the possibilities clearly indicated by the present trends, urban environment design is slated to provide better living conditions for the masses of tomorrow's urbanites than were ever dreamt of.

An environment pregnant with wonderful amenities may be rapidly despoiled, however, and turned into an unpleasant location if, under modern pressures, its use is not planned and regulated with full understanding of the needs and the means of the ways of life this environment will hold. Urban studies and design are now focused on two categories of concerns: on the one hand, how to provide the "industrial society" with an idyllic pre-industrial kind of frame of living; on the other hand, how to use, in the best, fullest, and most efficient manner, the new tools of our ever-expanding technology. The former concerns are chiefly aesthetic and ethical, but most of the solutions considered are aimed at the problems of a society which has been vanishing in the last half century. As to the technological tools, they offer today the prospect of an extraordinary set of gadgets to equip the homes, the arteries of traffic, the means of communication, and many other services.

Adorning a world of garden cities which would have suited the needs of 1900 with the gadgeteering which will be generally available by the year 2000, such as [sic] the fascinating game of perhaps too much of modern urban design.

While technological progress makes production easy, and the equipment and redevelopment of the environment relatively simple to plan, understanding the new ways of life and providing satisfactorily for them remain very difficult. It used to be accepted, even twenty years ago, that people wanted to live in garden cities of an optimum population size of 30,000 to 50,000 each. What has been called the "New Town Blues" and other considerations have forced even the enemies of the large city idea to revise the optimum figure to at least 100,000 inhabitants, and in some countries to 200,000. New cities are being planned which are designed to reach soon a population of 300,000 or even 500,000. The half a million figure may have been first advanced in very large countries with great expectations (for instance, in the case of Brasilia), but the 300,000 figure has been given as a target for the new town being planned in the vicinity of Zurich. A large agglomeration can better afford the expensive array of services and amenities now required to make urban life pleasant.

In the competitive world of a transactional and affluent society, people want excitement, and they do not want that excitement to be provided by wars, revolutions, and other catastrophes, as was too often the case in the past. In Roman times, Rome had to provide its people with bread and spectacles. As the Renaissance dawned in Italy, L. B. Alberti warned the urbanists of the 15th century that cities should be designed to offer both *commoditas et voluptas,* that is, physical and cultural amenities in addition to functional quality. In the welfare state and the affluent society, bread is provided; some amenities too, in some locations. The planners are working hard to make use of all the possible gadgets within their reach. A great deal of concern for making the environment more and more functional is stressed. In the midst of it all, the urban ways of life are undergoing a deep and widespread mutation, largely caused by the new technology but reaching far beyond the use of all the gadgetry and labour-saving devices made available. The individual, living in conditions of higher density of human contact, liberated from hard physical work by the sweat of his brow, develops new and yet vague needs for his body, his emotions, his intellect. Society evolves with the individual and must provide for the new needs of more abstractly minded people.

ENVIRONMENTAL REQUIREMENTS FOR THE MODERN METROPOLIS

The metropolitan structure cannot rest on the concept of a system consisting of housing, green spaces, less pollution, and decent transportation from home to work. The modern metropolis requires much more. It cannot assume that the places of work will take care of themselves and will locate wherever transportation, housing, and some green spaces are available. The transactional nature of the work that is increasingly occupying most of the labour force in the metropolis calls for certain rites to be attended to. Entertainment of customers is an old rite of business transactions. Recreation of a collective nature is an old rite of metropolitan communities (the "spectacles" of Rome). Modern society is thirsty for such rituals in addition to those traditionally provided by religion, local custom, and the walk in the park. As more time becomes available for the leisure of the masses, sports develop on a huge scale, some reminiscent of the Roman arena sessions. Still better, the 20th century has revived the long-forgotten Greek tradition of the Olympic games. The liveliness of a modern metropolis may perhaps best be measured by the size and rate of occupancy of its symphony halls. It was recently reported that the number of symphony orchestras in the United States had doubled in the last ten years and had reached about 1,000 by 1965. Despite the progress of radio and television and musical recordings, the collective rite of the symphony is gaining in an urbanized society.

Perhaps what people are coming to seek in the modern metropolis is, first, the economic opportunity offered by expanding employment in the transactional activities; second, the greater social fluidity resulting from both the economic opportunity and the physical proximity of the powerful, the wealthy, and the scholarly; and third, the excitement and spiritual satisfactions derived from the recurrent collective rites dense urban life offers. Such a concept of the metropolis appears more realistic in the 1960's than those rooted in the industrial landscape of the 19th century. For about two centuries the industrial revolution made the term "urban" synonymous with blue collar labouring masses. H. G. Wells, in *When the Sleeper Awakes,* saw the masses of an enormous London of the 22nd century still dressed in blue overalls and spending all their day at hard physical labour. The present evolution of society was little foreseen; it is not yet widely realized. The metropolis which is now being built or renewed must fit a demanding society, largely occupied in abstract transactions, thirsty for collective rites, increasingly educated to higher aesthetic and ethical standards.

With production becoming easy, one can hardly settle for a design of cities made of monotonous and endless repetitions of mass prefabricated elements. It must have variety, beauty, and prestige. It is not being too optimistic to assume that the public will ask for more and more of all these. It would be very pessimistic to assume that these cannot be provided.

Many cities continue to live the life of the industrial revolution, centering their raison d'être on local factories, warehouses, and a regional market. But these do not qualify for the category of modern metropolis which is now shedding the impact and the forms of two centuries of industrial development and returning to the ancient functions of the city, those of the acropolis, the agora, the forum, and the academy. The concentration of transactional and ritual activities clothe the metropolis in new attire. A new crop of urban centres, rapidly achieving the status of a metropolis, rises in unexpected locations as the demand for amenities develops massively and imperiously, and as the mobility of urban masses increases. Transhumance is nowadays an urban rather than a pastoral way of life.

Many authors have been constantly reminding us that cities are people more than they are buildings. Now that material structures and facilities are so easily erected or transferred, the design of the material environment must be subordinated and inspired to express the rapid and too often neglected evolution of the population's ways of life.

DISCUSSION

THE WHITE-COLLAR CITY

Bloomberg: Professor Gottmann, you seem to put special emphasis on the contrast between the predominantly blue-collar city of the past and the emerging white-collar city of the present and future. You note the difference in amount and character of leisure as one distinction between their respective ways of life. Are there other forms in which we can see a new way of urban life emerging?

Gottmann: Certainly. But first of all, this white-collar predomi-nance is emerging very rapidly as the occupational structure changes. Thus we have another kind of people becoming much more numerous in the city and having different requirements and different interests. Let me give just one example. Recently I ascertained that probably during 1968 (unless it had already happened by the end of 1967) the number of people employed in mining establishments in the United States would fall below 600,000, while the number of people on the faculties of higher education establishments would rise to 600,000; hence we will soon have more professors, or at least instructurs, than miners. Possibly we already live in that society. Such is urban society in contemporary terms. This example of occupational change illustrates what I mean by the shift from the blue-collar to the white-collar city.

Obviously, a city with more professors than miners is going to be very different from one with more miners and operatives of manufacturing establishments than professors. This specific example is rather symbolic, but we can formulate its meaning. Our cities will be predominantly inhabited by people who will have had more and, we hope, better education. In any case, what they think of, or accept, as requisite modes of conduct will be different from what has prevailed into the present. More and more city-dwellers will express new requirements with respect to the recreation they will seek, the kinds of homes they want to have, and so on. There will also be a difference in terms of the discipline they will be prepared to accept.

Bloomberg: What do you mean by the kind of discipline they will be prepared to accept?

Gottmann: Well, I believe it is clear that we are going to have a considerable increase in the density of people in a given space, whether you take that space as the whole of the nation, as the restricted city community within municipal boundaries, or as the larger metropolitan region. Most importantly, we are heading towards a concentration of population in a small proportion of a country's total land area. As a consequence, a great many more people will be living and working at much higher densities than they formerly did.

In 1851, the density of the City of London was 185,000 inhabitants in the one square mile which constitutes the financial quarter. This kind of density in the middle of the nineteenth century was not extraordinary among the larger cities of the West. But it was terrible overpopulation, the kind of thing against which Dickens wrote, and against which the revolutions of 1848 developed; and it's one of the things that Engels wanted to strike at in the *Communist Manifesto*. Yet one may say that society needed such a high density and that people worked successfully in it.

Today, with greater affluence and greater mobility, we can spread that out so that Manhattan has a residential density of only about 75,000 inhabitants per square mile, although the density is higher, of course, in certain areas. No doubt it is much lower per square foot of floor, even in the ghettos, than it was in the overpopulated cities of the Industrial Revolution, 120 years ago. So, we could say that one kind of residential density is decreasing, which means diminished physical pressures and less social promiscuity in the home. In the predominantly rural era of the nineteenth century, however, it was only a small proportion of the total population that was crowded together. Now a majority of the people live in metropolitan areas that house between 5,000 and 30,000 per square mile. And these are densities to which most of our people are not accustomed.

Densities are, of course, higher in the Old World than in the New World, because in America you have had more cheap land at your disposal. But now you, too, have an increase of density, which at work time reaches between 100,000 and 200,000 per square mile in the business district of any substantial American city. You attempt to release some of this pressure by building vertically in the form of skyscrapers. Although this lowers the number of persons per room, you still have very high density in terms of square miles of ground. The individual must accept his "neighborhood," whether it is the 10,000 per square mile residential area or the 200,000 per square mile central business district. This situation requires a certain amount of discipline. One has to behave in a way that is appropriate to a world full of people. We are not supposed to smoke a cigarette in certain situations because we can inconvenience or endanger another person—inadvertantly burning his clothing and so forth. So there is a rule, "No Smoking in the Elevator," and we accept that. We don't feel that it is a terrible imposition on our freedom. There are also, for example, rules limiting the number of people in a meeting place. And there are many others, such as those concerning our daily behavior, which are not written in red signs on the walls, but which we learn to abide by and which reflect the densities of the modern age.

I think that educated people can accept a complex structure of rules more easily than people who are engaged in straight production activities—the blue-collar workers. They also accept a private life that can be perfectly private without having to be necessarily separate in terms of geographical neighborhood. This is very important.

EVOLUTION IN PLANNING TECHNOLOGY NEEDED

Schmandt: Recognizing these changes, do you feel that the conceptual tools and the technologies of the planners are capable at the present time of devising a spatial arrangement to accomodate these new factors and new forces so that the environment itself can remain functional?

Gottmann: I have, of course, the reputation of being rather badly disposed toward the family of planners, and I'm afraid that there has been a great deal of misunderstanding of my position. I do believe that planning is necessary, that planners can attend to, and can understand, this process of change; but I've been very worried by the fact that planners, as I've seen them, essentially have stood by the same principles and concepts that had evolved at the end of the nineteenth century. The concept of the small new town—for instance, a la Ebenezer Howard—was formulated to alleviate the terrible housing conditions of the workers at a time when densities in the city of London were 185,000.

It is true that we have to become more and more specialized; but we also have to be able to understand what the specialty connected with ours is doing, like the urban geographer and the planner, for instance. And it is not enough just to read one another's publications, nor to meet from time to time. It requires a great deal of discussion and debate to understand one another's viewpoint, or to be able to follow one another's evolution of thought. This is true whether between chemists, physicists, and biologists, or between people in research in information, in publicity, in management, in "R and D." In other words, we are all in a situation where the future of our society and the success of our work, either collectively or individually, depends very much on our capacity to communicate with other people with whom we have to work. And I think that this is one of the reasons for the skyscraper and the concentration in the big cities. And this is why it is so difficult to break up such concentrations. You haven't had as much planning legislation in

America as there has been in the big European cities. For the past forty years, there has been legislation to limit or forbid large new office buildings in the central areas of London, Paris, Amsterdam, and other major cities; yet despite such legislation the construction of these buildings has continued, circumventing the regulations by way of exceptions and special rulings. Obviously there must be a very good reason why this happens.

Bloomberg: One gets the feeling that in this case the planners were playing King Canute and telling the tide not to come in.

Gottmann: People, believing the big obstacle to be just size, say in effect, "Let's break the megalopolis up into a hundred smaller communities and in this way overcome the difficulty because the city then will be only one percent of what it is now." They fail to realize that in such case we will have a hundred obstacles to get over. I'm afraid that the technology of today, despite the progress of communication, does not favor the dispersal of work; nor am I sure that the social revolution favors such dispersal. The majority of those people who do not already belong to the several upper strata, who are not yet white-collar, want to benefit by the present fluidity and get into, and belong to, the new arising city. It's not easy; they're usually not trained for it. A great many of them will be frustrated and will not succeed. But they want to try, and they are entitled to do so.

The response of the upper strata, however, is "Let's get away from them." I'm not speaking only of the racial problem of the American cities; a similar problem existed during the Industrial Revolution in the big cities of Europe. This is probably again, but on another level, the problem of the large cities of Europe and Asia. Too many of those who support breaking the metropolis up into small towns are actually planning to flee from the mass entering the city to participate in its opportunities. The possibility for such participation exists for those who can help relieve the main shortage of the labor market today, that of well-qualified, well-trained, and well-educated workers. But instead of devising accommodations so that by living together with the newcomers, at least at work time, we could help accelerate their evolution, we say, "Let's break the city up into small towns."

THE QUESTION OF DENSITY

Schmandt: In connection with this question of density, some of the experiments with animals seem to indicate that when a certain stage of crowding is reached, dysfunctional and even pathological consequences begin to be observed in the behavior of the animals, I wonder if possibly there is a stage in the density patterns of human beings, such that once we cross this threshold, we begin to have, if not clearly dysfunctional consequences, at least radical changes in the behavior patterns of individuals.

Gottmann: There is, it seems to me, a certain variety among human beings as to their capacity to live and work at higher density. I am convinced that we can perfectly well have very high geographical densities while providing enough privacy for those people who want it. The techniques of construction allow for building perfect ivory towers within very densely occupied space. Right now we may hear what is going on in the next room—the typewriter; but architects know perfectly well how to design buildings so that we wouldn't hear anything that happens out there. As a matter of fact, such arrangement probably does not require a high expenditure. I have noticed in Purin Hall where I have stayed now for a few months (it's not a luxury building but it's apparently well constructed) that I don't hear at all anything that occurs below me or next door to me; but I hear noises in the corridor. Apparently there are certain regulations about how doors should be made—possibly fire department rules—that cause the doors to be permeable to sound.

But these are trivial technological problems. We can produce the kind of environment, man-made environment, that our instincts (the "animal" side of us) would require. But we also know, from studies of the social behavior of apes and monkeys, that certain disturbances in the family structure occur with the appearance of an outsider—a bachelor outsider, for instance—in an area populated with families that have already their accepted "gentlemen's agreement" way of life. The appearance of an outsider produces great turmoil in the established community. A city, on the other hand, knows how to deal with bachelors that come from the outside! But we are still, to some extent, disturbed by this kind of thing when it happens in great numbers. Again, I think that it is not a matter of "disorganization," but of "inorganization" of society.

You can see the same kind of problems being discussed in the

Bible at a time when nobody conceived of agglomerations approaching the size of modern urban areas, or even those of the last century. In fact, we are still fighting about the kind of moral principles that the Bible was concerned with and that concerned the old Greeks—Aristotle, Euripides, and Aristophanes. I don't think it's a matter of size and of density; it's a matter of our values and of the organization of our society.

SOCIAL COMPLEXITIES OF THE MODERN METROPOLIS

Bloomberg: I would like to proceed from that and at the same time go back to a couple of points you made before. You pointed out, dealing specifically with the world of work, that we now have much more specialization than in the past—many more categories of vocations and careers—but that we are also forced to communicate a great deal more within this great complex matrix of specializations than we had to in the relatively simpler sets of specializations in the past. But hasn't the same thing happened generally in urban life? It seems to me that in the past we had rather neat categories of people, and it was fairly easy to know where class lines were. Ethnic groups were quite well defined and lived in their own enclaves. And the communications between these relatively simple groupings were also fairly minimal. They didn't talk much with one another. I get the impression now that there are many more different, at least marginally different, styles of life than ever existed before; there are multitudes of choices: you can be a suburbanite, a semi-suburbanite or an exurbanite, or you can be a city-dweller who has a backyard or a city-dweller in an apartment. You can hobnob with many kinds of people or certain sects, or you can go off to Haight-Ashbury. There are just so many different ways of life; and yet all of them seem to be in communication with one another, through television, through the mass media, through the fact that we are all spatially so mobile. It seems to me that it is very hard for an individual to insulate himself anymore from at least an awareness of other options, which he may or may not like. And I wonder if this, then, isn't something that is happening, not just in the vocations, but in urban life generally—that we're becoming both more differentiated and yet at the same time involved in a richer kind of communication across these lines of differentiation.

Gottmann: I would agree. The richness of that communication between groups, whose differences often are mainly shadings, is an essential part of the urban evolution. It is probably one of the great problems of our century. First, physical mobility is much greater; and that puts us and the others into contact with all kinds of people with whom we previously had very little chance of having contact. Do you remember Wendell Wilkie's famous book, *One World,* which asserted that the world was shrinking? I think the world is expanding for each of us, getting many times bigger every year. We have to deal with many other people from distant lands or with specialities and interests previously unrelated to us, and they, in turn, deal with us for their work and their way of life. At the same time social mobility is also considerably increased. The result is that we have to live in a world which is full of exotic elements. That is something which society has always resented. An exotic element is fine if it is something bizarre which you can put in a museum on a pedestal, and touch, and ask questions about. But when it is a constant, hourly—not just daily—development in our life and in our work, a few of us may like it (the more adventurous kind I suppose) but to most it becomes really disturbing.

YOUTH IN THE METROPOLIS

Bloomberg: You mentioned the family, and I want to get to that in terms of the new or emerging urban way or ways of life. Isn't an amazingly large part of the younger generation now, to some degree, international? We see it, for example, in the areas of consumption: fashions, music, travel back and forth, not only across the Atlantic but also across the so-called Iron Curtain. Our young men are much more easy about going to Canada to get away from the draft, it seems to me, than at any time before, though this is still a difficult decision. I was in Canada recently and some Canadians I met were reacting by saying, "It's good to have so many of your activists coming up here. We need someone to stir us up." In both of these instances there's a kind of almost bland acceptance (as we accept the Telstar satellite communications link) of something which only a decade or two ago would have seemed absolutely outrageous or unimaginable.

Gottmann: If I may add this; one of the reasons why the big city is attractive to the young from other parts of the land, is that it permits them to break away from a community in which they are very localized, and in which everybody knows all the virtues and vices of their family. And if they want to break out of this world, they have to get away into the big mass in which they will be just newcomers, and where those with whom they mingle will not know their precise backgrounds.

Bloomberg: One more question on this plane and then I'm content to leave it. Among the youth particularly (and this goes back to something else you said) you get demands for more privacy than they've been allowed in the past, particularly in the sexual sphere, but also in terms of such things as their political choices, "doing your own thing," as they say; and at the same time you get a demand for more "community." They say they want to have more authentic relatedness to one another. They want people to be less impersonal: "Talk to the guy sitting next to you on the bus." "At least when you live on the same block, you should know each other." Is this paradoxical? Is this just the naiveté of youth believing that somehow you can have everything? Or is there some way that the urban life of the future may resolve, or could possibly resolve, the high mobility, the lack of localism, the changefulness, the privacy—resolve it with the demand that people relate in ways that the word "community" seems to suggest?

Gottman: Well here you are going beyond my competence. I have only a layman's feeling about the point you raise. At a meeting in Paris about a year and half ago—an international meeting of people concerned with urban matters—the American delegate, Buckminster Fuller, said he felt that the younger generation didn't want to hear about settlement. When he was asked, "What do they want to hear about?" he replied, "About unsettlement."

SPATIAL DISTRIBUTION:

Schmandt: This discussion raises an interesting question that is relevant to planning. It strikes me that the planner is basically oriented towards space, towards the distribution of activities, cultures, uses, in space. This suggests that as specialization increases and, as the major relationships between individuals become less spatial and more aspatial, the concepts of the planner may be incapable of dealing with these new relationships. As society becomes more highly specialized, my relationship is more to others of similar speciality in other parts of the city and other parts of the world than it is to my neighborhood; thus the space in which I live and even work becomes less relevant to me and many of my relationships. If planning remains oriented toward spatial distribution, doesn't it run into difficulty at this point in devising the kinds of arrangements that are statisfactory in terms of emerging living patterns?

Gottmann: I feel very deeply that spatial distribution alone is not enough, although, of course, you can't plan without putting things on a map. You have to make a design on paper and immediately it takes a spatial form; but the spatial form is not the whole design. Many times I've had a rather humiliating experience when trying to lecture about the matters we have been discussing. After emphasizing the social and political process which I think is the essence of our problem, and stressing the need of communication with both the other occupations that are related to ours, and the exotic elements in our society, at the end the question comes up: "But don't you think that with better communication, the better highways, and the better televisions and telephones, it isn't necessary for us to be all together, we can be farther away from one another?"

Schmandt: Propinquity becomes less important.

Gottmann: But I'm not sure it does. Isn't it incorrect to think that because you are reluctant to accept certain propinquity, this is possible for the whole society at its present stage, and therefore everyone can escape it with the help of technology?

Bloomberg: What you are saying is that in the emerging urban society you are going to be close to people spatially whether you like it or not. You now may choose to find ways of relating to them, or you may try to channel your relationships only to those people who are like you occupationally, or in some other ways, and try somehow to insulate yourself against all these other people around you.

URBAN SOCIAL RELATIONSHIPS

Gottmann: I don't think you can remain in the main flow of that society, where the action is, if you choose an ivory tower and specialize yourself very narrowly.

Bloomberg: Even if it's connected with all kinds of phones, television, and so on—with all the other . . . .

Gottmann: It's very possible, certainly, that we each have ten satellites at our personal service and still feel lonely and frustrated; physical communication does not necessarily give a feeling of belonging.

Schmandt: Does this mean, then, that community, in the sense of belonging and shared values, is much less important today than, say, the concerns of the individual and his family with their practical interests in the environment in which they live? Is this concern now more important than a sense of shared values with one's neighbors? Does "community" for the average individual mean, in other words, simply a sense of security, a sense of order, a sense of certain standards of conduct to observe—so that his own life-style is not threatened?

Gottmann: I'm afraid that you put the place of order too late in the sequence of concepts, when you say security and order. Let me recall to you that in his *Laws,* Plato outlined his conception of the ideal community. He put it on an island, and kept the whole population in the interior at some distance from the seashore. Relations with the outside, he said, should be handled by a small group of specialized civil servants. The population itself should not be in contact with the outside—you see, no exotic visitors! Everybody would keep on doing his job producing as he should; then you get good political behavior and a good society. Of course, the question arises as to what should be done when the population increases and pressure on the sources of raw materials mounts. You have to provide more land (of course, at that time they weren't thinking of skyscrapers and white-collar majorities). The answer then was that you have to get the surplus to emigrate and set itself up on another island. This strategy could provide for an infinite future assuming that there was an unlimited supply of such unoccupied islands.

This kind of theory was largely influenced by the constitution and structure of Sparta. The people were settled in relatively small towns. There was no big city and the one port that depended on Sparta was far away. This was very different from the metropolis of Athens where there was resentment of the influence of the merchants, mariners, and outside forces in the politics of the city-state. But we know, after all, that Sparta disappeared. And who remained to chant the greatness of Sparta? Only Athenians! All the virtues of Sparta come to us through the teaching of the Academy in Athens, which never had any roots in Sparta. Athens was a large city; Sparta just another village among many others. Yet the Platonic and Aristotelean philosophy has molded our thinking about what the ideal life should be. And the philosophy of planning is still impressed with it. I don't think the view it represents is any longer compatible with our modern society, where production does not have to take place in the immediate vicinity of the working and consuming population, where the whole matter of opportunity is different, and where the stability of the community over generations cannot be maintained. Perhaps after we have completed our present transition and have established a new white-collar society, with production well organized and mechanically directed by a few men outside of the city, we shall again be able to have stability through generations in the city. But this will not occur in the twentieth century, and probably not in the twenty-first.

Bloomberg: What you are saying seems to write off in large part the kind of advocacies that, for instance, Erich Fromm has made when he argues that the mentally healthy, the "sane" society, would really be a society of many small communities. And I think also of Herb Gans, who appears to be asking whether there isn't some way we could preserve, for those who want such a way of life, the little Italian enclave, the little Polish enclave, the "urban villagers" as he calls them. We seem to be writing off both of these viewpoints.

Gottman: That's true; I'm afraid I can't agree with them. And I can't agree, not because it isn't my taste or because I'm anti-Plato. It's simply because, as I study the present evolution of society, I don't see any possible way to be at peace with that evolution and still hold to their philosophy.

URBAN GOVERNMENT AND PLANNING

Schmandt: This also relates to the view that planning should seek to understand and influence the actual developmental forces at work in the urban community rather than to aim for some theoretical future metropolitan form as a goal.

Gottman: If you mean that in terms of governmental structure, then I think that there are a great many different cases. There are some city areas that grow in a way that probably will need a metropolitan form as a goal—even as a means to a different governmental goal. This doesn't mean that we eliminate the other tiers of government below. I don't think that the very big metropolis can go on annexing surrounding territory as it has done in the past. But metropolitan planning and even some consciousness of the interweaving of metropolitan interests is something that in certain cases is very important. I'm not sure it's important for every city and in every case.

Schmandt: What I'm thinking of, more than formal governmental organization, are the dynamic changes that are taking place in the development of urban and metropolitan societies. Since we know so little about the probable effect of current plans on the future, is it wise to superimpose a form or grand design on an urban agglomeration, when such action may in many ways decisively affect, either for good or for bad, the development of what then takes place? If we must seek deliberately to create a form, perhaps it would be a better idea to keep the spatial pattern so flexible that the developing forces within a society would have maximum room for emerging free of any superimposed physical arrangements that the planners devise.

Gottman: This is a very important question. I must confess that I don't have a firm view on it. I feel very strongly, however, that, on the one hand, growth and evolution, particularly at the rate, depth, and breadth we are experiencing them today in the world, needs some chaos; and it would not be wise to think that we can control and organize this chaos fully even before we have understood all the factors. And we are far from having even adequate studies of many of the important factors. Indeed, only as we go along do we discover the importance of the various elements. On the other hand, however,

I don't think we can abandon this evolution to the chaos it develops because the results would be too uncomfortable and too ugly for us. Consequently we have to try to control it; we have to try to put some order into the evolution. But we have to do this with enough humility to understand that we are trying to suggest costly solutions which will probably improve things only for a while. I don't think we can promise that we have *the* solution once and forever, or even for fifty years. For instance, I am very much in favor of new towns insofar as we recognize that such communities, planned well ahead, represent one of the best ways we now have of putting a certain amount of order in the growth of rapidly mushrooming metropolitan areas. Otherwise, if we let this sprawl develop indefinitely, it is anarchy; it is chaos; it is one mistake after another; it's ugly. Still, we have to accept the idea that probably these new towns will be outdated very soon. Meanwhile, let's still try to create them. So we accept that kind of planning which has to be done as if it were definitive, but with the idea that in fact the final solution is beyond human capacity in our time—which is more or less, if you wish, a humility to which the physical scientists have come in their concept of the universe. Even the mathematicians have now a theory about the inconsistency of the mathematical systems. We also have to accept this in planning, which is a much more complicated matter, I'm afraid, than a mathematical system.

Bloomberg: I wonder if I could suggest a "rural" analogy, just to see if I grasp the point. What you appear to be saying is that we have to talk, perhaps, about trying to domesticate the evolutionary process rather than completely controlling it. You may begin by saying that the cow is an important source of food but it would be nice if it had shorter horns, a bigger udder, and more meat on it. And you may seek to breed it to acquire those characteristics. Later on you may find that changing food tastes and new food stuffs makes the cow itself seem somewhat obsolete. But in the meantime you haven't created a whole new animal. You haven't tried to play God. You haven't tried to control evolution; but you have, within the evolutionary production of this kind of animal, made some adjustments and changes in the character of the beast.

Gottmann: Very much so. Very much so.

Schmandt: Let me give you an example of this. In this country public policy has been pretty much committed to the revitalization and preservation of the central business district and oriented strongly to the concept of central place—the aggregation of cultural and other activities in a civic center. As time passes, the more we invest in this pattern, the more we perpetuate it. Suppose, however, that the developmental forces—social, economic, and political—seem increasingly to contradict the concept of central place and central business district. And suppose there is growing pressure for the dispersion, let's say, of cultural and other forms of activity, to provide greater opportunity and greater freedom of choice to more individuals. As these developments occur, would we not find it extremely difficult to change because of the heavy capital investment we have placed in patterns which may, from an evolutionary point of view, already be obsolete?

Gottmann: Of course, with all that is happening now, we have to ask our financial specialists also to accept the idea of taking greater risks. They accept the risk very readily in terms of weaponry that becomes obsolete, knowing that in some cases it lasts for a century, while in others it may become obsolete in five years. I think we have to develop the same attitude for our cities. I think that some of the attempts of cities to preserve certain images of what kind of places or communities they are, amount to fighting against a definite trend in such a way that the struggle itself almost threatens to erase those values most worth preserving. In Scranton and Wilkes-Barre, Pennsylvania, for example, local people are still trying to revitalize these cities. And this seems to me little justified. However, they want to have their place in the competition, in the race. Do we have the right to tell them that they have no chance, because we have seen many trends developing that seem to indicate that such is the case?

I don't think that every old center is doomed. Some may still be able to survive possibly for a century, or even two or three. Which ones? It's hard to predict because it depends in part on their consciousness of the fact of competition. And it depends also on how much their rivals will try to get ahead of them. You recall the situation following World War II—in the years 1946 to 1948—when people were saying that New York was old, had too much congestion, and so on. And what was the interest of being on the seashore in the air age? You could just as well be in a central position on the continent. People were saying that Chicago, possibly Denver, were the big cities of the future. Several New York corporations bought land around Chicago and Denver to build new headquarters there since "Manhattan was too congested" and "belonged to another time." Yet we all know what happened: New York has stood its ground fairly well until now. And if it has not stood its ground well enough, it's because of Los Angeles! Perhaps some of the old predictions will be true given another fifty to seventy years of attrition of New York's dominance, chiefly by Los Angeles—and perhaps by Detroit, although in the latter case the possibility is much more fragile. Certainly nobody foresaw this pattern in the 1940's. Chicago really did not come up to the hopes based on its central position and already acquired organization of large interests. These are recent trends and it's still difficult to foresee how patterns will develop. I'm sure you could find similar indeterminacy on a lower regional scale between much smaller towns or cities.

Bloomberg: This raises the question of some of the kinds of critical policy alternatives that face the leaders in the various institutions in these great megalopolitan settlements we're developing. One of them is, of course, do you stay or do you leave? This conversation has brought another to my mind because I'm particularly struck by your example of obsolescent war equipment. It does seem true that, if we construct a building for a few million dollars and it doesn't get fully utilized or something else about it turns out wrong, we treat it as an error so massive and so terrible that we never can recover from it. Yet, we can take one missile of the same cost and if it goes off halfway up and explodes, we say, "Oh well, that's part of the game!" Isn't this a kind of policy alternative that people aren't even talking about as yet? One alternative would be, say, to build for risk, to build knowing that you may be wrong, knowing that in five years or ten years it may be obsolete, but asserting that it's still worth the investment of a lot of resources because maybe it will last, too. Can we set this against the policy alternative which insists that somehow we act as if we were always building for permanency? Are these really basic policy alternatives: do we build for permanence or do we build for risk and perhaps change? Do we stay or do we leave? What are some of the other critical policy alternatives that we face as we move on into this future?

Gottmann: Well, I'm afraid there is a great variety of alternatives. But may I reverse the question and put it in this form? What is it that limits the variety of choice? I think that what limits it returns us to something Professor Schmandt said a few minutes ago about the security we are all looking for. Then, there also is the need to provide for the opportunity to get ahead, in addition to being secure and safe. The difficulty which limits us is that we have to strike a certain balance between security and opportunity. This is very basic in political geography and probably in politics in general. For the planners to recognize this takes a very big, massive step. In all their years of work they have not accepted the competition between cities and between individuals to the same extent that they have accepted competition between two armies, or baseball teams, or competition between corporate firms. We have to accept this fact of competition and allow it opportunity to function. Obviously, if we permit emerging competition, a great deal of opportunity, the established structure will be frightened and will want to defend itself against disestablishment. The debate about reapportionment is one of the facets of this. Why hadn't there been more reapportionment? Because the political structure was trying to keep secure and it felt frightened.

Reapportionment opens the way to a new distribution of political weights throughout the country. Curiously, recent analysis has found that the results of reapportionment, particularly of state legislatures, work quite differently from what was expected. Instead of having the clear-cut dichotomy of urban and rural, there is also a major suburban element, which often aligns itself with the "country boy" districts against the urban. Moreover, reapportionment has apparently benefited the Republican party more, on the whole, than the Democratic party, which is contrary to what was expected. If it is so difficult to plan that kind of simple arithmetic, think how much more difficult it is to foresee the evolution of the city as a whole. We know its potential in all its diversity. But the balance between opportunity and security is a stumbling block; and perhaps it has to be explained not only to a few experts but to a mass of people yet unconscious of it. Then people may decide they don't want to sacrifice all or most of their opportunities for the sake of security alone.

Schmandt: The discussion about security and opportunity raises a question that is being increasingly talked about here. The opportunity of the citizen to participate in the planning and decision-making process. The movement in this country is towards what some refer to as advocacy planning or the attachment of professionals to such groups as neighborhood associations. With the aid of these professionals, community organizations are enabled to confront the established planning profession, at least that portion attached to the public sector, with demands, ideas, and propositions that may not accord with what the more conventional elements in the profession feel is sound planning, and which may even threaten the security of the planning establishment as such. I wonder what your feelings are in this respect.

Gottmann: May I, again, answer by a little detour, because it's a very difficult question on which I don't have a clear feeling. The last two years or so, I've been consulted in some capacity, but not always by official governmental authorities, about problems connected with the planning of New York, Paris, London, and Tokyo—four of the biggest cities which have known both successes and problems in planning. I think it is true that where a new setup in the planning structure was permitted, the best progress developed—not only in terms of the city generally but even for the solution of a given problem. In other words, I believe that the planning structure, like many other structures, needs some kind of rejuvenation. And by introducing either a new institution or at least a new method, by creating a little revolution in the structure, some improvement is likely. This is, I suppose, because any established structure has a tendency to believe in concepts that are already outdated and may even be very backward in a changing society—and also because it allies itself with the other structures, such as political and welfare establishments, that are trying to defend themselves against needed change.

THE POSSIBILITIES OF PARTICIPATION

Bloomberg: So that, in a sense, one source of rejuvenation would be a positive reaction to the demands for direct representation in policy-making of what were once thought to be purely recipient populations—poor people, students, newcomers—who now begin to organize and begin to insist that the planners not only listen to them, but maybe that they ought to have their own planners. Instead of a single planning agency you would get a competition between planners or planning agencies which have their bases in different kinds of constituencies. This would be something quite different from what we've had in the past, wouldn't it?

Gottmann: I think so; and I also think that it's important to get people involved with the problem, because there clearly is an evolution of planning underway and there should be more talk about it. It's not enough to get only the experts involved. It is essential to get the people involved, to have them discuss it. Then things become easier for the experts to do and for the politicians who do see need for change, because at least there are different opinions expressed. There is debate.

Bloomberg: But what usually happens is that, when the plan is finished, the planning agency says, "Now we will have some public hearings." Anybody who wants to object to this plan can show up at a given time in such-and-such a room. That's what they call debate and discussion. It turns out to be very minimal, of course, since the plan is already prepared. The poor citizen with no extra knowledge comes in and gets showered with statistics and charts and graphs. Schmandt seems to be proposing that the discussion begin at a much earlier state and that the citizen, if he is willing to join a group, have his own planner to generate his own statistics on behalf of his values, and his own charts and his own graphs, very much the way labor unions do in time-study and wage negotiations. Management used to assert that they were the only ones who could set wages, not only because they had the "sovereignty" but also because they alone possessed the expertise. And then labor unions began to hire their own economists, their own time-study experts; and the bargaining became equalized. We've never had any kind of equalization in the bargaining between the planners and the politicians for whom they worked, on the one hand, and all kinds of people who were affected by the plans and programs. Do you really think this could now happen?

Gottmann: It's very difficult to get the people, the whole body of the people, to participate. But you can get people to become more consicious of things and get at least their own leadership to participate. You can get the newspapers and the television and the radio to begin discussing such matters even if they don't do it in a completely disinterested way. But at least from the discussion, things will arise. The assertion that the big city has to be evil is a great, sick assumption of modern society. The major cities are growing more numerous and even bigger, and obviously the whole trend is toward the very large city. Yet we continue saying: "It's evil. Oh, my God, keep us from it!"

Bloomberg: A very ancient complaint.

Gottman: Of course the ancient complaint was against Babylon, which never exceeded, as far as we know, 200,000 people. And planners today say that a city of 200,000 people is the ideal size!

THE COMPUTER AND THE CITY

Bloomberg: That's a good point, too. Is the computer another source that might be a rejuvenating change?

Gottmann: I've been systematically asking people in management and people in the advanced electronics field whether they feel that the computer and some of the other advances—also the improvement of the switchboard for closed circuit television and so on—will help us a great deal to decentralize the decision-making process. And until now the usual answer is "no." On the contrary, the capacity of concentrating more data and having it processed in a few minutes has itself helped to concentrate management. I have heard the same on both sides of the Atlantic. I think it is a great illusion to believe the planning process will be transformed because we have such improvement in the technology of what we call information storage and retrieval. You know, it's very interesting: the first cities arose at the same time that recorded history began, that written records started—the tablets, the inscriptions, papyri. And it looks as if, today, a new dimension of organization in the twentieth century also is starting, with a completely new basic revolution in the technology of records, as all the inscriptions and writing go now on tape or into the computer. It's a curious coincidence—twice in history.

Schmandt: It seems to me you are saying that there are really two diverse forces at work today: one embodies technology and information, retrieval systems, the use of the computer, and so forth, and actually tends towards centralized decision-making; the other is a strong countermovement towards greater involvement, greater participation, greater division of labor, which in turn leads towards (or which logically would lead toward) greater emphasis on decentralized decision-making.

Gottmann: It's decentralized in terms of the number of people participating in the decision. But it's not decentralized in terms of their spatial distribution because, if they have to participate, they want to be closer to those who command.

Bloomberg: Let me just suggest, though, that the technological revolution is still going on; and it is true, for instance, that with the new off-line technology, neighborhood organizations, conceivably ten years from now—not even twenty or thirty but ten years from now—will have access to highly sophisticated computer systems. We're about to establish in Wisconsin a state computer system which will serve all the universities simultaneously, as well as all of the state government and the major local governments. Presumably its capacity will be that large. Now it seems possible to me that a large neighborhood organization, that is busy trying to generate its own counterplans to those of the city or country government, would, through its own off-line equipment, be able to utilize such a central computer if it had on its staff, even on a part-time basis, the necessary expertise—perhaps a university faculty person and a computer expert. Then it could come into the hearings with its own printouts. I think this may be what Professor Schmandt has in mind. I don't know if this is contrary to the impression that you have.

Gottmann: No. I think we agree on that.

Bloomberg: That still means that the decision is finally going to be made in one place sometime, somehow; but it certainly makes decision-making much more complex. It adds more sequences to the process.

PLURALISM IN DECISION-MAKING

Gottmann: And there, again, you see the competition aspect. Your decision is no longer independent of the decision of a competing organization. It is not a monolithic structure. This pluralism in decision-making then means that you want to have the very best advisors. And they ought to be such that you will get at least as good advice as any possible competitor will get. And that builds up, one might say, into the skyline.

Schmandt: This whole trend seems to indicate that basic changes must occur in the institution of planning and in the planning profession itself. Planners must be willing to accommodate themselves to these new developments; they must be willing to become involved in the competitive arena of ideas and to step beyond the narrow ideational and conceptual confines of their profession or discipline.

Bloomberg: We're back to humility in a way—in a very profound sense. You know, I'm still reflecting on some of the things you said earlier, Professor Gottmann: that urban man for a long time was thought of as the controlling being who no longer believed in God, no longer necessarily believed in faith, who had great powers to control his destiny, the great machines of the city. It seems to me you suggested that, like many intellectuals, the decision-makers increasingly are beginning to see that urban man, if not at the mercy of some god, or some fate, is nevertheless still very limited in his capacity to determine and control his own history, even for relatively short periods of time. And this suggests a kind of "urban virtue," if you want.

Gottmann: I think so. That is the basis of freedom. And it is also, I think, the reason why we want to move so much in the modern city, both as individuals and as groups—because of our search for an inner life that we doubt can be planned for us but which provides security. That is probably what people are asking for when they talk about privacy, community, and so on.

Bloomberg: This suggests that *1984* is a demonic version of what is potential in the city, the totally controlled, monolithically established social order. But what you see emerging is really not . . . .

Gottmann: Much more in the sense of plurality and complexity; and out of that plurality comes the need of more inner life and some kind of inner security.

Bloomberg: We have, then, an alternative that is neither Huxley's *Brave New World* nor Orwell's *1984* but a kind of urbanism which in the experience of human beings will embody a lot of very ancient, long-lasting convictions we've had, at least in the Western tradition, about what constitutes the virtuous human community.

Gottmann: And we need a good novelist to portray in contemporary and futuristic terms what living in such a community might be like.

REFERENCES

ELIAS, C. E. et al. [eds.] (1964) Metropolis: Values in Conflict. Belmont, Calif.: Wadsworth.

GOTTMANN, JEAN (1966a) "Why the skyscrapper?" Geographical Review 56 (April): 190-212.

———(1966b) "The ethics of living at high densities." Ekistics 21 (February): 141-145.

MARTINDALE, DON (1958) "Prefatory remarks: the theory of the city." PP. 9-62 in Don Martindale and Gertrud Neuwirth (trans.) Max Weber, *The City.* Glencoe: Free Press.

Ministry of Housing and Local Government (1964) The South-East Study: 1961-1981. London: H. M. Stationery Office.

New York City Planning Commission (1964) The Future by Design: A Symposium on the Considerations Underlying the Development of a Comprehensive Plan for the City of New York. New York: City Planning Commission.

Premier Ministre (1963) Avant-Projet de Programme Duodécennal pour la Région de Paris. Paris.

Révue Française (1966) Special issue on urbanism. Vol. 56, No. 2.

ULLMAN, EDWARD L. (1954) "Amenities as a factor in regional growth." Geographical Review 44 (January): 119-132.

U.S. House of Representatives, 89th Congress, 1st Session (1965) "Message from the President of the United States relative to the problems and future of the central city and its suburbs." Washington, D.C.: Government Printing Office.

VERNON, RAYMOND (1959) The Changing Function of the Central City. New York: Committee for Economic Development.

4

Urbanization and Social Character

Notes on the American as Citizen

SCOTT GREER

□ ALL OF US ARE aware of the immense increase in the proportion of the population living in cities. Indeed, if we have paid attention to the literature on urbanization we have probably memorized the figures by now, realizing that the United States is over 70 percent urban, that the 64,000,000 new citizens expected by 1980 will be urban dwellers, and 80 percent of them suburbanites (Hauser, 1960: ch. 4). But the importance of this piling up of people in cities is by no means clear and the whole concept of "urbanization" obscures as much as it reveals the nature of our society. For our urbanization is not the dense concentration of the nineteenth century factory city—its growing edges are the open neighborhoods of suburbia. Nor is the way of urban man sharply dichotomized from that of his relative in the country—often he has moved directly from a village to a suburb, visually and socially little different from his previous home. "Urbanization" is a loose summary word, absorbing many meanings and pointing to different things. Let us see how it can be made into a working tool.

Cities are possible because of a geographical division of labor. Briefly, urban dwellers have not extracted the raw materials from the soil; they have bought, stolen, or taxed these resources from countrymen. Thus cities depend upon the agricultural hinterland. In the past they were never dominant, for the countrymen could not produce enough surplus; urbanites were a small class released from bondage to soil and the seasons by the work of their rural cousins. Our enormously increased urban population today rests upon the increased productivity of the extractive industries; this has allowed us to reverse, in less than one hundred years, the proportion of the American population dwelling in cities.

The building of cities does not, in itself, change the basic structure of human life (Greer, 1962). It is change in the basic political economy of the society which allows—indeed forces—the creation and growth of cities. Two meanings wrapped up in the term "urbanization" can be separated in view of this fact: the first is simply the "concentration of people in cities." The second, and more basic, is the social transformation that releases manpower from an agriculture which remains adequate to sustain large and growing urban centers.

The release of labor from extractive industries is made possible by two intertwined developments. The first is the use of nonhuman energies harnessed to machines as substitutes for the human beast of burden. The second is the coordination of human behavior in larger and larger networks of interdependence. The two interact. Nonhuman energy makes possible the release of men from manual work and, equally important, the constraints of distance: it allows rapid communication and the coordination of behavior over greater spans than was possible before. This, in turn, allows for the development of giant armies, corporations, nations, and other functional groups. The latter hasten the tendency to exploit nonhuman energy and the machine. Thus, the two developments of technology, physical and social, result in an extension of the network of interdependence, the social interdependence of large groups cemented by a common dependence upon the man-machine-resource nexus sustaining the entire society. So close is this dependence that none of our great cities could survive more than a few weeks without the national market, while few (and deviant) are the individuals who could get a bare subsistance from the earth on their own.

These radical transformations of society have been given the summary name, "increase in societal scale" (Wilson, 1945). They have two major consequences for the ordinary citizen: they increase his dependence upon a large-scale system and they free him to a large degree from dependence upon the near-at-hand, the neighbors and neighborhood and parish of his immediate surroundings. But his freedom is not from the boundaries of social groups per se; it is simply a shift from the constraints of Main Street and the neighbors to the control system of the bureaucracy. For, with increasing scale the major work of the society is delegated to large task forces. These, formally organized and rationally oriented towards production, distribution, and survival, uniformly take the shape of bureaucracies. In them the citizen holds office, affirms norms, receives punishment and rewards, and achieves (or fails) his destiny. Upon them he is dependent.

Thus, underlying the process which concentrates people in cities is the major process of *organizational transformation*. This is a process which can best be measured by its effects upon a total society: cities and hinterlands are equally affected in the process. In setting the conditions within which one can earn a living, enjoy citizenship, procreate and be protected, the organizational structure is a major clue to the way people will behave. From interdependence among ourselves we learn to organize our actions, controlling ourselves and each other. The organizational system tells us, in short, what men must do if they are to hold a job in the society (Greer, 1955). (Thus we require of American workers a rigorous attention to chronological time, to efficiency, to the subtle cues of interpersonal blandishment and bargaining, with a weather eye to the "rules.") But beyond the requirements made upon the individual citizen are other constraints derived from our social organization: the mutual dependence of the economy, the polity and the family system, separately and interacting, set the limits of what we can and must do *collectively*. The Great Depression indicated the force the polity will exert upon the economy when unemployment grows to disastrous proportions, while the present "war on poverty" indicates the ways in which the economy *forces* issues upon the polity. At the juncture of the great organizational segments of the society, where the President treats with the directors of big steel and the United Steelworkers of America, the interdependence of giant organizations (and our common dependence on their coordination) is dramatic and clear.

Urbanization stands, then, for the concentration of people in cities and the processes which lead them there. Concealed in the term, however, and frequently assumed without being made explicit, is another meaning—a certain kind and quality of living which we call "urbanity." Derived from the same linguistic roots, "urbanity" indicates a life-style which was once automatically associated with the dwellers in cities. It meant that those who dwell in cities become more sophisticated, accustomed to variety and wont to see the world in a larger perspective. Contrasted with the peasant or countryman, the urbanite was free from the automatic demands of custom and habit, the brute pressure of "the nuzzling herd society of the village." Literate and exposed to literature, he was both more "root-less" and more "cosmopolitan" than the yokel. Urbanism, in this perspective, is the cultural parallel of increasing scale: the urbanite was envisioned as a citizen of a larger universe because he lived at the meeting place of many disparate and distant worlds.

In societies of the past, energy-poor and dependent upon the peasant village for surplus, the three meanings implied in "urbaniza-tion" were closely related in the empirical instant. The concentration of people in cities required a sufficient scale of organization and surplus energy to provide dependable fodder for Caesar's herds, and the people of the cities (the "urbs") did indeed become different in their culture and point of view from the daemonical universe of the peasant villager. They worked at the structure of controlling the larger area of city state or empire through the army, the state bureau-cracy, the church, and the market. Their cities rose and collapsed with their varying success in controlling the peasants. But meanwhile, everywhere underlying the urban society, was the vast majority of mankind still chained to village, province, *latafundia*—a low energy source but the *sine qua none* for the entire system.

Today the matter is otherwise. As new sources of energy surplus have been harnessed to our machines, farming has become our most highly mechanized industry. The number of people actually working at agriculture is about the same as the number of the unemployed in the United States today, and only a small minority of the workforce stands for the tradition of village and country. The massive mode of the society is the employee of some exclusive membership organiza-tion, living in a giant metropolitan area and moving (or hoping to move) his residence towards its outskirts. We cannot define his way of life as "urbanity," for his country kin live in very much the same fashion as he does. All we can say is that the culture of urban man is radically transformed. All men are exposed to the communication flows of the urban society, a necessity flowing from their depend-ence upon the market and the state. Thus the city is no longer either the source of increasing societal scale, nor a prime shaper of the dominant culture. It is part of the more inclusive natonal system, and we must bear in mind its limits within, and its dependence upon, that system.

SOCIAL CHANGE AND SOCIAL CHARACTER

The culture of urban man is a major clue to his social character. For culture, that general agreement within a group concerning what the world is and what should be done in it, limits and determines the vocabulary in which he can communicate with others and formulate his own course of action. Its sources as diverse as dreams, images, logic, and misunderstanding, culture is a set of definitions which can be expressed symbolically with a common meaning among men. Of all the traits which could enter into the common storehouses, various and exotic as the findings of the anthropologists, only some can be accepted and used in any given cultural group. That which is accepted will pass through a history of natural selection. Favored by its resemblance to the existing culture, it will survive insofar as it is useful to the members of the group in playing their inescapable roles. As new traits are incorporated into the common culture, they add to the bases for communication and the ordering of individual behavior oriented to the rewards and punishments provided by the group; they become new limits.

The individual's character as a social actor will be intimately related to this preexisting common culture. For social character is nothing more than the probable behavior of an individual when confronted with a given kind of life situation. The Balinese go to sleep in the presence of overwhelming danger; the crew of the Texas City refineries remained at their station to maximize control for the common safety, though their entire environment might explode at any moment; the Hindu prays before the bayonet. However intertwined in the life and nervous system of the individual, each of these courses of action is a learned response to a kind of social situation, a statement of fact and of moral imperatives. Thus social character reflects the common culture, and can best be identified through social behavior, including verbal behavior. It is always a constructed pattern inferred from observations over time—as abstract as the id and as concrete as the prejudices which limit and frustrate our aims.

The basic importance of social character to the collective and its fortunes is this: it is one set of limits upon what men can understand and find it necessary to do. This means that the average social character of an aggregate tells us what can be assumed as common before clustered people form into a structured group. It provides the beads which can be strung on many different sorts of organizational wire. Social characters tolerating or condemning the game of gossip are major considerations in forming the kaffeeklatch groups of Levittown, while the most scientific public administration practices fail in a society of feudal sheiks, whose nepotism is morally prescribed. Moore and Feldman have recently dramatized the consequences for economic development of a social character "uncommitted to labor" while those attempting to change fertility rates in traditional societies are well aware of the characterological barriers to sterility. Social character is a "lurking variable," an old palimpsest on the supposed tabula rasa which frequently makes nonsense of our efforts to rationalize the snarled affairs of a society.

While social character is a major basis for commonality in a social collective, it is more. Unlike the common culture, it points to patterns of *individual variation* in definitions and goals. In any collective large enough to be called a society there will be variation in the common culture, hence the character, of the important subgroups. Even in hunting and gathering societies the differences by age and sex are marked: the general unifying culture is subtly modified as it applies to the warrior, the old woman, the child. That which is accepted as general social fact and that which is a categorical imperative for action separate: the latter is specific oftentimes to the role. Thus a woman in New Guinea or the Tennessee hill country may know that blood revenge is required in a certain situation without feeling any compulsion to execute it. Social character varies enormously within a society—by one's subgroup, and by one's role within that subgroup.

This variation is a major resource for the society, for it supplies an array of alternative possibilities for social action. Small group research projects have made clear the variation in social types which take leadership roles under routine as against crisis conditions, while Pareto and Machiavelli speak of Lions versus the Foxes. And from another perspective, Levy points out the brilliant coup of the modernizing Japanese, who enrolled the Samurai estate, en masse, in the police force. This preexisting social character fit nicely the requirements of the function, and the dangerous Samurai had a job in modern Japan.

Beyond the differentiation of social character brought on by variations in group and role, there are other processes tending to differentiate. It is important to remember that we have defined social character as the probability of an individual's acting a certain way *in a certain situation.* But the social situations any actor faces will vary enormously, and what holds for one in a given situation may not hold for another. Thus the "role-set," as Merton has called it, may include for an individual the roles of leader, follower, father, son, provider and client, ally and competitor—the list is nearly endless. Under the circumstances, there are many opportunities for error if the roles are not clearly defined and the situations clearly identified. The literature of the American frontier is replete with the pioneer father who treated his household as a factor of production, while Hughes underlines the temptation for French-Canadian foremen to treat their kinsmen as kin and *not* as factors of production (Hughes, 1943). The interaction of roles in the life-space of the same person may result in all kinds of "sports," while his effort to organize behavior with consistency from situation to situation may be still another dynamic. Fundamentalist in politics and religion, the small businessman is often the victim of consistency in his business policies, harvesting his marginal existence in the form of bankruptcy.

These are, then, some of the causes of variation in social character. As societies increase in scale, variation increases. The different kinds of work, of life-style, of ethnic background, tend to maximize the number and variety of subgroups which interact intensely among themselves and have little to do with other such groups. Massive migration into the larger society and within it, the fractionation of the job structure (from a few dozen titles to 40,000) and increasing choice resulting from leisure, wealth, and education, allow immense opportunities for variation in social character. At the same time, however, the increasing inclusion of the entire population in a large-scale national state and national market increases the uniformity of communications via the mass media and mass education. From variation among small groups segregated by region, class, race, and religion, we move to variations among segments of one integrated system. Integration is important; it is the necessary condition for the persistence of the entire social system.

THE DISTANCE FROM PAST TO PRESENT

Robinson Jeffers says in one of his poems that he would have men keep apart from each other in small groups, so that occasional madness would not infect the whole. And for most of mankind, during most of its career on earth, this has been the pattern. The hunting and gathering societies that have dominated our history perforce had to stay apart; grazers, they literally lived from hand to mouth. With an energy source providing much less than one percent of that available to the average American today, they stayed in motion most of the time; like hummingbirds, metabolism was their master. Only with the agricultural revolution of a few millenia past was it possible for man to settle down with a modicum of security and the resources to develop, eventually, city-states and empires.

It is important, therefore, to remember that most men, through most of our recent history, have spent their lives in what Turner calls "the daemonic world of the peasant village." They had, in a sense, a long childhood. Tied to the soil, illiterate, receiving the common culture by word of mouth from the older members, the individual was born, lived and died in a tiny circle of earth. With powerful traditions and a frozen technology suited to little more than a subsistence economy, he acted with fatalism, piety, and a superstituous fear of the unknown which hovered around him. Investing the world with magic powers, he believed implicitly in immanent justice—the automatic punishment of transgression by the very nature of things. When his limited expectations were fulfilled he was, presumably, content. When famine, war, and plague descended he accommodated, to persist or to perish.

To be sure, empires were erected upon the economy which the peasant maintained. Armies marched and countermarched, cities were built and looted, and high cultures were developed within their walls. But this involved only a few persons, at the most perhaps one out of ten. For most men the capital city was as far away as the moon. Neither the literate world of the high culture nor the role of responsible citizen and leader meant much to the traditional peasant villager. His character was to maintain the older traditions of work and family, and to accept, with piety or fatalism, the hardships imposed by armed men, believing it beyond his powers to control or destroy them. His religion justified some of the world's way with him (Turner, 1941, especially vol. 2, ch. 20).

This state of affairs persisted, with minor variations, among all the "higher civilizations" until a few hundred years ago. It is still the norm in much of the world—the 60,000,000 Indian villagers who live on no road are the thing itself, as are millions of others in Africa, Spain, South Asia, and Latin America. But in certain parts of the world nonhuman energy resources of enormous surplus, the fossil fuels, were discovered and harnessed to machines (Cottrell, 1955). Increased power allowed for the rapid extension of networks of interdependence; social organization increased in its extent and the intensity of control; the modern world emerged in western Europe and America. As it did so, the common culture changed from that of the peasant village to that of the rational, productive, society.

From the peasant village to industrial society the distance was, for the ordinary man, probably as great as that from the Shoshone Indians to Imperial Rome. Among the western European nations where the change has been best documented, we see the complete destruction of the earlier world in a period of less than two hundred years. Beginning with the creation of national markets and the rationalization of agriculture, the "great transformation" destroyed the subsistence economies of the peasants through improvements in production which drove them from the land. Arriving in the cities, they provided immense pools of cheap labor useful for the rapidly developing factory and market system. The products of the latter, in turn, were traded back to the hinterland, joining city and country in an indissoluble link of economic interdependence. The bifurcated society of city and village was fused in a national whole.

The peasant villager was poorly prepared for this transition. It is doubtful that anything less violent than hunger could have driven so many into the chaotic social world of the cities, for that world took an immense toll of life, family security, health, and happiness. The casualties provide Marx's most vivid documentation of capital and its evils. But the destruction of the village as a social security system forced the peasantry to make their "great leap forward" into the world of rational production. It forced them to subsidize, through their weak bargaining position, the rapid expansion of European and American industrial society.

In the process the social character of the average man was forced to change, for he had to learn the common culture of city and factory in order to survive. He learned that work was instrumental, negotiable, and contractual. Beyond the bargain he had no protection. He learned that the public order was problematic; in the stews and sinks of London he learned to take account of the law and of that peculiarly urban social form, the police. (Not that he worried so much about lack of police protection; like the bottom dogs of our contemporary cities, his was the opposite need—protection *from* the police.) He learned to focus upon the market and its prices; his home, livelihood, family, and future depended upon the prices for labor and commodities.

In short, men became rational producers. The process of rationalization is, in this context, the separation of means from ends which allows one to estimate the fit between the two. Men moved from traditional, habit-bound, time-honored acts which guaranteed a livelihood, to rationally designed and prescribed acts which had an end reward—a paycheck. The reward of pleasure in the act itself frequently disappeared; what could be gotten from a tense and near-hysterical fourteen hour day of loom operation except the end reward, the paycheck (Frank, 1948)? This separation of the task into disvalued means and split ends was the source of Marx's concept of *alienation,* for it left the worker without pride or satisfaction in his work. From Durkheim's point of view, the rationalization of factory labor had another consequence; in destroying the value of the task as action in and for itself, rationalization destroyed the norms for behavior. "Anomie" means, precisely, the "lack of norms," or limits. The limits that remained were not visible in the rules for behavior during the performance of the task; they could only be inferred through foresight and planning. The net result was the devaluation of work as a good thing in itself and the definition of work as a *means* to an *end.*

The resulting social character of industrial man was one which fit the task forces of production and a life in the city maze. From the functional requirement that men be coordinated in production developed the limited work role, the job. This required self-control, for one must be at work on time, must carry out his role on time and at given levels of accuracy, and must remain at work while the steam shafts rotate. From the requirement of balancing labor market and commodity market developed the household budget: week by week and year by year the necessary costs and income must be balanced on pain of hunger, disease, or the debtor's prison. And from the generally low level of income (one which rose little or not at all for decades) came the most rigid requirement of all, the necessity for parsimony. The norm of efficiency deified in mechanical production was thus translated into the everyday working rules of the household, exalting the cheapest articles of consumption. In brief, the social character of early industrial man was one emphasizing self-control and abnegation.

Much, however, did not really change. There is, after all, more to life than the acts of producing and consuming, the major spheres where new learning and a new character were required. There remained such various realms as public affairs, religion, the high culture, kinship, and social class. In these matters the social character of the peasant villager could and did persist in many ways.

Urbanization of his residence and mechanization of his job did not elevate the political role of the ordinary citizen. Illiterate and accustomed to compliance, he merely learned the self-controls necessary to avoid trouble with the police, the army, the wealthy, and others who dominated the urban scene. Having no tradition of public responsibility or participation, he became only a consumer of spectacles and a spectator of consumption. Nor was he able to participate in the high culture of the society. Uneducated, without security or leisure, he found his high culture in religion, where he was a poor consumer. Only in the deviant sects and cults did he find an opportunity and reward for active participation, and the reward was less intellectual than social. With his family he remained traditional and authoritarian, struggling to maintain the older kinship norms even though the network of kin was attenuated by migration and child labor. Nevertheless, his escape from the village probably facilitated his conservative authoritarianism in many ways, for he was free from the pressure of a close community and its public opinion.

The foregoing is a highly simplified view of social character in the early phases of industrial society. Emphasizing the mode, it does not do justice to the increasing differentiation of the population which followed upon the increasingly complex division of labor—a differentiation which resulted in the proliferation of social types. The peasants remained, a dwindling minority oriented to traditional, non-contractual, and habitual social roles, along with their reciprocal types, the landed gentry and the rentiers. In the cities the free professional man, expert and small businessman at once, became important, along with the salaried professional of government and private business. But cutting across all roles were the dominant concerns of the dominant organizations, the rapidly expanding empires of production and the market (Williams, 1960: 382).

The norms of production—self-control and enterprise—had some surprising consequences for the behavior of the workers drafted from farm to factory. Segregated in the working-class slums, thrown together in the factories, sharing a common lot and a common fate, the urban working class in every industrializing country began to experiment with organization and collective bargaining. The very demands of rational production encouraged such a development, for literacy and technical competence, organization and organizational skills, were taught in the processes of work. The new common culture of working-class man was thus created, to be articulated by the emerging working-class leaders, the craftsmen of physical and social organization. It was a culture which incorporated the norms of self-control, parsimony, planning: it accompanied a social character oriented towards work, security, and equality. Its consequences were apparent in the increasing control the workers exerted over the conditions of production. Out of common fate had evolved a common culture; with common enemies came the creation of a social class. All of the differences in life situation came together under the generalizing term "the working class," and Victorian polemics are studded with references to the "labor problem." Economic weakness, common enemies, the life of factory and slum provided the organizational leverage; a family culture based upon traditional kinship, religion, and thrift provided common folk values; individual weakness and inability to play public roles provided the basis for a corporate protest (Perlman and Taft, 1935; Greer, 1955).

The effects of the emerging industrial working-class culture were not confined to the situation on the job. As the society was increasingly committed to the complex interdependence of the urban factory and the urban market, it became increasingly dependent upon the orderly performance of tasks by the workers. Their technological monopoly was translated into political pressure through the strike; their organizations, resting on common fate, brought increasing pressure on the party organizations. They fought for two resources beyond trade unionism: formal education and the vote.

In their victory they achieved the most important social revolution of all time. They forced the society to treat them as citizens, through their achievement of the political franchise; and to treat them as responsible actors, through their achievement of the intellectual franchise, universal education. But their success was complementary to the inexorable demands of developing industrial society, a society that could not be operated without the assurance of proper behavior from its vast work force of manual workers whom the political franchise involved organically in the status quo. Nor were illiterate workmen capable of the increasingly complex work roles that the productive system kept throwing up; mass education was a necessary working tool for the entire economy.

The end result was a vast increase in the "internal inclusiveness of the society," as noted by Wendell Bell (Bell, forthcoming). The various forms of partial citizenship decreased in their actual distribution and were condemned in principle, though they persisted in the differential handicaps of the poor, the illiterate, the rural, and the ethnic. At the same time, other blessings and burdens of citizenship were more widely distributed. Universal manhood conscription may not seem an increase in citizenship—until one remembers how many regimes could never have dared trust weapons in the hands of the entire male citizenry. Universal taxation is another ambiguous gain, yet it forces the political beneficiary to look at the price of benefits—just as it ties the taxing power more closely to the consent of the citizens.

These immense changes were hastened along by crises in the over-all structure of the large-scale societies. In the twentieth century, modern warfare twice demonstrated the intricate interdependence of the polity and the economy, increasing that interdependence in the process. The great economic collapse of the 1930's made clear to all the political vulnerability of a large-scale society which could not provide economic security for its population. The Full Employment Bill of 1946 was a major symbolic act, for it stated authoritatively the principle that every worker had a right to a job, the basis for economic citizenship. The trend is clear, from the struggle for popular education and the franchise back in the early nineteenth century to the fight for the Civil Rights Bill in 1964. It is in one direction and that direction is towards the equality of citizens before the state and the public order.

THE SPEED OF CHANGE

The society of the United States was a relative latecomer to Western history. Emerging precisely at the point in time where scale was rapidly increasing, it was itself one aspect of increasing scale in western European society. From the beginning it was a dependent part of international markets and empires. Its closest approximations to a peasantry, the chattel slaves of the South and subsistence farmers of the frontiers, were really a far cry from the remnants of feudal agrarian society. Its class system, weakened by distance from Europe and the opportunities of the frontier, never resembled the harsh outlines of government based on feudalism and conquest and, with early independence from the British system, the feudal concept of estates evaporated. The society of the United States thus maximized those aspects of increasing scale which destroyed the age-old agrarian society and exalted the rational producer.

"Americanization" has been a derogatory term for the changes which have occurred at a rapidly accelerating rate in the western European countries. The process of increasing internal inclusiveness is not, however, a characteristic unique to the United States. It occurred here earliest, and progressed most rapidly, because fewer aged structures constrained the process. The lack of hereditary ruling and taxing elites elevated the role of the entrepreneur in the cities, while the availability of land on the frontiers allowed millions the opportunity to become free and rational producers. The political norms invoked by the Declaration of Independence and the Constitution led easily and logically to the political and intellectual enfranchisement of the total population. This, in turn, led to the economic enfranchisement won, at least in principle, in the 1930's and 1940's.

But such is the irony of history, that the society of rational producers was hardly realized before its radical transformation into something new and strange became evident. World War II energized and activated the American economy to an enormous extent. In the process the technologies of production, given a chance by the unlimited demand and the shortage of manpower, raced ahead towards the maximum application of nonhuman energy to work. In the postwar years the results became obvious: American society was not only the wealthiest in the world, it was also far wealthier than it had ever been before. That wealth, distributed through collective bargaining between corporations, unions, the free market, and the government, raised the resource level of millions of American households. Leisure increased, as the forty-hour week and the three-week vacation became the norm. The average income was more than double that of the early years of the twentieth century. Higher levels of formal education, made easier by a late start at work and increasing educational facilities, marked each succeeding generation. The goods of the market, the pleasures of leisure, and the rich and novel symbolic worlds of the mass media became part of the "American way of life."

The society of rational producers accepted all this as its due. The enormous spurt of prosperity was a just reward for self-control, parsimony, and labor. The American people expressed its values through the major investments of the postwar years: the deification of the family was evident in the rapid spread of home ownership. Appetite for movement and privacy was satisfied by the flood of automobiles which carved new concrete channels through fields and city neighborhoods. Commitment to the children was evident in the increasing "child market" for commodities, the "adolescent culture" of the mass media, anxieties and plans for higher education, and, always, the preference for the single-family house in the neighborhoods of the likeminded (Greer, 1965).

The suburban move did not create these values. The suburbs have been felicitously and accurately called "new homes for old values." What occurred was the result of release from those constraints of poverty and space which had prevented the average American from living as he had wanted to do all along. Now with new consumer freedoms the old working-class culture could express itself in many new ways. That culture was evident in the conservatism of the "new middle classes"—their focus upon the child population, their pious religious participation, their concern with earning, buying, and planning to buy.

David Riesman has called them the "nouveau riche" of leisure (Riesman, 1950). He has pointed out their great new opportunities for learning, for development, for the creation of a new social character appropriate to the economies of plenty. But it is important to note certain concomitants of the new society which actually prevented learning, in old ways, important norms required for life in the society of increasing scale and internal inclusiveness. The knowledge, skills, and norms of good "consumership" are hardly adequate for a round of life; taste is no substitute for morality. And what was happening to morality?

Morality is that aspect of the common culture which emphasizes duties rather than rights. But the new affluence provided little in the way of guidelines to responsible public action. Indeed, through its cushioning effects, it tended to pull the teeth of older protest movements. Labor unions, their jurisdictions firm and their members' paychecks increasing, found little cause or profit in militancy. The socialist movements, committed to a reevaluation of work and social class, found the workers content with their "fair share" and uninterested in radical change. Feminism virtually disappeared, as the female became the captain of consumption and director of child-rearing in a child-oriented society. Aside from the continual pressure of colored ethnics for fuller inclusion, little action in the public domain reflected the duties, as against the rights, of citizenship.

In reality, the inherited social character of the rational producer had never had more than slight relevance to the demands of the public interest. Oriented to the maximization of profit at the entrepreneurial level, to parsimony and security at the level of the rank and file, the public purpose was assumed to be automatically fulfilled through individual aggrandizement—even the protest movements were powered by a concern with equal opportunity to aggrandize. No wonder the vast increase in energy, leisure, and symbolic communication has not resulted in the education of adults to a concern with public affairs and the high culture.

The Gross National Product is shared out among millions of households, where it is consumed in private. Much is undoubtedly being learned in the process—styles of dress and furniture change; child-rearing and sex practices are exposed to questioning and conscious thought; many of the peripheral crafts and skills of the society are diffused in the do-it-yourself manner. But the hard core of responsibility for the public affairs of society and one's action in them is not an area where adult education has flourished. "Public affairs" programs on the mass media remain public charges because there is little effective market for the dramatization of dilemmas and promises in the larger society.

This weakness in the political culture is maximized by the processes which are transforming the organizational shape of the society. The continuing changes in the space-time ratio (the cost in time of traversing space) result in a continuing process of national integration. The corporations safeguard market and resources through continuing mergers. As this occurs, the effective center of control shifts from the city, where the major industry is now a branch plant, to the national headquarters, where real decisions occur. Parallel changes take place in labor unions. (A labor leader described, ironically, a dozen bargaining sessions around the country waiting the word from New York—where the management of the telephone company and the president of the union were settling the issues.) Increasingly, even at top levels of local control, the industrial-commercial empires are manned by salaried workers.

The local political community is also changing, *para passu*. While the nation becomes overwhelmingly urban in its residence, the city loses its governmental form. The increasing command of space incident to the diffusion of the automobile has allowed the development of endless tract housing on the peripheries. These settlements, separately incorporated, live within the ecology of the metropolis but are not subject to any public powers at the local level. Typically, small residential enclaves and suburban municipalities are stewards for highly local interests; they are little concerned with the great issues of the metropolis as a whole and, like dogs in the manger, prevent any broader power from acting. Thus the governmental fragmentation of the metropolitan area prevents the generation and solution of significant issues at the place where they occur—the local political community (Greer, 1962).

Public action is learned through participation in the near-at-hand. It is learned through membership in a group, where one's private vocabulary is measured against the common culture, and where one's individual enthusiasm or grievance can be turned into social fact through communication and collective action. But study after study has shown that Americans are poor participators in the social groups accessible to them. They are most apt to participate in the voluntary organizations of suburbia—those groups devoted to the problems of household, family, and neighborhood. And of those who do participate, some must learn at least the ABC's of public action. But they are a minority and their public affairs tend to the trivial and the parochial.

For the rest, true significance is nolonger local. The shifts in space-time ratio have radically changed the meaning of space and it is more useful now to speak, not of statute miles, but of what John Friedman calls "interactional space." The very concept of space is reduced to its effects as a barrier to and a channel for social inter-action. In this sense, many of the most significant leaders of American thought and action are not primarily citizens of any local com-munity. With national and international roles, the nation is their community, O'Hare Airport their crossroads, and the airstrips from Boston to Norfolk their Main Street. Even at the bottom of the occupational pyramid, where men engage largely in routine jobs seldom taking them beyond a plant or neighborhood, some dimen-sions of the world are supralocal. Each evening when they turn on the Big Screen their attention moves through interactional space to the national society. It is the locus of true significance, but it is far away and it is not experienced as personal action in a concrete social group.

While the organization of work and distribution of consumption goods have been radically changed, a more basic dynamic has con-tinued to revolutionize the *process* of production. We have seen the origins of intensive urbanization in the progressive mechanization of the extractive industries, which released workers from the rural areas. We have noted that agriculture is the most automated of our indus-tries. But the process has not stopped there. The "secondary" industries of fabrication are increasingly staffed, not with men, but with machines operated and controlled by other machines and powered by nonhuman energy resources. Production soars while the blue-collar labor force declines. As this occurs, more labor has moved into the white-collar jobs of sales and clerical tasks (the processing and diffusion of orders), into service jobs and the professions. Men increasingly work, as Riesman points out, not with "hard" materials but with "soft" symbols and social structures (Riesman, 1950).

Many of these jobs, however, are now in danger. The development of electronic data processing has provided a new kind of machine, one which can process orders at lightning speed with a far lower rate of error than the fallible human being could ever achieve. Digital computers with "canned" programs perform analyses in one minute which took, only a few years ago, several hundred man-hours on hand calculators (and let us recall that the computer is only about a decade old). Meanwhile, the service industries are increasingly applying new machines and sales activities are being revolutionized by prepackaging. Even the teacher is experimenting with teaching machines—devices which minimize error and improve learning through the abolition of the "human relationship" between pupil and mentor. The revolution which began with agriculture bids fair to permeate the traditional worlds of human work.

The world of the rational producer is working itself out of a job. The society based upon the economy and logic of scarcity becomes a monster cornucopia of production; the world focused upon cities, hinterlands, and regions, becomes involved and almost dissolved in the giant national system; the economy exalting the value of hard work and self-control produces a set of tools which threaten to make work as we have known it obsolete. The results are a set of massive problems, a fierce tension between the common culture, the social character, and the politico-economic situation of Americans.

THE PROBLEMS OF PLENTY

Each year production in the United States increases by two or three percent and personal income moves steadily upward with it. The vast array of consumer goods spreads further: to color television, pastel sports cars, power-driven reels for deep sea fishing. The number of millionaires passes 100,000. Yet the unemployment rate hovers around 5 percent; it is 14 percent for the youngest entrants to the labor force, 30 percent for those who did not finish high school, and 50 percent for those in the latter category who are colored. "Silent firing" becomes the rule. The author recently visited one of the major steel plants, economic resource for a populous valley, where no one had been hired for a couple of years and no one was going to be hired for several more. A large proportion of the stock of conventional jobs is disappearing.

These are the results of substituting machines for routine human work, for *drudgery*. Norbert Wiener believed that anything which could be done better by machines should be, for machines allow "the human use of human beings." The difficulty is that human beings can be used in American society only in a job, and jobs are not multiplying as fast as would-be workers. Then, since the job is one's claim to economic citizenship, the society dichotomizes into the relatively skilled, honored, and prosperous holders of increasingly technical jobs, on one hand, and the unemployed on the other. Subsisting on inadequate doles, the latter are "social basket cases," casualties of increasing scale.

The unemployed are a minor proportion of the citizenry at present, but there is no assurance this will remain true. By Adam Smith's definition, less than half the labor force is now employed in "productive" work, and some economists believe that in two decades it will be more like 15 percent. The accumulated labor force, meanwhile, is accelerating, as the child crop from the war and postwar years makes ready to marry and procreate. The society faces more than a temporary problem: as one in which the right to economic citizenship is affirmed it must generate millions of roles to confer that citizenship. The political rights of the unemployed (as well as the vulnerability of complex orders to sabotage) will force that duty upon whomever holds political authority.

In a global perspective, it does not seem a grim or tragic predicament. When one considers the problems of nations whose wealth per capita is less than one-twentieth of ours—the energy-poor and inefficient producers who dominate world society—it seems no problem at all. And, indeed, certain solutions to the problem seem clear and feasible. These solutions would require either a change in the assignment of work, a change in the nature of the role which guaranteed one economic citizenship, or a change in the nature of work. Given our certainty of production, any of the three would have the major effect of distributing economic citizenship to the increasing population of the republic.

The sheer number of human hours required by our productive plant is decreasing faster than our standard work week. (And probably faster than we can see from statistics, for featherbedding is a concomitant of silent firing.) It is not difficult to see that shortening of work weeks and the outlawing of "overtime" could easily effect a redistribution of jobs. The cost of production would increase somewhat in the short run, but the long-run human costs of multiplying the unemployed would be greatly minimized. Further, the resulting increase in leisure could be counted as a social good.

To go further, the very requirement that economic citizenship depends upon the job could be amended. The guaranteed household income is a distinct possibility and, as the authors of *The Triple Revolution* point out, the total cost of guaranteeing $3,000 a year for every American household would run less than two percent of the Gross National Product (Perrucci and Pilisuk, 1968). Such a minimal level would not attract many people who had the opportunity of earning the average income for the nation—over twice as much. But it would have the advantages of shoring up broken households where middle-aged women are busy raising a significant proportion of our next generation, of removing some of the stigma of failure from the older men whose skills are obsolete, of providing some kind of base on which the unemployed youth of the poor might lay the foundations for a rise to useful employment. In short, those bred to a society which once demanded menial, casual, and heavy physical labor would not have to bear alone the burdens of transition to a world where such labor was of decreasing value. Even though they might remain net charges upon the society, the effects upon the middle-aged miners in the worked-out coal fields of Illinois or West Virginia could be counted as a suitable social investment.

The foregoing are means of adjusting to the decline in our conventional stock of jobs relative to our burgeoning population. More important, however, is the long-run possibility of using labor, released now from manual drudgery, in a wide variety of spheres where the chief resource demanded is simply human action itself. From a strictly economic point of view, what is needed is work which is highly labor intensive, which produces negotiable goods, and produces goods which do not compete with themselves in future market situations. From a social point of view, what this writer prefers is work which enhances the quality of human life, for producers and consumers alike. The kind of work that is immediately apparent is that focusing upon the human services: education, health, the arts and sciences.

Indeed, we can already see a shift of investment towards these areas of production. The recreational industries, from tourism to fly-tying, are booming; the most important generator of new jobs in the economy today is education; even health is an increasingly big business. But the shift is not fast enough to accommodate the new generations of job-hungry youth and it is unlikely to accelerate fast enough to provide work for the millions of new households that will form in the next decade. This is in large part because the fields of health, education, welfare, and the arts, at best supplied by a very imperfect market, have been wards of the government. As such, they have been as stunted as most charity cases.

The obvious solution is a massive increase in governmental investment in these areas of production. Where the market is incapable and the task is crucial, the public sector must act, as even the late Senator Taft acknowledged. Education is a public task by general acknowledgment, and so, increasingly, is research. The massive increase of resources in these areas could produce a rapidly professionalizing labor force with increasingly powerful tools for solving human problems.

Research, justified by a concern for all of mankind as well as the national interest, could face other crucial problems than that of exploring space. These include, at least (1) the discovery of new sources of nonhuman energy, to take the place of our depleting reserves of fossil fuel and to make this power available to people now the victims of the erratic distribution of coal and oil; (2) the delineation of the social and physical causes of overfertility and the implementation of population control in the variety of human societies; (3) the discovery of the basic principles of learning and the development of a variety of educational techniques, suited to the varieties of persons and cultures and the different capabilities required; and (4) perhaps the most important of all, the delineation of the nature and constraints of social organization in our rapidly expanding and increasingly complex society, with a concern not only for inexorable constraints on collective action, but also for the conditions of increasing freedom to choose. In short, our abundance of material goods and services could subsidize an increase in knowledge and techniques which would help the entire global society to achieve "the American standard of living," a goal impossible at present.

The above policies have all been suggested by reasonable men, yet they have a wildly utopian ring. Emphasizing the interdependence of the total society and, in the process, the necessary interaction of government and the economy, they violate the common culture of the American as a rational producer. They sound immoral to him and therefore, impossible to the observer. To pay people more for not producing, to pay an adequate livelihood whether people work or not, is anathema. Further, to increase the amount of the wealth distributed by the federal government and then to invest that wealth in the "nonproductive" work of health, education, welfare, reserach, and the arts is to violate both the norms concerning who should have power and those concerning what should be done with it.

The social character of Americans is based upon the discipline of poverty and the religion of work. Their moral armature depends upon the mortification of the flesh through toil, a debt paid to the self-controls which constitute conscience. Leisure itself is a problem for such a character, and the prevalence of moonlighting on the one hand and do-it-yourself tasks on the other testify to the use of make-work as a means to security. The harsh character suited to the subsistence job or the subsistence farm, inappropriate though it is to the soft life of routine employee and aggressive consumer, continues full force in the absence of a replacement. For the threat that one's character is obsolescent is also a threat that one's self is obsolete: hence the anxiety over work, the threat in leisure.

Leisure for others, particularly for those who cannot find a job at productive labor, is the same threat doubled. Used to privatizing his economic life, the rational producer has only a personalistic vocabulary for achievement or success; the failure of others is proof of their incompetence and hence their lack of character. Even as his own life is organized around the armature of threat, scarcity, and work, so he cannot believe men will work hard without these whips—and he cannot accept the morality of men who do not work for a living. His harsh judgments of others are the mirror images of his own concept of self.

His privatization of public issues is simply a continuation of his heritage, the common culture of the merchant, farmer, and laborer. It is evident in his alienation from public responsibility and in his suspicion of the political process and public affairs. Since he is involved only peripherally, through the ritual of the vote and his spectatorship of the big screen, government is unreal and far away. Where he lives dwell only consumers of goods and compliers with order. Only in the small scale world of the local polity does he have access to a concrete political process; its trivial issues generally fail to arouse him to public action, save for sporadic efforts to impose a veto where his minor interests are considered in danger.

Democracy is an aristocratic concept, and the ideal political norms of the United States imply a common culture stressing noblesse oblige, the duties as well as the rights. The duties begin with the obligation to understand, take positions, and contend. They continue through the obligations to serve the public interest and to pay the price in the form of goods, services and powers surrendered to the public purpose. Democracy is, then, the socializing of political responsibility.

Judging from our data on American politics, the common culture has diffused these duties only as vague desiderata. The average citizen has not integrated them in the essential armature of his social character. His knowledge of the high culture, including his own political history, is typically vague and inaccurate. His knowledge of public affairs, even the burning issues of his politics, is hardly any better. (In one recent referendum for metropolitan government the slogan of the opposition was: "If you don't understand it, don't vote for it!") His education is typically already obsolete by the time he takes his place in the ranks of householders and citizens; it is not kept up by lifelong learning. Under the circumstances, it is easy for him to lose all perspective on the vast issues of the continental nation. He has no idea of the relative share of the Gross National Product going into governmental expenditures, or the relative importance of armaments in those expenses, much less the share spent by local, state, and federal governments. Illiterate of statistics, he is unable to read the alphabet, much less the language necessary to understand the world of national policy.

He translates these issues, then, into his own vocabulary of everday action. The federal government is defined as a small business man, forever on the edge of bankruptcy. With the imperative of thrift and parsimony held sacred, he hardly understands, much less appreciates, the importance of research and development, risk capital, and the long view that attempts to predict and control the future. It would help if he could improve his economic notions to take account of, at least, the imperatives of big business enterprises, but he does not understand big business. His economic folk thought omits the most important aspects of the contemporary society: rapid technological change, the complexity of organization, and the interaction of political and economic goals.

He confronts opposition, from the world without and from ideological opponents within, in the crude dichotomy of chauvinism. All that threatens is inherently evil. His ethnocentrism is so pronounced that he continually underestimates the power of other nation states and overestimates the importance of his domestic enemies. His political illiteracy is perhaps most dangerous in the areas where he should be competent—the values behind the Bill of Rights and the history of governmental oppression which gave rise to that bill (Mack, 1955-56). His inability to accept the problematic nature of the real world blinds him to much of the reason for controversy and, therefore, to the basis for protecting the person with a minority opinion. Life to him is simple in principle: thus the anxiety generated through his blurred awareness of the Communists as radical critics of his way of thought. His response is a religious reaffirmation of what he takes to be his "way of life"; the deification of small business (though 200 corporations produce 80 percent of the economic values in the society), the decrial of the federal government's spending (though most of it goes for the wars of past and future which he supports in principle), opposition to "growing federal power" (though most of his security is a by-product of the welfare programs and economic stewardship of the national system).

But he is not really given to political contention. In his everyday social relationships, he avoids conversation about politics if there is any genuine difference of opinion, for it is "controversial." (He imagines a politics without controversy.) His chief source of political information is in the mass-media figures who agree with him, and his outlet for political opinions is the circle of kin and friends upon whom he can count to share his assumptions and pieties (Greer, 1963). But political questions do not really limit his friendship circle, for he tempers the winds of conversation to the company, excising if necessary all subjects of general import. (In an unpublished study of a middle-class Oregon population, Robin Williams and his associates found no relation between friendship and similarity of values.) While such behavior probably reflects, rather than creates, the lack of significance found in public issues, each opportunity lost perpetuates the existing situation.

Half-in and half-out of the large-scale society, the "typical" social character sketched in above is a changeling. One can see traces of the peasant villager, the vague fear of the unknown and the lack of confidence that the world is in any way subject to one's choice and efforts. One can see, in many cases, a "daemonic universe" in which the President of the United States is a "conscious agent of the Communist conspiracy," while fluoridation of the water supplies is a plot by Communists to weaken the bone structure of American bodies. One can see also the hard carapace of the rational producer's social character: the suspicion of political leaders and the public enterprise, hatred of positive government of all sorts and especially taxes, contempt for controversy and the political process. The common culture has not produced a social character adequate for a broadly based democratic polity; it has produced a society of employees, spectators, and consumers who think they are free enterprisers.

Yet so constituted is the American polity that its ability to act rests finally upon the consent of the average citizen. The local political community is based upon their agreement and must return to them for changes in taxes, in powers, and in its formal structure. The rules for the local community are, in turn, frozen into the constitution of the states, while the latter are given conditions for the integration of governmental action in the society. At each level the political leadership must have the assent of the citizens. Only thus can government action be legitimated through the common political culture, and the law which is that culture be frozen into statutes and charters.

One might well ask how, under the circumstances, the country is governed at all? And indeed, the history of the United States includes some vast hiatuses in which little governmental direction was evident—whether we speak of the lawlessness of the western territories before statehood, the "lynch law" of the South and the Indian "wars," or the uncontrolled development of our giant corporations and our vast and ragged cities (Wood, 1961; Josephson, 1934; Kroeber, 1961). In the past, the society, simpler in its organization and protected from its enemies, could live with a modicum of governmental responsibility. Today this becomes less possible by the minute: the anxieties of the cold war with its threat of thermonuclear holocaust, and the anxieties of economic change with its threat of disastrous depressions inevitably increase the responsibility of the federal government and the national political leadership.

National leaders dominate the big screen and the issues they confront are the most salient in modern life. But they are far away from the social presence of the citizens and they achieve support only through the delegation of trust to party or person in the forced choice of a two-party election. But indeed, this distance from the voters is probably the only reason they can act at all. When we look at the states and their polities, we are struck with their incapability to act with respect to their legitimate problems. This inability of state governments to organize consensus probably rests, in turn, upon their very exposure to the rank and file of their constituencies. We hear much about the overrepresented rural vote as the cause of state incapacities, yet the only careful studies of the problem indicate that urban delegations to the capitols get what they want— they simply cannot even agree among themselves on what it is. The doctrine of states' rights has foundered in our complex society upon the inability of the states to act as semi-sovereign governments. They have been, instead, veto groups—brakes upon the policies generated at the national level.

Equally rigid has been the hold of popular democracy upon our city governments. Limited in their abilities to tax, their abilities to expand their jurisdictions and powers, and their abilities to reformulate their structure, they have been passive victims of the powerful transformations associated with expanding scale. Their limits are, in turn, due to their exposure to the population through the referendum, that plebescite which assumes competence and concern on the part of everyone aged twenty-one or more. Unable to act under these limits, unable to reorganize themselves (so they could act) under the state constitutional limits, cities have increasingly turned to devices which evade the vote of the electorate. Special district governments with their own taxing power can integrate jurisdictions and increase governmental control—at the expense of being virtually invisible to the citizens. State and federal grants-in-aid for new purposes break the deadlock of local indifference and stalemate, but at the expense of being administered by professional bureaucracies hardly exposed to the citizenry. Positive government evaporates from the local democracy in the process, leaving the mayors of our cities in the role of "caretakers." As such they try to update the services legitimatized in the past and to increase the flow of nonlocal funds into their bailiwicks. Prisoners of the common political culture, they attempt to "wire around" the dead cells of the local democracy (Greer, 1963).

THE UBIQUITY OF CHANGE

To recapitulate, social character has been identified as typical ways of acting in important social situations. Based upon a culture created to handle the everyday situations that must be confronted, social character perpetuates that culture as individuals attempt to adapt the new to the framework of the familiar, applying wonted behavior to the emerging world. Thus the social character appropriate to the rational producer, the laborer, the small businessman, and the peasant, has been applied to the rapidly urbanizing society of the twentieth century. In the process, its inadequacies are evident; it fits neither the situations of the individual actor nor the demands of the large-scale system. It is perpetuated in the rhetoric of politics, the folk theory of economics, and the cliches of popular morality. Yet one suspects that it can give lip service in these areas only because of their lack of salience to the ordinary citizen (as burial rites are among our most conservative customs largely because of their unimportance to the ongoing society). Let us examine, then, the changes in those situations where Americans *must* be competent and concerned with outcomes.

The typical American worker is becoming a white-collar employee and his work is changing to the control of symbols and people. Further, he is apt to work within a large-scale formal organization, one which integrates the behavior of hundreds of men with that of hundreds of machines on one power grid and one time schedule. This means that the work he does is demanding, for error can cumulate geometrically. What is more, the action demanded is *rational choice* in respect to the larger system. It rests upon a precise knowledge of abstract theory applied to the concrete facts at issue; it encourages rationality in the formulating and resolving of the question at issue.

The increasing scale of the organization within which the individual acts has other consequences. At the bottom of the skill hierarchy the workers are conscious of the sheer complexity and integration of behavior in the corporation. At the top, the managers are aware of this complexity as it persists, thrives or fails within a complex organizational environment, national in scale. The very scope of their roles is conducive to wider horizons; the North Carolina textile industry is integrally related through the world market to those of Japan and Hong Kong, while many major American corporations do a large minority, or even a majority, of their business overseas. With jet aircraft, the executive life can literally be lived in an international network of organizations. This is a far cry from the squires of the Pennsylvania mill towns still described by nostalgic novelists.

Meanwhile, back at the plant, the squires have given way to the experts at social engineering. David Riesman (as so often) first underlined the significance of the "human relations in industry" approach to the problems of authority in economic organization (Riesman, 1950: 302-337; 330-338). The solution of brute mechanical problems of production has been accompanied by an increasing awareness of the problems of social coordination, integration, and morale. Perhaps a decline of the social character called the "rational producer" among routine workers has accompanied the post war prosperity and the increase in security. At any rate, "labor commitment" is a problem not only in the underdeveloped countries (Moore and Feldman, 1962). The role of the organizational engineer is increasingly one which requires an understanding of the varieties of humanity which, together, constitute "the firm." Labor leader, personnel man, arbitrator,—all are in the same profession, and provide major reference groups for management.

The changing nature of everyday life is evident also in the world of leisure. Here it is important to remember that, for the first time in history, the whole range of men has the right of *choice,* and in many different areas. Increased formal education and access to the mass media means a broadening of cultural equipment, for many different vocabularies run through the collage of noise and images on television and, when they are new, allow alternate definitions of known events. The increase in household incomes means an improved power to act; it is evident in the growth of what some have called the "ardent amateur" (Foote, 1960). This wealth is invested in fields as diverse as sports, movie-making, and the arts and crafts. Perhaps most important of all, the increase in leisure has increased our life-space, allowing action in a variety of realms.

In a recent national poll it was reported that six percent of the adults interviewed had taken an active part in a theatrical production within the past year. This finding is to be read along with a recent statement that there are now over 10,000 little theaters in the United States. Galleries for the arts multiply in the metropolitan areas and spring up in the smaller cities, while the sale of books—both soft and hard cover—has increased faster than the population.

All in all, it seems possible that the release of man from the iron role of the laborer is beginning to produce an efflorescence of man as a maker and man as a social actor. It is, to be sure, a phenomenon broadcast throughout the population. It is most evident in the hobbies and crafts, pursued and consumed in the privacy of the household. It is least evident in social action, the significant role in public space. Yet the consistent finding that a large minority of the citizens in the suburban municipalities is involved in their local community at a number of levels may indicate an increasing preference for social and political action as leisure pursuits. Equally important is the finding that public life in the suburbs is accessible to and engaged in by women as frequently as by men.

Another realm of everyday life that is not amenable to the old norms of the rational producer is that of child-rearing. The authoritarian norms of the past assumed a clear congruence between the world of the fathers and that of the sons; only the most obtuse can believe that this holds true today. Instead, children are given the most general norms and discover for themselves how they are applied in the large-scale, formal group such as the junior high, the high school, or the university. Nor do the norms of self-control, parsimony, and work fit very well in the relations between parents and child; there is so little real work for the child in the ongoing round of the household, that it penalizes the parent more to find it than it does the child to perform it. Indeed, the work of the household is not the proper task of the child. With the widespread realization that formal education is the key to the future, the pressure on children is calculated to ensure commitment to the goals of the educational system. With this achieved, the child is rewarded with his "fair share" of the consumables, a share calculated by comparison with his peers. For the parents, this is an expected cost of rearing children.

The major cost in rearing children, however, is the burden of uncertainty. With the breakup of the scarcity economy and its ways, parents are involved in the enterprise of inventing a rationale for child-rearing. The thorniest problems in learning theory, authority relationships, and depth psychology are innocently posed and somehow settled by these amateur social scientists who struggle with the unknown. Grappling for certainty, they swing between the discarded norms of their own childhood and the efforts of social scientists to generalize about the complex and delicate process which transform the neonate into the American citizen of tomorrow. Their effort to be rational and their consequent uncertainty are, again, not likely to reinforce the norms of the authoritarian social character.

The everyday life situations which the contemporary American must handle show certain recurring attributes. One is the pervasiveness of the *problematic:* on the job, in the children's bedroom, or facing the big screen. The intellectual answer is demanded and it cannot be intuited or remembered; the social answer which resolves the impasse is required yet it does not come "naturally." Both kinds of uncertainty can lead to what Riesman calls "the other-directed character," the person without any intrinsic standards for the good or the true. It can also lead, however, to belief in the possibility of rational consensus, a solution to the problems of the intellectually uncertain. And it can lead to a belief that one must take the standpoint of the other in order to understand him and solve the problems of the socially uncertain.

Perhaps a new American social character is emerging from these recurrent problems of everyday life. It would be one organized around the armature of intellectual control and socially created agreement. It would emphasize the necessity to think, in abstract terms, of a larger and more problematic world than was ever assumed by the rational producer. Particularly in respect to facing, on television and elsewhere, the variety of social worlds, pluralism and scale would require that an immense value be placed upon taking into account the role of the other person. Projecting these demands upon the events revealed by the big screen, the public world in which one is totally involved (through the bomb, conscription, taxes, the danger of depression or inflation), such a social character would require a general education and source of information far beyond what has ever been available in the past to either common men or privileged elites.

Such an education is possible in the United States today and is approximated for some of the oncoming generation. Should the above speculations be correct, that generation would be the first to realize and formulate a new American character type. It would be congruent, not with Americanism but with the human world of large-scale society. Freed from the older constraints of poverty, parochialism, and authoritarianism, our first postwar generation is still an enigma to its parents. It has not known the depression of the 1930's, a catastrophe which unnecessarily reinforced and prolonged the dominance of a scarcity culture, nor has it known the appalling experience of watching a Hitler dominate the proudest societies on earth. At the same time it has been educated in schools immensely superior to those of older generations (schools which Dr. Conant says have much that is right about them and, generally, only minor arrangements which are wrong). Finally, it is a generation which has grown up with the world in its living room, with the White House, Little Rock, Kruschev and De Gaulle before it on the big screen. Its perspectives should be broader, its pieties more relevant to reality, and its questions more significant.

In the meantime the American society is not filled with either the "ideal type" of rational producer nor with the new character type sketched in above. What seems the likely case is a complex mix of social characters, with old styles, new styles, and hybrids galore. (Anyone's social character may vary greatly between the realms of work, politics, leisure, kinship, and the high culture.) One would expect the new style to predominate where the task was most inexorable and the new norms most requisite, while the older style would be common where it could be maintained without difficulty.

The fundamentalists of religion, politics, and work should be found in the remnants of the smaller-scale society and in the roles of small scope within the larger world. The small towns of the rural areas, the open country neighborhoods, the South, should be heavily loaded with the older types; they should be common among small businessmen, farmers, ranchers, miners, and common laborers. They should also be found among Negroes and other ethnic groups cut off, by segregation, from full participation in the wider society.

The new social character should be most common in the metropolitan areas, in the most rapidly expanding and large-scale economies of the Middle Atlantic and the West Coast. They should be common among such workers as those employed in mass communications, entertainment, education, government, and distribution. They should be most common in those jobs requiring broad decisions and exceptional technical expertise, though one would expect them to appear in force among the ranks of the service workers— maintenance men, restaurateurs, landscape gardeners, and the like. For it is in jobs where the intellectually and socially problematic are most crucial that one would expect a major alteration in the ethic of hard work, parsimony, and self-control.

But both are extreme types, and it is not likely that either will be found to dominate the social positions indicated. Between the old style and the new, lies the tolerant middle, the rational producer whose carapace of habit and character is softening in the sun of prosperity and leisure, the new-style American who still commits himself erratically to the grim conscience and religion of work. The middle range is made up of the changelings: arbitrating between the extremes, it is the cushion which prevents the war of norms from utterly hamstringing the society. It is sorely pressed today in mediating between the southern sheriffs and the Student Nonviolent Coordinating Committee. It is that part of a generation torn between mutually irreconcilable commitments. Its bad conscience and its overconstrained leisure, its fetishism of commodities and privatization of action are alike, symptoms of a social character no longer adequate to its world. The tensions will probably not be resolved in this generation, however; for that resolution we must look to the next crop.

Meanwhile, the mix is changing. The present generation controlling this country is one which remembers the great depression, a bench mark of poverty. Like children who never had enough to eat and thus become, as adults, compulsive eaters, this generation has been obsessed with the consumption goods of the private market. But as new generations come on, we shall find a differing hierarchy of values and a new set of norms. The passing of the depression generation will be accompanied by the declining importance of those who spent their earliest years in small towns and open country. (The true meaning of urbanization is obscured by their presence for, so rapidly have our cities grown, that a large proportion of our "metropolitanites" is really made up of country boys in the suburbs.) Increasing quantities and qualities of formal education should radically alter the cultural equipment of these new, city-bred, post-depression citizens.

Perhaps, from all these trends, one could hazard the prediction that the people of the United States are busy inventing the social character appropriate to the large-scale society. As they do so, they will be laying the foundations for the first truly democratic civilization in the history of mankind. No effort is more important than this, for we are committed, willy-nilly, to a democratic solution to our problems. This means that our public policy cannot long remain far above the common culture and the social character of the usual citizen.

REFERENCES

BELL, WENDELL (forthcoming) The Democratic Revolution in the West Indies. Berkeley: Univ. of Calif. Press.

COTTRELL, FRED (1955) Energy and Society. New York: McGraw-Hill.

FRANK, LAWRENCE K. (1948) Society as the Patient. New Brunswick: Rutgers Univ. Press.

GREER, SCOTT (1965) Urban Renewal: A Sociological Critique. Indianapolis: Bobbs-Merrill.

––– (1963) Metropolitics: A Study of Political Culture. New York: John Wiley.

––– (1962a) Governing the Metropolis. New York: John Wiley.

––– (1962b) The Emerging City, Myth and Reality. New York: Free Press.

––– (1955) Social Organization. New York: Random House.

FOOTE, NELSON N. (1960) Housing Choices and Constraints. New York: McGraw-Hill.

HUGHES, EVERETT (1943) French Canada in Transition. Chicago: Univ. of Chicago Press.

HAUSER, PHILIP M. (1960) Population Perspectives. New Brunswick: Rutgers Univ. Press.

JOSEPHSON, MATTHEW (1934) The Robber Barons, The Great American Capitalists, 1861-1901. New York: Harcourt, Brace and World.

KROEBER, THEODORA (1961) Ishi in Two Worlds: A Biography of the Last Wild Indians in North America. Berkeley: Univ. of Calif. Press.

MACK, RAYMOND W. (1955-56) "Do we really believe in the Bill of Rights?" Social Problems 3 (Winter): 264-267.

MOORE, WILBUR and ARNOLD FELDMAN (1962) Labor Commitment and Economic Development. Princeton: Princeton Univ. Press.

PERLMAN, SELIG and PHILIP TAFT (1935) Labor Movements (History of Labor in the United States, 1896-1932). New York: Macmillan.

PERRUCCI, ROBERT and MARC PILISUK (compilers) (1968) The Triple Revolution: Social Problems in Depth. Boston: Little, Brown.

RIESMAN, DAVID, REUEL DENNY and NATHAN GLAZER (1950) The Lonely Crowd. New Haven: Yale Univ. Press.

TURNER, RALPH E. (1941) The Great Cultural Traditions. New York: McGraw-Hill.

WILLIAMS, RAYMOND (1960) Culture and Society, 1780-1950. Garden City: Doubleday.

WILSON, GODFREY and MONICA WILSON (1945) The Analysis of Social Change. London: Cambridge Univ. Press.

WOOD, ROBERT (1961) 1400 Governments. Cambridge: Harvard Univ. Press.

Part II

CONTROLLING THE URBAN ENVIRONMENT

Introduction

ACHIEVEMENT of the "good life" is little evident in the wave of criticism directed at the modern metropolis. Hostility toward the big city is not a new phenomenon—it has been with us since ancient times—but the complex problems of today have intensified the hue and cry and stimulated renewed emphasis on the need to control our urban environment. Yet the very ideological beliefs, images, and theories of planning which underlie and feed the criticisms militate against the formation of a public consensus sufficient to permit decisive action.

What strikes one about the current literature in the field of planning, however, is its increasingly comprehensive outlook, a broadening out from its former narrow physical base to encompass the full range of programmatic development concerned with all aspects of the quality of community environment. Equally important is its increasing willingness to acknowledge that planning is not a scientific and neutral process divorced from politics but one shot through with value conflicts and policy choices. The planners' world of today—much to the consternation of some—is no longer circumscribed by the drawing board, the master plan, the zoning and subdivision codes, and the pleasant dialogues with the lay planning commission of genteel businessmen and professionals. Like other public officials and administrators, the planner and his expertise are now being challenged by the social and economic realities of the urban communities and by the demands of groups heretofore excluded from effective access to the system.

The four essays which comprise this portion of the volume consider the relationship of planning and technology to the question of control over the urban environment. In Chapter 5, Hans Blumenfeld discusses the criteria which must be kept in mind by those who wish to plan the physical city with a real sense of its function as a container for social and psychological behavior. Distinguishing between the qualities relating to the urban environment as a place for making a living (economic) and those pertaining to it as a place for living (social), he examines the association between the two and the factors which characterize each. His principal emphasis is on the physical arrangements which serve as a basic condition to both economic and social functioning. Although the root causes of violence, crime, racial strife, drug addiction, and juvenile delinquency lie in social, economic, and political conditions, these maladies are exacerbated, as he points out, by certain aspects of the urban physical environment. The quality of the latter must therefore be assessed in terms of its impact on the social realm.

The decisive operational criterion of quality, according to Blumenfeld, is the range of choice accessible to the urban resident. In the economic sphere, this means the spectrum of mutual choice between employer and employee and between buyer and seller. In the social sector, it relates to the diversity of choice available in the location and type of residence, in education, in cultural and recreational facilities, in medical services, in voluntary associations, and in personal contacts. However, as is apparent from this and other essays in the volume, the question of quality must also be considered against a backdrop of contradictions or counterforces present in the urban milieu. These include centralization and decentralization, concentration and dispersal, separation and integration, continuity and change, all of which call for judicious balancing if effective control over the urban environment is to be achieved.

In Chapter 6, Peter Self continues the examination of the physical and social dimensions of the city with particular emphasis on the relation of size to the quality of urban life. Noting that the problems of the large cities are increasingly seen as social rather than physical, he demonstrates how these are not to be attributed to size but to the pace of social change and the obstacles which change encounters. The new towns were started as an attempt to develop a way of life superior to that of the large cities but ended up as modernized suburban components of the metropolis. The residents of these towns simply could not be denied the opportunities—the range of careers, jobs, and specialized facilities—which were available within the larger urban region. And for Self, as for Blumenfeld, the opportunity structure is a major criterion of urban quality.

The potential scope for the management of urban development is far greater today than in previous eras because of the technological conditions which now permit increasing freedom and flexibility over the location of factories, offices, residences, and other physical facilities. As a result, the choice is no longer between a concentrated or spread city; it is between an unstructured or structured region. Achievement of the latter, Self observes, will depend upon the creation of sufficient subcenters at appropriate points to accommodate the social and economic needs of an expanding population and provide a broadened range of opportunities. The basic need in this regard is to nucleate a set of facilities which will reduce dependency on the major center and at the same time prevent the haphazard scatteration and development of uses and activities throughout the area. This goal, as the author sees it, calls for the creation of a regional-local system of government in which an areawide authority would be responsible for macro-planning and overall development while local management units (each large city would be broken down into these) would perform the micro-planning and administration for enhancing the environmental quality at the very localized level. In this fashion, some of the contradictory forces in urban functioning could be resolved.

Chapters 7 and 8 both have something to say about the limitations of planning capabilities even in an age of computer and atomic power. Harris' examination of the development of cities in Latin America usefully reminds us of some of the forces at work which may be less visible in the more fully urbanized and industrialized societies but which nonetheless exert their pressures and constraints on the historical shaping of cities, often in ways contrary to the planners' prospectus. Terrain itself has had a powerful effect on the spatial layout of urban settlements in Latin America; but surely rivers, valleys, mountains, waterfronts, and geological substratum continue to have their influences on metropolitan patterns in our own society in spite of our ever more massive and sophisticated construction technology.

Latin American cities, even though they are the homes of a powerful ruling class in each nation, have been the relatively helpless targets of migration from the countryside which regularly swamps all efforts to bring about a rationalized and controlled residential development. Harris notes that only a national development policy—which obviously would call for sacrifices beyond what the ruling strata thus far have been willing to make (a point he leaves implicit)—could begin to slow this incoming demographic tide. But is not the same true for the troubling population movements in this country, including the still cumulating exodus of the more affluent citizenry into a fragmented and often civically irresponsible suburbia? And the contrasts between migration and resettlement patterns in Latin America and in the United States remind us of the immense force of historical background and cultural predispositions which planners so often still ignore.

A central theme for Harris is the polarization of urban life around affluence and poverty, especially evident in the restructuring of residential patterns in the city and their attendant densities, facilities, and amenities. But is this really so different from the American scene, or has the presence of a larger set of middle strata and less distinctive gradations up and down the class ladder only softened and somewhat concealed the same kind of underlying social substratum? In any case, Harris does not suggest or even imply that planners can do much about the class structure of a society or the proclivities of a ruling strata to promote their own interests while accepting only minimal sacrifices; rather, he sees Latin Americans as more realistic about the time it will take to cure the city's ills, and more patient—a mood he recommends to his countrymen.

Nieburg sets the theme and the mood of his chapter with the opening assertion that there is "a vulgar, self-serving, and optimistic myth that new technology and science offer the best and only solutions to the problems of high-level economic stagnation and rigid social cleavage." Reiterated affirmations that solutions to urban problems can be produced by research and development programs, promulgated by systems analysts armed with ever larger and faster computers, are seen as essentially a political ideology protecting the established power relationships which often lie behind the production and continuation of many of these very problems. So powerful is this doctrine in the United States today, according to Nieburg, and so well supported financially, that university professors as well as planners and politicians are easily co-opted into its service. But he finds little evidence that the claims of the "tech-fix" ideology are justified; rather, he sees an endless output of false quantification and self-confirming hypotheses which have little visible effect on the existential problems besetting the society beyond the walls of computing centers and the horizons of plush, technologically preoccupied "think tanks."

Nieburg sees a great many of our urban problems originating in new power and political participation on the part of previously exploited and excluded segments of the population. But "tech-fixers" are seldom allies of or aligned with the restive elements of the urban social understructure; rather, they serve programs whose purposes in fact combine a soporific manipulation of surface conditions with control and suppression of the political potential of deprived minorities. The result, he argues, is a vicious circle of futilities in which blind faith in science and technology generates an inevitable self-defeating development of mechanical and social inventions that never bring the promised "final breakthrough." If there is hope, Nieburg sees it in the restoration of the political wisdom and pluralistic participation which democratic statecraft requires to ensure that meaningful priorities are set and that the courage exists to enforce them.

Whatever their differences, these four chapters share in a respect for the complexity of urban society, an awareness of the limitations of our capabilities to plan and control its environment, and an understanding of the basic importance of the social, economic, and political patterns which lie behind the physical and technological surfaces that have too long preoccupied too many of our planners.

—W. B. Jr. and H. J. S.

5

Criteria for Judging the Quality of the Urban Environment

HANS BLUMENFELD

☐ A SUDDEN AND general awakening of interest in "the urban problem" has taken place in America during the past decade. Some critics have said that the greatest threat to the future of mankind, next to nuclear war, lies in "the problems of the cities." Statements of this kind, however, mistake problems *in* the cities for those *of* the cities. Poverty, unemployment, racial discrimination, crime, alcoholism, drug addiction, and other social maladies certainly are to be found in the urban communities, but they are problems of society, not of the cities. They happen to have their locus in the cities because that is where people are. Since these problems are more visible in the cities, because of their concentrated populations, they force themselves on the consciousness and conscience of society. Were the same problems scattered in many small pockets, public attention would not be so forcefully directed to them; they could even be swept under the rug more easily.

While most of the so-called "urban problems" are, in fact, those of the larger society, there are some which refer to a smaller unit within the total urban environment: the "city" as a political entity in contradistinction to the "suburbs." A good deal of semantic confusion has arisen from the widening gap between socioeconomic fact and legal fiction. Until the turn of the present century, the legal boundaries of the city followed its socioeconomic boundaries fairly closely. Philadelphia, for example, expanded its corporate limits in 1854 from two to 130 square miles; and half a century later, five counties united to form the city of New York. Since then, the local body politic in the United States has suffered a hardening of the arteries. Canada, as well as several European countries, has, on the other hand, given recognition to the transformation of the city into a much larger and looser metropolitan area.

The expansion of cities into metropolitan areas, usually seen merely in quantitative terms, has resulted in a profound qualitative change, in the emergence of an entirely new form of human settlement which is indeed, as it is frequently accused of being, "neither city nor country." It differs profoundly from the city as it has been known throughout history. Although its peripheral sections show some of the aspects traditionally associated with a "rural" environment, low densities and extensive open areas, it is nonetheless "urban" in its totality. The entire area forms a single functional unit, a single labor market and a single housing market. The division into "city" and "suburbs" is obsolete and obscures the reality.

URBAN GROWTH

There has been, and continues to be, considerable hostility toward urban growth. "The intellectual versus the city" has become a fashionable theme of scholarly studies in contemporary America. This hostility, however, is by no means exclusively American nor is it new. It appeared in the Western World with the first appearance of the big city: in Anakreon's shepherd poetry in Alexandria and in Cicero's praise of "rustic" Tusculum in Rome. It disappeared temporarily with the dissolution of urban life in the "Dark Ages," only to surface again during the Renaissance. Both Elizabeth I and Cromwell tried to stop the growth of London by establishing a "Green Belt." The French kings left Paris for Versailles and the kings of Prussia left Berlin for Potsdam.

At the other end of the political spectrum, socialists like Karl Marx and anarchists like Peter Kropotkin, were even more emphatic in their condemnation of the big city and in their call for dispersal of industry over the countryside. Both they and later critics advocating "decentralization" were alarmed by the increasing centralization of decision-making and the consequent alienation of the individual from participation in the determination of his own fate and that of his fellow beings. However, given a complex economy, integrated and interdependent on a national and even international scale, it is inevitable that many decisions vitally affecting the local community will be made elsewhere—and the smaller the community, the more this will be the case. Moreover, it is evident that spatial decentralization can be brought about only by a powerful central decision-making body. Only such a center can decide to locate the many various elements of economic and cultural activities simultaneously in one place—be it a "new" or an "enlarged" town. Where decision-making is decentralized, as it largely is in the United States, each individual decision-maker of necessity locates where he can find the supplementary facilities and institutions which he needs—in existing large urban areas.

Louder than the warnings of the decentralizers that "bigger is worse"—and more typically American—have been the "bigger and better" barkings of chambers of commerce and other spokesmen of the business class. What both groups have in common is the emphasis on quantity. What ultimately matters, however, is the quality of life in urban areas, be they large or small.

THE "GOOD LIFE"

Few writers on urban affairs fail to quote Aristotle's dictum that "men come together in cities for security; they stay together for the good life." Considering that many people in American cities fear to walk on the streets and few dare to walk in a park at night, one might feel that we have come full circle and that our cities fail by the most basic and elementary criterion. The modern city is far better protected than its predecessors against "acts of God"—fires, floods, earthquakes, and hurricanes; but not against acts of man.

Do people remain in the city because of the good life? It is not too difficult to define the good life in general terms: health, happiness, wisdom, and virtue. An environment may be defined as "good," if it produces healthy, happy, wise, and good men and women. But it is difficult to measure these general qualities. Perhaps life expectancy can serve as a yardstick of health, but what are the indexes of happiness, wisdom, and virtue?

Mortality is no higher in our cities than in the countryside. This is in itself a quite remarkable achievement, because for centuries the cities have been the breeding places of deadly epidemic diseases, in particular of water-borne diseases. Indeed, the universal provision of an ample and safe supply of water—also decisive for the control of fires—is an essential precondition for the very existence of the modern city. However, as Frederic L. Osborn, the head of the British Town and Country Planning Society, remarked, after having been treated to many glasses of chlorinated water on the occasion of his visit to the United States: "The American engineers seem to be more concerned that I should live as long as possible than that I should enjoy the time that I am alive." Increased life expectancy means an increased quantity of life; it still tells us nothing about its quality and the quality of the environment.

SOCIAL AND PHYSICAL ENVIRONMENT

The old school of environmental geographers attempted to explain human characteristics and human behavior as direct effects of the physical environment, of climate, soil, and nurture. To some extent this notion still persists. The crumbling buildings of slums are supposed to "breed" crime and vice. But serious study leaves no doubt that a far stronger influence is exercised by the social environment, by the totality of human relations into which an individual enters in his lifetime—in the family, in work and business, in school and church and neighborhood—formal and informal, from person to person, and mediated by various modes of communication. This all-pervasive human environment is determined by the structure of the total society, by its "culture." The man-made or man-modified physical environment is the effect rather than the cause of the quality of the life of human society. But the physical environment reacts on the social, indirectly influencing it by limiting or facilitating human relations. In addition, it has a direct influence on health and may have, through its aesthetic aspects, an influence on happiness.

THE CITY AS AN ECONOMIC MACHINE

Of all the social relations into which men enter, one with another, probably none are more basic and more variable than the economic, the arrangements for the production and distribution of wealth. "Wealth" was not mentioned among the four aspects by which we tried to define "the good "life." This is at variance with the predominant American attitude which tends to regard the Gross National Product as the be-all and end-all of the life of society, by which all other aspects are measured. Many products forming part of the GNP may be useless and even positively harmful to health, happiness, wisdom, and virtue—as is the strain of accumulating them. Yet, it remains true that there can be no life without the material means of life. There can be no living without "making a living."

The millions of people who have been and are still pouring into American cities from the countryside and from all over the world have not been attracted primarily by the prospect of "living" in the city, but by the hope of "making a living." In the context of a fully developed money economy this means "making money." Essentially American cities developed as money-mining camps, with the mentality characteristic of such camps: a sense of impermanence and an indifference to the despoiling of the environment by the waste products of the mining process.

The negative aspects of the industrial city should not obscure the profound significance of the fact that for the first time in history the urban areas have become the main producers of the wealth of society. Pre-industrial cities were largely, sometimes exclusively, consumer cities, places in which the ruling elite consumed the wealth which they extracted from the countryside. It is, therefore, hardly realistic to expect in the working cities of today the kind of "urbanity" which graced the life of the Kaloi-k-Agathoi in classical Athens, of the Princes of the Church and their retinue in Papal Rome, or of the gentry in Georgian Bath.

Before discussing the quality of the urban environment as a place for living, it may therefore be appropriate to consider its qualities as a place for making a living; to evaluate the contemporary American urban area on its own terms, as an economic machine.

THE PRODUCTIVITY OF THE URBAN AREA

Urbanization is the product of the interacting processes of rising productivity and increasing division and specialization of labor. As productivity increased in "primary" production (agriculture and mining), and as more and more goods and services previously produced on the farm were supplied from the outside by specialized factories and institutions, an increasing portion of the total labor force was shifted to "secondary" production (manufacturing and construction). And as productivity in the secondary sector increased, as more and more functions previously carried out by the manu-facturer—research and development, selling, accounting—were per-formed by outside specialialists, and as many other services were similarly transferred from the household to other specialized enter-prises, an increasing portion of the labor force was shifted to the "tertiary" or "service" sector.

The highly advanced specialization and division of labor character-istic of modern society requires a complex and many-sided cooper-ation between an ever growing number of specialized establishments, as well as a growing number of specialists within each establishment. Moreover, changes both of product and of production process become more frequent as technical progress accelerates and as a rising level of living leads to more differentiated and variable demand for consumer goods and services. These changes can proceed effec-tively only if the enterprise can draw on a broad and deep assortment of specialized goods, services, and skilled workers—resources usually found only in the larger urban areas.

From the point of view of the urban worker, these developments mean a wide range of job opportunities. In a small community an employee who leaves his job because of a change in product or process is generally faced with a choice of pulling up stakes and seeking employment elsewhere, or to accept being "a square peg in a round hole," working at a job which does not utilize his best skills and talents. For the total economy, this means a loss of productivity; for the individual, a loss of income.

There is considerable evidence that productivity and income per person do indeed increase with population size. As census data show, per capita income tends to be highest in the largest urban areas. This relationship is based in part on the concentration of high-income occupations in these areas. However, wages and salaries for the same occupations also tend to be higher. A recent study of the Stockholm

Region (Kristensson, 1967) concluded that the size factor alone accounted for a fifteen to twenty percent difference between Stockholm and the rest of Sweden. This variance must be ascribed to higher productivity resulting from the wider range of choice.

A wide range of mutual choice between employer and employee and between buyers and sellers is certainly the decisive criterion for the urban area as an economic machine, as a place for making a living. But it is hardly less important for the city as a place for living: a wide range of choice for location and type of residence, of shopping and consumer facilities, of educational, cultural, and recreational opportunities, of medical services, of voluntary associations, and, last but not least, of personal contacts.

ACCESSIBILITY AND TRANSPORTATION

Without accessibility, however, the mere existence of a wide range of choice within an urban area is only an empty promise. This was dramatically illustrated by the fact that the unskilled workers of Watts who did not own a car could not avail themselves of those jobs in the Los Angeles area which might have been open to them.

Accessibility can be brought about in two ways (which are not mutually exclusive): by locating points of potential origin and destination close together, and by decreasing the friction of space between them. It is possible to reduce the need for commuting to work by providing within each section of an urban area an approximate balance between the number of residents in the labor force and the number of jobs. A purely quantitative balance, however, is not sufficient; a combination of residences in a price range accessible only to white-collar workers with an equal number of blue-collar jobs in manufacturing plants (a situation found in some large real estate developments misnamed "New Towns") will be of little value. A fairly broad range of types of jobs as well as of types of residence is required. Similarly, a distribution of a broad range of shopping and public and private service facilities in balance with the demands of the surrounding residential population can substantially increase accessibility. Minimizing the need for travel is a valid criterion for the arrangement of the physical environment.

Ample data on travel in urban areas confirm that the percentage of residents of a given area travelling to a given number of potential destinations for work or other purposes decreases with increasing travel distance. Other things being equal, people prefer to locate their residence as close as possible to their place of work. But other things are not equal, and are becoming less so. The more work becomes specialized, the more some jobs become more attractive than others. The more the level of living rises, the more some types of residences and of neighborhoods become more attractive than others. In other words, people avail themselves increasingly of the wide range of choice, which is the very hallmark and raison d'être of the large urban area.

For these reasons, the second method of improving accessibility, "decreasing the friction of space," is of growing significance. Even more important than minimizing the need for travel is the criterion of maximizing the possibility of travel; or, expressed differently, increasing mobility. It is, therefore, quite natural that in the public mind demands for "improving transportation" outweigh all others.

CRITERIA FOR URBAN TRANSPORTATION: MOBILITY

Mobility or mutual accessibility is as important for goods and for messages as for persons. The problem is less with messages since for the most part they are equally accessible throughout the urban area. Goods transportation, however, deserves more attention than it is usually accorded. To a greater extent than is generally realized, it has been relieved by substituting movement by pipeline and, for fossil fuels in the form of electricity, by wire for surface movement. Minimization of surface transportation by wider use of such substitution, as well as by proximate location of establishments which generate substantial interchanges of goods, is an important, if secondary, criterion for the arrangement of the urban physical environment.

The main problem, however, remains the reduction of the friction of space for the movement of persons. Friction of space cannot be adequately measured in terms of miles, but only in terms of time, money, and inconvenience. It is possible, though difficult, to express these three aspects by a common denominator—time or money—by a systematic analysis of the "trade-offs" between these different aspects, in which people actually engage. Although quantification of inconvenience has not yet been achieved, translation of time into terms of money—at present usually $1.55 per man hour—is common practice. It is on this basis that proposals for investment in road and transit facilities are considered justified. "Saving time" is generally regarded as the goal and product of a good transportation system.

There is considerable evidence that, in the long run, the benefits of increased mobility are taken out not so much in terms of less time, but of more space and of wider choice. It seems that the time people are willing to spend on a trip for a given purpose is fairly constant. Even in large urban areas the median duration of the journey to work has been found to be no more than half an hour. If walking at a speed of 3 m.p.h. is the predominant mode of movement, about half of all people will live within 1.5 miles of their place of work; if streetcars at 9 m.p.h. are universally used, within 4.5 miles; if car driving at 18 m.p.h. is predominant, within 9 miles. The area within which destinations can be reached within a given time is enlarged thirty-six times. If the density of potential destinations were the same, the range of choice would also be increased thirty-six times. In fact, the overall density of American urban areas is about one-fifth to one-sixth of the density prevailing a century ago, when walking was the predominant mode of transportation. It seems that the benefit of increased mobility is being taken out about equally in terms of more space and in terms of wider choice, each being increased more than five-fold.

The direct result of increased mobility is an enormous increase in the number of person-miles (and ton-miles). Once this new pattern is established, many trips require more time, money, and inconvenience than is acceptable. In particular, any unusual circumstances lead to congestion and delays. At many places and times movement is far slower than "normal." Thus, the "traffic problem" constantly reproduces itself; it is never "solved." It might appear from this that all improvements of transportation are futile; the expected saving of time does not materialize. However, if people take out the benefits of increased mobility in terms of more space and wider choice, it means that they value these higher than they value savings in travel time, cost, and inconvenience. The benefits of mobility are real, both for "living" and for "making a living."

It may, however, well be asked: How much space? How much choice? Is there not a point where "enough is enough"? Even if there are millions of jobs in an area, a man can only occupy one at a time; and he can have social contacts with only an infinitesimal fraction of the persons living in the area. The New York Regional Plan Association raised exactly this question: "Isn't there a maximum beyond which it becomes more of the same?" Its answer was: "Apparently not" (Regional Plan Association, 1967b: 15). The appearance may not be entirely conclusive; there may well be a diminishing return. Certainly many people come to the New York region, but many others live happily in smaller cities. Yet whatever the population size of the urban area, broadening the range of choice by increased mobility remains a valid criterion.

The value of "more space" is not so easily resolved. Its desirability insofar as manufacturing and warehousing are concerned raises little doubt. The modern landscaped one-story plant with ample space for loading, parking, and expansion is superior to the multi-story loft building, in terms both of productivity and of environment for the life of the workers. Nor is there much doubt as to the desirability of more space for schools and playgrounds, parks, and sport facilities. Somewhat more questionable is the rapidly increased land absorption by residences. Some of this increase is forced by zoning requirements which many municipalities enact in order to keep taxes low by making it impossible for low income families to live within their boundaries. Some is due to the fact that each individual, in making his own decision, does not consider the combined effect of similar decisions made by all other individuals. People move to the "suburbs," in the expectation of being close to both "the city" and "the country." But the more people who make this choice, the further they have to move away from the city and the further the country moves away from them. Viewed in this light, the general move to the low-density urban periphery is self-defeating.

The "city-country" location is not the main reason for low-density suburban development. The desire for ample private outdoor space is deep-seated and the satisfactions derived from its various uses are real, as surveys all over the world have shown time and again. Thus, making an ample and growing supply of land per capita accessible, not only by road and rail, but also by pipeline and wire, is an important criterion for a good urban environment. However, with the presently prevailing methods of ownership, control, and development, much of the land made accessible is not used. The "overall density" previously referred to relates population to *all* land within a given area. This includes not only that actually used for any urban purposes (including parks), called "developed" in the planners' language, but also "undeveloped" or "open" land. Much could be gained if both types were held together instead of being scattered haphazardly. The cost of accessibility—of all types—within the developed areas would be reduced and the open spaces would be far better suited for recreation as well as for agriculture and forestry. The criterion of more accessible land must, therefore, be supplemented by that of compact and continuous urban development, at whatever density.

CRITERIA FOR URBAN TRANSPORTATION: SAFETY

Reducing the friction of space not only has costs in terms of time, money, and inconvenience, it also exacts its price in accidents. The toll in human life and personal injury on the streets and highways of America is rightly a cause for alarm. It is important to realize that whatever may be achieved in reducing this toll by improvement of the road, the vehicle, and the driver, the individually directed fast movement of many heavy bodies is inherently dangerous. If 60 persons travel in two buses, the probability of a collision is one. If the same 60 persons travel in 40 cars, the probability is 40 x 39, or 1,560. Considering that the buses have been assumed to carry 20 times as many persons than the cars, the danger to persons is still 78 times greater in cars than in buses.

For reasons of safety alone—aside from many other considerations which will be discussed later—minimizing automobile and maximizing transit usage is an important criterion. Collisions between vehicles, as well as delays, occur almost exclusively because of two kinds of friction, cross-friction and side-friction (weaving). Provision of a separate, grade-separated right-of-way eliminates both types completely in public transportation; for vehicular movement, freeways eliminate the former completely and greatly reduce the latter. Hence, minimizing movement of all types of vehicles on surface streets and maximizing use of grade-separated facilities, subways for transit, and especially freeways for cars and trucks, as well as buses, is an additional criterion for a safe environment.

CRITERIA FOR URBAN TRAFFIC: INDIRECT EFFECTS

So far this discussion has dealt only with the benefits and costs to the user of transportation; that is, with the bargain between seller and buyer which, according to classical economics, is supposed to result in maximizing the benefits of both. However, in an urban environment, any "deal" affects not only seller and buyer, but third

[148] THE QUALITY OF URBAN LIFE

persons on whom it confers benefits—and malefits.

Third-person benefits of vehicular traffic are few; third-person malefits are many and overwhelming. The latter may be divided into two groups: interference with other movements, and dangers to pedestrians. The first includes the interference of private cars with the movement of transit vehicles and of trucks, and in particular the interference of all vehicular movements with that of pedestrians. Pedestrian movement suffers considerable delay by being forced to make detours and to wait for traffic lights. But delay is not the only and not the most serious malefit inflicted on the pedestrian. He is in almost constant danger. Noise and vibration, as well as the glare of headlights, disturb rest and sleep in adjacent houses. Exhaust fumes pollute the air. These factors, together, increase nervous tension. It has been rightly said that in our eagerness to "go places" we are in danger of destroying places worth going to.

The fact that traffic interferes with a good living environment has long been recognized. Even before the advent of the automobile, separation of residential streets and traffic arteries was an accepted principle of city planning, a principle which has developed further into the concept of the "superblock" of "precinct." The most complete and systematic statement of this concept was developed by the famous "Buchanan Report" (1963) which proposed to establish standards of amenity for each built-up precinct. These standards, designed to preserve a "civilized" living environment, would determine the number of cars which would be permitted to move within the boundaries of each precinct.

PROTECTION OF THE PEDESTRIAN REALM

Traditionally the public domain of streets and squares has been a communal outdoor living room. It served for the play of children, for informal chats and meetings, for assemblies, demonstrations, and processions, and simply for sitting, standing, and walking around to enjoy the sun or the shade, to see and be seen. More and more the upper- and middle-income groups have transferred most of these activities into off-street enclosed and open spaces, at some loss of contact among neighbors. The lower-income groups, however, still rely on the street as their outdoor living room.

Large volumes of speeding automobiles have destroyed the age-old amenities of this pedestrian realm. It should be noted—and was recognized by the Buchanan Report—that a limited volume of cars moving at moderate speed is quite compatible with street life. Indeed, it may enhance it as one can appreciate by watching the use made of cars by boys and girls on Saturday night on any small town main street. In many cases, however, complete separation of cars and pedestrians is desirable. This separation can be achieved in three ways: in time, horizontally, and vertically.

Separation in time was practiced in Imperial Rome and is now being successfully applied to Copenhagen's main shopping street, Strøget; but its applicability is limited.

The first systematic attempt at complete horizontal separation of vehicular and pedestrian movement was the "Radburn Plan," developed around 1930. Radburn also uses vertical separation in the form of pedestrian underpasses under arterial streets. However, its most discussed and copied feature is the horizontal separation between cul-de-sacs for vehicular "service" access and walkways leading to the front door on the other side of the houses. Ironically, separation has here been applied where it is definitely not needed, because the number of cars serving the dozen or so houses on each cul-de-sac is minimal. As a consequence, the pathways and front doors are little used and most pedestrian life, including the play of children, occurs on the "service" roads.

Separation is needed where traffic volumes are high, particularly in city centers and shopping districts and in concentrations of high-density apartments. Shopping centers owe their success largely to this separation. Strangely, the example they provide has been followed more widely in Europe than in America, by transforming parts of the city center into pedestrian malls, sometimes extended to whole pedestrian precincts.

Horizontal separation can cover only a limited area, because access from the rear must be provided and because the network of streets cannot be interrupted for too long a distance. Over a large area complete separation can be achieved only by establishing separate levels. Ideally, pedestrians should move on arcaded sidewalks above the vehicular levels. In built-up areas it is, however, generally easier to accommodate them on a lower level, as has been done in Philadelphia's Penn Center and in Montreal's Place Ville Marie.

A combination of horizontal and vertical separation is also possible by a system of walkways in the interior of blocks, with overpasses or underpasses crossing the streets. Whatever the method, protection of the pedestrian realm by total or partial segregation of vehicular traffic is an increasingly important criterion of the quality of the urban environment as a place for living.

While separation eliminates most of the malefits inflicted on the environment by the motor vehicle, it cannot overcome what may be the most serious one: the pollution of the air by the waste products of the internal combustion engine. This by-product of the motor age affects not only the immediate environment, but entire urban areas extending over hundreds of square miles.

WASTE AND POLLUTION

The mentality of the money-mining camp led its inhabitants to dispose of waste by dumping it in the easiest and cheapest way, whether on the soil, in the water, or into the air. But production and consumption have increased and are daily augmenting the volume of wastes of all kinds, to such a degree that life itself is endangered. The situation has been aptly described by a West German scholar in the following words: "No satisfactory solution has yet been found for disposal of the accumulated waste resulting from living standards brought by industrialization. Despite purification systems, rivers and canals can no longer eliminate the pollution; they are becoming biologically dead—sluggishly flowing cesspools. . . . The atmosphere all over the world is contaminated by certain technical robots that are part of the equipment of all civilized states and are regarded as indispensable." And he continued: "While on the one hand the findings of a number of research groups in the field of biology and geophysics have given us an added insight into this liaison between human existence and natural preconditions, on the other we are using technical means to abuse nature in a horrifying way. But, by destroying vital biological systems we are contriving our own destruction" (Manstein, 1968).

Too many attempts to dispose of waste merely shift its incidence from one element of the environment to another. If garbage is dumped, it pollutes the earth; if it is incinerated, it pollutes the air; if it is run through grinders and flushed down the drain, it pollutes the water.

Water pollution may be the most immediate problem. Great concern is caused by the proliferation of algae. This is ironic when it is considered that many nutritionists see in the growing of chlorella algae the best hope to supply mankind with an adequate protein diet. The algae in our sewage effluent are, unfortunately, not chlorellae. But their exuberant growth testifies to the wealth of plant nutrients contained in sewage. Plants are the ultimate source of food for all animals, including humans; and the excrements—and ultimately the bodies—of animals are food for plants. The Chinese have become the most numerous nation on earth by carefully husbanding this natural cycle—at the risk of the spread of waterborne diseases. Modern technology should be able to reestablish this natural cycle without this risk and on a larger scale.

With a chemical technology able to make almost anything out of anything and something out of everything, the entire concept of "waste" is becoming obsolete and should be replaced by the concept of recycling. Rather than thinking in terms of "disposal" of waste, we must learn to think in terms of the reuse of by-products. The by-products of industry and transportation which pollute the water and interfere with its natural biological cleaning process, and also those which pollute the air such as particles of fossil fuels resulting from incomplete combustion, consist of usable materials. However, the cost of recapturing them generally exceeds their market price and their re-use and is, therefore, not worthwhile for the enterprises which dump them. But such action may be very much worthwhile for the community as a whole. A difficult problem of cost allocation arises here.

Although water pollution may be the most immediate problem, in the long run air pollution may be even more serious. Here, too, a natural ecological balance is involved. Plants absorb carbon dioxide and produce oxygen; animals absorb oxygen and produce carbon dioxide. But so does combustion of fossil fuels, and in vastly greater amounts. Concern has been expressed that continuation of the rapidly rising trend of combustion may reach a point where the volumes absorbed and produced may adversely affect the earth's climate.

A more immediate danger lies in the pollution of the atmosphere by the poisonous exhaust fumes of motor vehicles. Some reduction can be achieved by measures to encourage a shift from private to public transportation and from surface streets to freeways (which minimizes exhaust-producing starts and accelerations) and by the installation of after-burners. But the only complete and satisfactory solution lies in the replacement of the internal combustion engine by a different type of power plant. Electric batteries with a greatly improved power-to-weight ratio or thermo-electric cells hold some promise. Some research and development in these fields is going on, but far too slowly. If an effort comparable to that of putting one man on the moon were made to allow all men to breathe freely on earth, results would quickly be produced.

The list of human actions endangering the biosphere could be extended indefinitely. They all are expressions of the mentality of the money-mining camp which in its "pragmatic" single-purpose pursuit of immediate gains brings about unforeseen and often catastrophic results. The floods which for the first time in seven centuries reached and heavily damaged the priceless art treasures of Florence certainly were the combined result of many such single-purpose actions in the watershed of the Arno River. What is needed is replacement of the single-purpose approach by ecological thinking. Preservation of the biosphere by maintaining and restoring the ecological balance of nature is the most basic criterion for the urban environment, as indeed of any environment for human life.

MICRO-CLIMATE

While the climate in urban areas generally differs from that of the surrounding countryside by higher temperature and humidity and a severe deficiency of ultra-violet rays, there are very great climatic differences within each urban area. Ancient Greek and Chinese city planning was far ahead of contemporary practice in paying attention to sunshine, wind, and humidity in the selection of sites. In our cities the wealthy, who have a choice, generally have occupied the west end where, thanks to the predominance of western winds, air pollution is less. They have also preempted the higher altitudes where the air is cleaner and the heat of summer nights less oppressive. Less favorable climatic conditions obtain in most areas occupied by the non--airconditioned dwellings of the mass of the population.

In addition to judicious selection of sites, many other means are available to improve the micro-climate. Water and plants, in particular trees and shrubs, mitigate extremes of temperature and absorb dust and soot, as well as noise. The placing and shape of buildings greatly influence the movement of air and can create or prevent both violent gusts and complete stagnation. Much of our present built environment adversely affects the micro-climate by shutting out sunshine and air and by radiating until deep into the night the heat stored in walls and pavements during a hot day. In narrow street canyons, this may even produce local inversion. Replacement of moisture-absorbing soils by the hard surfaces of pavements and roofs often leads to a harmful lowering of the water table in the immediate area and to flash floods downstream. Creation of the best possible micro-climate in all parts of the urban area, but especially in its residential sections, is an important criterion for a physical environment conducive to health and happiness.

RECREATIONAL FACILITIES

Health and happiness require not only passive freedom from disturbing effects of the environment, but, even more importantly, opportunities for active exercise of the body and mind. Although many American cities can boast of large parks, there is still a dearth of parks and playgrounds where they are most needed: in the densely populated low-income areas. Moreover, facilities and programs are often not geared to the real needs and demands of the population. Playgrounds, for instance, generally attract only a small section of the age groups which they are intended to serve. There is also a lack of small outdoor spaces in the immediate vicinity of the dwellings, easily accessible to preschool children and to the aged and the physically handicapped. Few immigrants from European countries, east or west, fail to comment on the lack of swimming pools to cool off in during our long, hot summers, and also of indoor facilities for physical exercise and for cultural activities.

Adequate provision of community facilities, geared to the needs of the surrounding population, is an essential element determining the quality of the urban environment.

IMPACT OF THE PHYSICAL ENVIRONMENT
ON THE SOCIAL ENVIRONMENT

The aspects of the physical environments so far discussed certainly have an impact not only on physical health and well-being, but also on mental and moral health. However, in this respect, their impact can only be peripheral; the human environment is decisive. Obviously, large sections of America's urban population are spiritually sick, tossed around between overactive aggression and passive despondency, expressed in a constantly rising rate of violence, crime, and delinquency, race and class hatred and strife, alcoholism, drug addiction, clinically definable mental sickness, and other symptoms of anomie. While the root cause of this state lies in social, economic, and political conditions, these are exacerbated by certain aspects of the physical environment. Before discussing these, a general analysis of the typical structure of our urban areas may be in order.

THE STRUCTURE OF URBAN AREAS

The structure of urban areas has developed—starting from the original and generally continuing center—in a fairly logical response to market forces. As an area increases at the square of the distance, the increasing supply of land with increasing distance from the center results in a rapid falling of land prices, and consequently of densities for all types of uses, toward the periphery. This density gradient is universal and inevitable, and in itself not unhealthy.

As Ebenezer Howard, the father of the "Garden City" idea noted, people are attracted by two magnets: city and country. Whatever one may think of Howard's therapy, his diagnosis is certainly correct. Hence, a basic criterion for the urban area as a place for living is the contradictory pair of requirements of accessibility to both the urban center and the country at its periphery.

While all people to some extent are attracted by both "magnets," the relative strength of the two varies greatly, not only with personal preferences, but primarily with the composition of the household. Single persons and childless couples, especially those in which both partners are working in white-collar jobs, are most strongly attracted by the occupational, educational, recreational, and cultural opportunities of the city center. They can satisfy their desire for "the country" by driving out to it. Preadolescent children do not have this mobility; they need open space right at their doorstep or in their neighborhood, and they have no use for the central city. Nor are their housewife-mothers generally attracted to the center for more than an occasional trip. The concentration of small dwelling units, generally in apartments, at the core of the urban areas, and of larger units, generally in single-family houses at the periphery, is therefore a perfectly logical response to entirely rational and voluntary decisions.

What is not rational and voluntary, but largely the result of arbitrary zoning, is the absence of dwellings suitable for those small households who want to live at the periphery; notably older couples and widows, whose children have left home, and who want, or have, to give up their homes, but do not want to give up their neighborhood. At the same time, many families with numerous children now live in and immediately around the core of cities, in areas called "slum," "blighted," or "gray." They live there not voluntarily, but because lack of money and lack of mobility leave them no other choice. Just as second-hand cars are cheaper than new ones, second- or twenty-second-hand dwellings are less expensive than new ones. This is the only housing that one-third to one-half of the American urban population can afford, and it is, with few exceptions, to be found only in the older inner areas of the central cities.

The result is an increasing territorial division by class. Separate residential locations for different income groups are, of course, not new. In the typical American small town the poor lived "on the wrong side of the tracks." However, the scale of the entire community was so small that all its inhabitants inevitably "rubbed shoulders." But the children growing up in the vast peripheral areas of our urban agglomerations never see "how the other half lives." In the United States this general class segregation is immeasurably aggravated by race segregation which confines the rapidly growing urban "nonwhite" population in overcrowded ghettos.

The remedy is evident: provide in all sections of the area dwelling units of various types accessible to all classes and races. This objective is an important criterion of a good environment. Its achievement will, of course, require the massive use of public funds to bridge the gap between the price or rent of a new dwelling and the amount low-income households can pay. It will also require the massive use of public power to break down the resistance of white prejudice.

The opposite approach, propagated under the fashionable slogan "bring the middle-income families back to the city," is utterly wrong. First, because the often repeated statement that "only the very rich and the very poor" live in the center is simply not true. Even in Manhattan, in 1959, over one-third (34.4%) of all families were in the $5,000-$10,000 income bracket (New York Regional Plan Association, 1967a: 121, table 39). Second, and far more important, "bringing the middle-income families back to the city" means displacing low-income families. Under such a policy more families would be pushed out than brought in, because the middle-income families demand more space, both inside and outside the dwelling. There are, moreover, better reasons for the poor than for anybody else to live close to the center. Their households frequently contain several persons looking for work, much of which is casual, part-time, or at unusual hours. The employment open to them may be either in the center—which does and will continue to contain the greatest concentration of such jobs—or at varying points at the periphery which can be reached by public transportation only from the center.

These observations are not meant to imply that no low-income families would live at the periphery, if dwellings within their means were available. Those with a steady job at a peripherally located establishment could live close to it. Those who own a car could reach employment in any part of the outlying area (as well as at the center) from any other part of the periphery. The number of families in these categories may increase; but at present, and for a long time to come, very many low-income families will be forced to live in the center because of lack of mobility as well as of income. To force them out by "slum clearance" is, in the words of Patrick Geddes, "a pernicious blunder." Yet, the practice continues. Though the verbiage has changed from "slum clearance" to "redevelopment" to "renewal," it is still the same old bulldozer, with rare exceptions, which is expected to "eliminate slum and blight." It is obvious that demand for low-rent dwellings greatly outruns the supply when rooms are overcrowded and families are forced to share bathrooms, and when landlords can find tenants for dwellings in need of major repairs. Slum clearance makes this condition worse by further decreasing the supply of housing. Instead, it should be increased both by new low-rent and low-price units at the periphery and by maintenance, improvement, and rehabilitation of old ones, without raising their rents or prices. This action should be combined with addition of facilities and services in the "blighted" areas.

"Urban renewal," as it has been and is being practiced in North America, is harmful not only because it decreases the scant supply of low-rent housing, but also because it completely destroys the social fabric of the neighborhood. The displaced persons not only are forced to pay higher rents or prices for dwellings often no better and sometimes worse than those that have been destroyed, but are also deprived of the support of neighborhood friends and families, stores, and institutions on which they rely far more than wealthier people. The time is long overdue to stop this criminal folly. Maintenance and constant improvement of existing houses and neighborhoods is one of the most important and most neglected criteria for a good urban environment.

PRIVACY AND NEIGHBORLINESS

It has already been noted that for the lower income groups the street is an indispensable outdoor living room, an extension, indeed an integral part, of their home. By contrast, the higher the economical, social, and educational level of the group, the more it identifies the home with their own dwelling and yard, and the greater the value it attaches to privacy, the protection from sight and sound. At the same time, its human contacts are selective over an ever widening area.

While both privacy and ease of contact are desired by everyone, rising income and education make more unrealistic than ever the dream of reproducing the "community feeling" of the old village by surrounding "neighborhoods" of 5,000 to 10,000 people with green belts. This unlikelihood does not invalidate the desirability of organizing residential areas in units of approximately this size, for which the Russians use the more modest term "mikro-rayon," with all community facilities of daily use, notably elementary schools, easily and safely accessible to pedestrians. But neighboring in any meaningful sense occurs only in much smaller groups of perhaps a dozen or so households. Actually, neighboring depends little on the physical environment. The widespread notion that it can be promoted by increasing densities is not borne out by the facts. A survey in the New York Region showed that far more neighboring occurred among people living at densities of two or three houses per acre than among apartment dwellers and even among inhabitants of two- to three-family houses (New York Regional Plan Association, 1967c: 52, table 48).

DENSITY

The current obsession with raising density by packing more people on top of each other in higher and higher apartment towers seems to be the result of a Leibnitzian prestabilized harmony between the urge of the land owner to squeeze the last square foot of rentable space out of his property and the compulsion of the architect to erect a steep monument to his prowess. It leads to such strange rationalizations as: "We wish more people would face up to the obvious fact that as our population doubles, the only way to bring low density close in is to develop much higher density at the center . . ." (*Nations Cities*, 1967). The obvious fact is that at a distance of three miles from the city center there is only one-ninth as much land as at a distance of nine miles. Yet we find proposals, such as the one reported in the *Nations Cities*, calling for the razing of most buildings inside of the present low-density belt, and the accommodation of all uses presently found in the intermediate "grey" belt by piling them up in the inner core. Even if this fantastic proposal were carried out, only a very small proportion of low-density dwellers could live closer in than they do now. It is impossible to believe that all of the very outstanding members of the panel, whose discussions this report purports to summarize, have checked this totally irresponsible statement of their reporter.

A wide variety of densities, between about 8,000 and 40,000 persons per square mile of residential area, can and should be achieved by a mixture of housing types of varying height. Beyond these limits excellent accommodation at higher densities can be provided for households of one or two adults who can afford to pay for airconditioned apartments in tall structures with underground parking, swimming pools, and other amenities. Excellent accommodation at much lower densities can be provided in single-family houses on large lots for people who can afford to pay for two or more cars and for hundreds of feet of streets and utilities.

Some observers believe that with rising incomes the proportion of these two extreme densities will greatly increase. This may well happen if the present extreme inequality in the distribution of the national income continues. However, should we develop sufficient economic rationality to maximize the utility of personal disposable income—which decreases exponentially with increase in volume—by distributing it more evenly, the intermediate densities will accommodate the vast majority of the population during the lifetime of houses presently on the drawing boards.

LIVELY CENTERS

While neither higher density nor green belts or other boundaries have any noticeable effect on human relations, centers at which people meet and engage in common activities can promote community identification. It is difficult to define the most appropriate population size to be served by such centers; probably a hierarchy is required. The traditional "neighborhood" size appears to be too small in a typically urban environment. Some have advocated groups of 20,000 to 30,000 persons (Carver, 1962). Those of this size will undoubtedly have a role. But at least in large urban areas of millions of people, spread out over hundreds of square miles, there is a need for larger centers, veritable "secondary downtowns," serving populations of 25,000 to 500,000 and more. The functions traditionally concentrated downtown—public, private, and professional offices, a great variety of retail stores and consumer services, hotels and restaurants, theatres and concerts, museums and exhibitions—are more and more to be found also on the periphery, but in scattered locations. By concentrating them in major centers, their attraction and accessibility would be greatly increased; and the infrastructure, notably public transportation, would be more fully utilized and could, therefore, provide a higher level of service.

Common use of such subcenters of various sizes and the contacts developed by their use may indeed provide identification with a unit which is closer to the individual citizen, more comprehensible, and more conducive to active participation in public affairs. Yet it should not be overlooked that identification with an ingroup implies a certain degree of rejection of outgroups. This may take such extreme forms as gangs of youths defending their "turf" with knives and guns against intruding outsiders; or the less violent, but far more vicious and harmful, zoning policies of exclusive suburbs.

Identification with a part of the urban area must, therefore, be supplemented by identification with the whole. Here we have another pair of contradictory criteria. The whole of the urban area is symbolized by its center, visually by the dominance of the silhouette of its skyscrapers and functionally by its uniqueness. Only in the largest center can those "highest-order" functions, which need the support of the entire metropolitan market, be located. In American cities this is often overshadowed by the second function, that of private and public management and their advisers, to whom easy accessibility to each other is even more important than accessibility to the entire urban area.

Because mutual accessibility is the very essence of the center, it is doubtful whether it is wise to dilute it by residential uses, although a concentration of apartment houses close to it certainly is desirable. In large centers the office towers, the seat of top management, tend to concentrate so highly that little space is left in or between them for other functions. The fact that such districts are dead and deserted after office hours has greatly alarmed some observers. Yet in itself this is no more alarming than the fact that a baseball stadium is dead and deserted when no game is on. What is important is that there should be at the center an area with a rich mixture of uses, vertically as well as horizontally, which attract life and movement in the evening as well as during the day and thereby support the infrastructure and those uses, such as restaurants, which operate during both periods. As previously mentioned, the center should be as far as possible a pedestrian realm, including plazas as outdoor living rooms, enclosed and shaped by the walls of the surrounding buildings, with the sky as the ceiling and with a sensitively patterned floor, furnished by plants, fountains, and sculpture.

SEPARATION AND INTEGRATION

Development of the finely grained mixture of uses which gives variety and interest to the urban environment is frequently prevented by zoning. Zoning is essentially a device to protect property values from being impaired by the vicinity of incompatible uses and building types. Such incompatibilities do exist and do require separation. At the same time, integration of complementary functions, such as residence and retail trade, and of different housing types is also required. Separation and integration are another pair of contradictory criteria which call for judicious balance.

All too often, the desire for protection, implemented not by an administrative device such as the British "development permit" but by the legal instrument of zoning with its inevitable requirement of "equality before the law," has produced a deadening uniformity and monotony in our residential areas. A well-designed mixture of buildings of different height, type, and use could produce far more attractive districts. Certainly the mixture should not be the same in each district; otherwise, the uniformity now found within each would be reproduced on a larger scale by uniformity among them. A strong individual character of each district is indispensable for identification.

CONTINUITY AND CHANGE

Nobody can identify with an environment which constantly changes its identity. Without a sense of stability and continuity, one cannot feel at home. But incessant change is inherent in modern urban life. Here we identify a final and most profound pair of contradictory criteria: continuity and change.

The realization of continuing and unpredictable change has led to the notion that "planning a product" should be replaced by "planning as process." Plans, it is said, must be flexible. But the artifacts which are the objects of planning and which constitute the physical urban environment just do not flex. They can be modified only in a minor way; the possibilities for "plug-in-cities," are in fact, quite limited. The only way to leave all possibilities open for future development would be to develop nothing at present.

The most that can be done in this respect is to leave some land open for future development, notably in transportation corridors. Leaving substantial areas open would mean scatteration, which has indeed been advocated as a means of providing flexibility and encouraging efficient adaptation to change (Lessinger, 1962). But such a policy would also mean an eternally unfinished environment burdened with severe costs in terms of both money and travel time. It might have the advantage of ultimately producing the mixture of old and new buildings so strongly advocated by Jane Jacobs (1961), but it would completely destroy the compactness which the same author values even more highly, and correctly so.

PERCEPTION OF THE ENVIRONMENT

The newness of an environment built all at one time is certainly one of the reasons for the "inhospitability" ("Unwirtlichkeit") of which the Germans at present bitterly and passionately accuse the new or rebuilt sections of their cities. Probably a more decisive reason is scale. Not only is the contemporary urban area as a whole of such vast scale that it defies visual comprehension, but in all of its parts there is a conflict of two different scales.

The pre-industrial city could be and was comprehended in two ways: from the outside as a silhouette and from the inside as a sequence of various enclosed spaces formed by streets and squares, all of them on a human scale, the scale of the man standing or walking on his own feet. To some extent, the structure of the modern urban community can perhaps still be expressed by the silhouette of the group of skyscrapers at its center and of smaller such groups at its subcenters; but these can hardly ever be seen together. However, the city's streets run on to infinity and its squares are torn apart by wide openings required for vehicular movement. Only within islands reserved for the pedestrian realm is the experience of urban space and of expressively detailed structure possible, an experience mediated not only by the eye, but by all the senses, in particular by the varo-motoric sensation of moving and turning, climbing and descending. Within these islands the rich heritage of the design principles of the historical town—not, of course, their specific formal expression—is still valid.

The vast urban area, however, cannot be experienced in this way. It can be comprehended only as what it is, not a city, but an urbanized landscape, a sequence of built-up and open areas and districts of different character. And it can be experienced only by using the means that have brought this vast new landscape into being—fast vehicular movement. Driving through it can be a rich and meaningful experience, as described by a sensitive Danish visitor to the United States: "On and on flows the traffic, across bridges and down broad ramps, farther and farther in sweeping curves out into the country, without stop, continuously rising and falling in time with the contours of the earth" (Rasmussen, 1962).

We have scarcely paid any attention to using our freeways and rail lines to make the urban landscape visible and comprehensible. Most of our freeways run in straight lines and, as the cone of vision of the driver is quite narrow, he sees hardly anything but the pavement and the sky (Tunnard and Pushkarev, 1963). Ideally a freeway should consist exclusively of large sweeping curves. In Germany, Hans Lorenz designed the Nuernberg-Aschaffenburg Autobahn on this principle. It is beautiful to look *at* and to look *from*.

In the urban environment these two requirements tend to conflict. From the point of the driver or rider looking *from* them, it is preferable that roads be above or at least at grade level. But the person looking *at* them (and hearing their noise) would rather have them below level in a cut or a tunnel. Occasionally, an escarpment offers a possibility to locate a freeway so that it will articulate, rather than violate, the natural form of the earth, and at the same time offer the rider a view over the lower part of the urban landscape.

Wide transportation corridors would open up further possibilities to reconcile both requirements.

The German term for the verb "to experience" is "erfahren." "Fahren" means "to drive"; and the prefix "er" indicates accomplishment of the purpose of an activity. The urban environment as a whole cannot be "seen"; but it can be "erfahren" and identified and thereby become a meaningful part of the citizen's identity.

CONCLUSION

In recent years, as the realization that urbanization is here to stay has sunk in, the old money-mining camp attitude has begun to give way to the insight that the city is not merely a place for making a living but for living. Care for the urban environment is growing rapidly. Freeways destroying the urban landscape are rejected and even literally stopped in mid-air. Protests are mounting against pollution of air and water, against noise and ugliness.

Kenneth Galbraith hit a responsive chord when he pointed out the glaring disparity between the plethora of—often useless or even harmful—goods and services, which are bought and sold at a profit for private consumption and the dearth of publicly consumed goods and services which can only be sold and bought "wholesale" with public funds. A generation ago an American president said: "The business of America is business." Strangely enough, this man, who saw the res privatae as the be-all and end-all of society, considered himself a Republican. But the sum of the res privatae does not add up to the res publica; accumulation of private wealth does not create a commonwealth.

Only if the res publica is given priority, can a good urban environment be created. This is not a criterion for its quality, but it is the basic condition, both necessary and sufficient, for the successful application of any criteria by which quality may be measured.

REFERENCES

BUCHANAN REPORT (1963) Traffic in Towns. London: Her Majesty's Stationery Office.

CARVER, HUMPHREY (1962) Cities in the Suburbs. Toronto: Univ. of Toronto Press.

JACOBS, JANE (1961) The Death and Life of Great American Cities. New York: Random House.

KRISTENSSON, FOLKE (Sept., 1967) "People, firms, and regions." Pp. 1-10 in a publication of The Economic Research Institute. Stockholm: Stockholm School of Economics.

LESSINGER, JACK (1962) "The case for scatteration." Journal of the American Institute of Planners 28 (August): 159-169.

MANSTEIN, BODO (1968) "Shaping the future in a rational manner." Perspectives 6 (June): 26-27.

NATION'S CITIES (1967) "What kind of city do we want?" (April): 17-47.

New York Regional Plan Association (1967a) The Region's Growth.

––– (1967b) Regional Plan News, No. 86 (October).

––– (1967c) Public Participation in Regional Planning (October).

RASMUSSEN, STEEN EILER (1962) Experiencing Architecture, p. 147. Cambridge: MIT Press.

TUNNARD, CHRISTOPHER and BORIS PUSHKAREV (1963) Man-Made America: Chaos or Control. New Haven: Yale Univ. Press.

6

Urban Systems and the Quality of Life

PETER J. O. SELF

□ THIS CHAPTER OPENS with a simple and rather naive question: Is there any relation between the size of cities and the quality of urban life? To many planners and governments, the answer to this query is clearly in the affirmative. This belief is reflected in the current worldwide movement to check and control the growth of the largest cities. Britain led the way in this movement with the Barlow Report (1940) which took the position that London had reached its maximum desirable size (8 million people in a built-up area of 720 square miles). The report led to a series of public measures (greenbelt, new towns, and controls over development, densities, and industrial location) which were aimed at stabilizing the size of London (Report of the Royal Commission, 1940; also Self, 1961).

Except in North America, the Barlow heterodoxy has now virtually become planning orthodoxy. Most European states accept the proposition that population growth should be switched from the largest centers to other areas, thus curbing metropolitan growth even where (as in the case of Paris) further substantial development is accepted as inevitable. In the Soviet Union and eastern Europe, similar ideas are often held in even more dogmatic fashion. In Poland, for example, current plans are already restricting the growth of Warsaw, still under 1.5 million, and expansion beyond this mark is viewed as definitely undesirable. By contrast, the Regional Plan Association of New York apparently sees nothing objectionable in the growth of that metropolitan region from 19 to 30 million by 2000 A.D. The Association holds that the problem is one of structure, not size (Regional Plan Association, 1967).

In Latin America and increasingly in Asia, the growth of the major cities is viewed with great apprehension and alarm by many planners, and also often by governments, even though adequate powers usually do not exist for tackling the problem. The rapid spread of Mexico City, Rio de Janeiro, and São Paulo is frequently deplored; while the possible prospect of a Calcutta complex of 50 million people coming into existence portends insoluble problems for the metropolitan planners.

THE BALANCE SHEET OF URBAN SIZE

In a purely schematic form, the balance sheet of urban aggregation is not difficult to summarize. On the credit side can be placed the expansion of most types of individual opportunity, including access to jobs, specialized services (public and private), cultural facilities, and freely chosen social relationships. On the debit side is spatial friction as shown in compressed forms of development, long journey times, traffic congestion, and certain social problems. This simplified picture, however, requires exploration and modification.

The growth of opportunities is clearly no simple function of urban size, since such development also depends upon the economic and social structure of the city. Quite small towns, such as Oxford and Cambridge, possess cultural facilities which do not exist in many far larger industrial cities. Job opportunities may be concentrated in a few very large plants, or may be scattered among a diversity of smaller industries and trades. Even in a large conurbation, for example the Birmingham concentration in Britain of 2.5 million people, service industries are sometimes seriously underrepresented, many kinds of facilities being available only in some still larger center (London in this case).

For those reasons, new town agencies in Britain have tried to secure a good balance of both occupations and social groups within the town. This effort has gone to the point where lists are drawn up of the desirable types and quantities of occupations which should be established in the town. Naturally enough, these efforts have had only limited success. Because of modern technology, large plants have grown up in some new towns where their dominant position as employers could only be overcome by substantial enlargement of the town. However, most British new towns do contain a fair diversity, for their size, of modern industries including industrial nurseries. The search for social balance has been less successful, inasmuch as many in the managerial and professional groups who work in new towns elect to live outside their boundaries.

On the other hand, the "ratchet" effect of urban growth is important for job opportunities (Thompson, 1965). Once a city has built up an adequate economic base, it becomes less vulnerable to economic depression than smaller places and is unlikely itself to lose population except to suburbs or exurbs. Simultaneously, a growing variety of occupations raises economic activity rates, bringing more jobs for married women and other dependents and lifting family incomes. This development is accompanied by greater expenditures for domestic goods and leisure activities. It also produces a more physically and socially mobile style of life, with more stress upon economic satisfactions and opportunities and weaker attention to traditional bonds of kinship and locality. These familiar social changes do appear to have a general correlation with the scale of urban aggregation.

Turning now to the drawbacks of aggregation, a very good case can be made for the view that the *physical* sources of friction are much more a function of urban structure and levels of capitalization than of size as such. The main determinants of the physical functioning of any city are: (a) the location of workplaces and social facilities, (b) the location and densities of residential areas, and (c) the transportation system. The relationships between these elements always involve a trade-off of advantages and drawbacks. The compression of residential development expedites access to workplaces and facilities, but produces housing congestion and a poor environment. Alternatively the spread of residences at low densities offers a better living environment, but increases the length and cost of daily travel. The central concentration of workplaces and facilities assists mutual contacts and permits the economic provision of public transport, but at the cost of central area congestion and laborious rush hours. Conversely, a broad scatter of destinations avoids these drawbacks but entails multiple private journeys and heavy investment in roads.

The various bargains struck between these considerations are reflected in the evolution of the modern city from a compressed and centralized pattern to the diffused form of Los Angeles and, potentially, to still looser versions of "spread city." Thus some of the indictments made against London in the Barlow Report, and often repeated, such as housing congestion and squalor and consequent poor health conditions, are not necessary conditions of urban size as such. These conditions can be, and in Britain increasingly are being, overcome through urban diffusion. As a consequence, transportation becomes the most obvious physical problem of the modern city.

Transportation could be solved, almost irrespective of urban size, if sufficient funds were available. As matters now stand, large cities often possess a considerably superior transport system to smaller ones, a situation which compensates for the greater length of journeys. Provision for private automobile traffic in big cities is far more intractable, but sufficient investment in multidimensional forms of traffic architecture would not only reduce road congestion, but also eliminate many of the blighting effects of automobile traffic upon the environment. The necessary investment, however, would be fantastically large.

Increasingly, the problems of large cities are seen as social rather than physical. However, many of the former should not be attributed to urban size but to the pace of social change and the obstacles which change encounters. The millions who pour into large cities in South America, Africa, and the Middle East transfer the location of their poverty, and thereby make that poverty more obvious and dangerous. For most of these people some slight betterment of material conditions is the result of the move, although those who fail probably suffer more severely in the city than in their traditional environment. The big city, in turn, tends to be blamed for all those social conflicts and tensions which result from the breakdown of traditional society and from economic development. Additionally, the urban concentration of large numbers of poor people who are experiencing the tensions of change represents a political danger to governments, a fact which influences official views upon urban growth (Abrams, 1964: ch. 1-4).

In the more developed countries, many institutional and social obstacles to the absorption of newcomers into large cities exist. Migrants into London, for example, find the housing market heavily rigged against them, because about one-third of all dwellings are publicly owned and allotted according to residential qualifications, and another large sector is rent-controlled. At the same time, the supply of moderately priced new dwellings is handicapped by green-belt policy and opposition from residents in surrounding localities. The existing size and congestion of London are blamed for these institutional obstacles to mobility and adaptation. Similar examples abound in the United States. For one, the congested conditions of American urban ghettos are not inherently incurable, but result in large part from the lack of economic means of the inhabitants, most of whom are black, and to the resistance of white suburbanites to their diffusion.

THE PROBLEMS OF URBAN RECONSTRUCTION

The simplified analysis of the last section has suggested that urban structure *rather than the urban size* accounts for many of the social disadvantages of large cities. This fact, however, does not make the disadvantages less serious. In the real world, large cities have to cope with severe problems of adjusting their structures to new demands and pressures. There is never enough money, skill, time, or political energy to achieve satisfactory transformations. The basic urban problems are those of reconstruction and adaptation, and the time-lag is such as to lend support to those who want to control the direction of urbanization in order to give large cities a better chance to catch up with their existing load of troubles before tackling new ones.

Neither modern technology nor social change will stand still. However the potential scope for the social management of urban development is far greater today than in any previous era. This is because technological conditions themselves allow increasing free-dom and flexibility over the location of plants, offices, and resi-dences. It is true enough that the economic advantages of urban aggregation continue in the shape of economies of scale and linkage, technical and educational services, pools of skilled labor, and proximity to large consumer markets. However these advantages need no longer be sought within a concentrated, high-density city; they can be realized, on the whole more efficiently, within broad urbanized zones which may extend for hundreds of miles and which need not be continuously developed. Considerable flexibility is possible in the structuring and size of these urbanized zones. In addition, new zones can be created much more easily and speedily than in the past to relieve the excessive pressure upon existing big cities.

A public policy of controlling urban size need not imply a Luddite-like rejection of urbanization, nor a commitment to some ideal size of city which must not be exceeded. Its aim can be twofold: one, to try to "regionalize" urban growth through spreading it more broadly over a wide area and reducing the pressures upon the central city; and two, to develop a greater number of urban regions, so as to spread benefits and reduce costs of urbanization more widely among the national or world population.

But even if these general goals are accepted, how are large cities to cope with their existing social problems? How are they to improve the quality of life for their inhabitants? Granted that urban reconstruction is a slow and lagging process, it is at least desirable to be clear about its goals.

URBAN BLIGHT

The problems of urban blight and decay represent a particular crisis point in the evolution of modern cities. The nature of the problem is not hard to understand in the case of those cities which have undergone continuous and accelerating outward growth over the last century or so. Given a steady increase in the range and flexibility of transportation systems, and given the costs of acquiring old high-density development and replacing it with acceptable modern buildings, the economic dice become overwhelmingly loaded in favor of *virgin-site development* at the city's frontier and against urban *redevelopment,* except in the city center itself and in a few favored islands of upper-class residences.

For all except the very rich, the quest for an improved residential environment, and for the social status which goes with it, involves journeying outwards through suburbs of ascending modernity and wealth. Left behind in the obsolescent areas are those too poor or too old to venture on this social odyssey. Private developers have little incentive for residential redevelopment in the old areas save where some particular attractions of open space or picturesque buildings can be successfully exploited. Finally, the processes of urban escape pollute the escape routes; the inner areas become "corridor communities" whose shopping centers are choked with through traffic and whose residential streets become commuters' shortcuts and parking places.

Theoretically, the market economy could also eventually provide remedies for urban blight. New peripheral development would spread so far and fast that population in the inner areas would substantially decline, land values would collapse, and urban redevelopment would become economically feasible. This was in fact the vision of big cities held by many of their detractors such as Ebenezer Howard. Why has this not occurred? The economic answer seems to be that work has not moved out from inner areas as rapidly as population. Manufacturing industry has increasingly been dispersed, and on the same basis as social groups, leaving behind those firms (typically small ones) which depend on low rents and cheap labor. The advantages of concentration have continued, however, to apply to many financial, commercial, and governmental activities, with the difference that a small number of very important city centers (the "world cities") have become increasingly important in these respects. The demand for unskilled labor to service these activities, and to cope with transportation and tourist demands, has consequently grown, perhaps only temporarily.

At the same time, the blighted areas around the city centers have become the chosen destination of large numbers of relatively impoverished migrants. Partly, they are impelled there by the decline of jobs in agriculture and the extractive industries, matched by the continuing possibilities of finding a variety of unskilled work in a big city. Partly, they represent simply a population surplus which must move somewhere. The attractions of a big city destination are increased for many people, despite the blighted living conditions, by the frequently radical complexion of big city politics (itself produced by the middle-class suburban exodus) which results in higher levels of public assistance and relief works and sometimes padded municipal payrolls. These blighted inner areas also have an obvious magnetism for criminals and vagabonds, as well as for peaceful rebels and nonconformists.

However, market remedies for urban dereliction are also retarded by public measures of control and assistance. The ladder of opportunity to the suburbs and exurbs is cut short by the conservative and restrictive planning edicts adopted by municipalities and counties. Large public subsidies spent on urban renewal have the incidental effect of maintaining land values in blighted areas, and thus increasing the costs of redevelopment. Heavily subsidized housing for big city workers helps to maintain a large pool of relatively cheap labor, and thus discourages the dispersal or mechanization of routine tasks done in city centers. A converse situation has arisen in the United States as a result of city employment's being dispersed to the periphery relatively faster while social and institutional barriers block the dispersal of the ghettos. There efforts are being made to pump more jobs into inner areas to avoid the need for moving the people out.

CONFLICTS IN PLANNING CONCEPTS

The social conflicts over the planning of big cities can now be simply understood. Ideologically, a conflict exists between the apostles of laissez faire (followers of Enoch Powell in a British context) who would rely upon the ladder of social opportunity, mediated through the market system, to empty and eventually rehabilitate the blighted urban ghettos, and those who favor strong public planning. Logically, the former group should seek to sweep away all institutional and discriminatory barriers to free mobility, including almost all local planning or zoning; but logic of this kind is not always or even usually practiced. Among the supporters of public planning, however, there is a further division: between those who advocate measures of planned urban dispersál as an essential precondition for the solution of big city problems, and those who would concentrate simply upon direct rehabilitation of blighted inner areas.

The former group of planning advocates is still relatively weak in the United States, but influential in Britain and many European countries. Their argument goes as follows. If sufficient population *and* employment can be removed from a big city, then elbow room will be available for creating an improved and modernized pattern of development. Those specialized activities which most require a central location would continue to exist and to grow there; but considerable amounts of industrial and routine office employment would be transferred elsewhere. The accompanying migration of many low-income workers would clear ground for the reabsorption into inner areas of some middle-class white-collar groups. The reduction of demographic and economic pressures would facilitate the redevelopment of inner areas so as to secure a much improved environment and a wider range of social groups. Transportation problems would also be greatly eased.

Public policies of this type have been followed in Britain since 1945, particularly in relation to London, but also for Glasgow and (more recently) for Birmingham, Liverpool, and Manchester. Eight new towns and a variety of smaller town development projects were designated to relieve congestion in London, and, to date, these have absorbed about half a million people. The new towns have also attracted over 500 firms employing more than 100,000 workers. (Town and Country Planning Association, 1968). Over 75 percent of the population of these new projects are ex-Londoners, and nearly all the firms are of London origin, having either transferred their activities or set up branches. It is true that the lowest income groups and those in greatest housing need are underrepresented in the new towns because the industries located there mainly require skilled workers; yet, release of accommodations by the migrants has certainly eased housing problems within London. A further and still larger program of planned dispersal was initiated a few years ago. It envisages sufficient new towns and other projects to accommodate more than 600,000 London migrants over a twenty-year period. Most of these new projects are more than fifty miles from the capital.

Critics of dispersal theories ask why London is not already effectively decongested. Actually, a considerable reduction of the population of most inner London boroughs has occurred, but this has been accompanied by a shift of social problems from the old working-class areas where public rehousing has been extensive to areas of decayed Victorian tenements which have absorbed a large influx of colored and other immigrants. Some rehabilitation of old dwellings for middle-class occupants has taken place, but of a very limited and selective nature. Dispersal policies, moreover, have not been consistently followed. Government and city policies seek to stabilize the population of greater London at eight million, although it is rapidly falling below this figure. A very substantial program of fairly high-density public housing is maintaining a large labor force to satisfy the demands of employers. It is also bulldozing much of inner London into a monolithic pattern, with destructive effects upon both architectural and social variety (though admittedly some of the housing projects have considerable merit).

By contrast with urban dispersal, public measures for urban renewal are almost universal. Their limitation is the intractability of the social causes of urban blight and decay. Since American urban renewal projects have done very little to help low-income groups, the new orthodoxy is to attempt a multidimensional treatment of the causes of urban poverty and squalor, rather than to rely upon physical forms of renewal. The difficulty with this policy is that it requires the emergence of effective community leadership, a requisite which is hard to achieve artificially but without which the program becomes merely a random handout of subsidies. In Britain, urban redevelopment does cater, primarily and specifically, to low-income groups, but it cannot deal with the numerous migrants who fail to qualify for public housing, or with all those who want some fuller cure for their blighted environment than a public housing estate.

These public policy conflicts are associated with certain ideological beliefs and images. The urbanist school deplores the destruction of the concentrated and centralized character of the traditional city, while the exponents of urban dispersal are guided often by anticity imagery. Both viewpoints are sometimes expressed irrelevantly or absurdly. It is hard to believe that the cultural values of a city center are preserved by surrounding it with twelve-story blocks of workers' dwellings. It is equally absurd to suppose any longer that a small town offers a viable alternative to the big city.

Yet there is real substance to the debate about urban form. One does not need to accept all the features of the traditional big city, such as high densities and a concentrated economy, to appreciate certain cultural and social aspects of its life. So long as the city center is not too gross or congested, and is sensitively adapted and protected, it provides a forum for multiple activity and intercourse which cannot be dismembered without cultural loss. As the major city center becomes more specialized and inaccessible, and secondary centers come into existence, each of these developments poses a similar issue. How far can facilities be scattered and dispersed without some loss, not only of convenience, but also of civic and social activities which have practical utility for law maintenance and other purposes, as well as intrinsic values?

Another feature of traditional city life is its opportunities for free association with like minded individuals or groups, a feature which has special appeal to the young and nonconformist. Essential to this life are the mixture of free contacts and the opportunity for the exercise of personal taste with a concentration of diverse facilities. It is possible that this type of life could be dispersed to a number of subcenters or spread throughout a mobile city region, but it is hard to believe that the results would be very satisfactory. Chelsea, Greenwich, and the left bank of the Seine are necessarily close to the heart of cities, and will take a lot of replacing.

This section has analyzed the social problems of big cities and concluded that these are certainly eased, although hardly cured, by planned policies of dispersal and urban restructuring. It has accepted some of the cultural and social values associated with the large city, while stressing that these are not dependent upon high population density or heavy commercial concentration. Many families who live near the center of cities derive no great benefits from their location to compensate for environmental defects; their removal would leave more scope for those whose tastes and social condition are in tune with these surroundings.

THE EVOLUTION OF THE NEW TOWN IDEAL

The new-town ideal started as an attempt to develop a way of life superior to that of large cities. It is sometimes said that the ideal has failed inasmuch as large cities are still very much with us, and that it is redundant, inasmuch as the peculiar degradations of the big nineteenth century industrial city are now preventable (see, for example, Jacobs, 1961). In fact these criticisms fail to observe the evolution or transformation of the new town concept. Certainly the "garden city," as Ebenezer Howard envisaged it, had a folksy, unspecialized, small-town character which is no longer widely admired or indeed realizable; yet this garden city is the prototype for a pattern of subregional development which has considerable contemporary significance. We will briefly trace this evolution.

Howard's garden city was a proposal for balancing and integrating the values of life, as they then existed, in town and countryside. The city was to be large enough to offer the basic facilities and amenities of urban life, but small enough to offer quick access for rural pursuits. It was to have a fair spaciousness of layout, a sufficiency of local employment, and qualities of community and democracy which would be nourished by the unity of the town, individual home ownership, and the sober, industrious, and fruitful lives of the citizens (Howard, 1946; Osborn, 1946).

Variations of these ideals have a philosophic history, including Aristotle and Jefferson, and have inspired community building in New England and many other places. Indeed they have their pale reflection in modern prosperous suburbs. Ebenezer Howard's vision clothed more universal ideals with the equipment of Edwardian England. Gardening and vegetable-growing were the surrogates for agricultural pursuits, bicycling and football-playing substitutes for gymnastics, and the leasehold system of land ownership, philanthropically administered, was an English device for combining responsible estate management with individual home ownership. The homes for alcoholics which would inhabit the surrounding greenbelt were Howard's answer to what was then a very great evil of big-city degradation: gin. Homes for drug addicts would be the contemporary greenbelt equivalent for the cure of big-city vice. Thus stated, one can see the element of small-town—big-city confrontation in Howard, an aspect easily intelligible to Americans but apt to leave a misleading image of the new town movement.

NEW TOWN DEVELOPMENT

Farsighted as he was, Howard himself, like any practical reformer, cannot be fairly lifted outside the circumstances of his times, but the development of new towns themselves is another matter. First came the evolution of management. The original two privately developed new towns, Letchworth and Welwyn, have already been followed by twenty-five public new towns designated under the New Towns Act of 1946. The public corporations building these towns were not to displace the normal mechanisms of local government. They were intended to hand over their assets eventually to the elected local council, in accordance with the concepts of community and democracy. The process has been complicated, however, by the fact that much of the housing is publicly owned and rented, and that transfer of ownership of large industrial and commercial freeholds to a small local government is politically controversial. In fact, while the management issues are debated, the assets of completed new towns are being entrusted to another public corporation, the Commission for New Towns.

The second evolution has been in the size of the new towns. Instead of the 30,000 or so people who were originally supposed to form a satisfactory community, the typical size of the planned new town has already grown to about 80,000, while one of the latest new towns to be designated (Milton Keynes) is being planned for 250,000 people. In part, this rapid expansion of new town size represents a political reaction to facts of population growth and urban dispersal. It is somewhat cheaper, because of economies of scale, and politically very much easier, because of local opposition to each new town designation, to build larger new towns instead of constructing still more of smaller size. Another part of the explanation is the increasing scale and specialization of both the employing industries within a new town and the services expected by its residents. For example, a department store such as Marks and Spencer's will rarely settle in a town of less than 80,000; specialized educational and health services (in modern Britain, even the provision of secondary or "high school" education) require or prefer units as large as this or larger; and all sorts of leisure pursuits (cinemas, theaters, dance halls, bowling alleys, professional football clubs) require substantial numbers to support them. Thus the population catchment area needed to sustain a modest degree of urban self-sufficiency is much larger than in Howard's day; how much so is matter for dispute.

The third evolution of the new town relates to work opportunities. It was a cardinal tenet of the founders that long-distance commuting was an evil and that the residents of the new towns should work locally. The development of the British new towns was deliberately engineered to achieve this goal. Prime emphasis has been placed upon the attraction of suitable industries from the big cities, and publicly owned housing in a new town is normally allocated in the first instance only to employees of these industries and local service trades. Contrary to some reports, these methods have been fairly successful in achieving their goal. Although all of the first batch of London new towns are only about thirty miles from the center of the metropolis, a substantial majority of their residents do in fact still work locally. Indeed the rapid growth of Hemel Hempstead new town was accompanied by an actual fall in the number of railway season tickets issued, as some previous commuters as well as new residents found local employment (Self, 1961).

The original no-commuting concept cannot, however, be maintained. As was earlier suggested, a great diversity of opportunities and facilities represents the strongest pull of the metropolitan region. New-town residents cannot be denied these advantages if they are within reach of them; and if not in reach, the new town itself seems likely to prove less successful and attractive. The residents of the original new towns around London enjoy both substantial local opportunities and reasonable proximity to the great range of careers and specialized facilities available within the metropolitan region. How they select and balance these options depends upon their personal skills, tastes, and ambitions. If a new town is developed so as to offer a large part of the ingredients of a satisfactory life locally, then it is a fair assumption that a majority of residents will be content, for the most part, with their own backyard. A minority frequently, and almost all occasionally, however, will seek other satisfactions, the availability of which will be a crucial element in the quality of new town life and in the ability of the town to draw and keep a diversified group of citizens.

The fourth evolution of the new town is the most difficult and challenging of all, namely the development of the community concept. Theoretically, a balanced community, such as the new town once aspired to be, is a microcosm of the entire population, containing due proportions of the old and the young, the married and the single, the rich and the poor, perhaps also the rebellious and the conformist. Yet to state these conditions is to recognize their incongruence with the modern dynamics of urban development which sort out and separate social groups according to income, marital status, age, and other factors. Moreover, the broader and more mobile the framework of urban growth, the more efficient and pervasive are these patterns of segregation. They are much more marked, for example, within a large and expanding urban region than in a relatively small and isolated town. New towns, however, cannot turn back the clock and align themselves with older traditions of social integration, since their obvious locations are within expanding and specialized regional systems.

The British new towns display a considerable imbalance in age structure and, to a lesser extent, in class structure. Attracting initially a preponderance of new and growing families, the bulge of young children becomes later a bulge of teenagers seeking jobs and pleasures which the town cannot easily provide. Old people remain scarce until the parallel bulge of young couples simultaneously become aged. Class distortions are shown in the scarcity of both low-income and high-income groups, the former because housing is too expensive, the latter because exurbs and rich suburbs have greater social prestige and appeal.

Some remedies for these problems are gradually being introduced, such as official encouragement for a much broader range of housing aimed at both richer and poorer groups, as well as at the unmarried and aged groups. New-town corporations are building old people's dwellings to encourage also the parents of young couples to move into the town. At the same time, the concept of a strict limit to the town's size is increasingly under attack because such a limitation would compel a large proportion of the next generation of young families, as these are formed, to move to new locations. The process of urban dispersal or "overspill" would thus become cumulative and perpetual, new towns begetting each other in a way that already occurs with large housing estates. There is a clash here between static physical plans and social pressures, with the issue still in doubt, but town plans should in future be flexible enough to cope with natural increase, especially since the alternative is the prolongation of a lopsided population structure. This is another reason why new towns must be larger.

The new town is necessarily moving a long way from Howard's vision. Is there enough left of the concept to dignify it as a separate urban form or even to distinguish it greatly from a superior suburb? (Within the British Town and Country Planning Association there was in fact at one time a considerable struggle to establish the primacy of the former concept and the unorthodoxy of the latter.) Obviously the differences in way of life between a modern suburb and a modern new town are much less than those that existed between London and Letchworth in Howard's time. Housing types and densities are similar as are residential specialization by age and income, the nature of leisure pursuits, and even the opportunities for access to countryside—now much more valued for motorized trips and specialized sports than for walking, bicycling, or vegetable-growing.

The main differences between big-city suburb and new town remain the limited size and more definite boundaries of the latter and its localized employment and greater self-sufficiency in social facilities. While in both cases, as in modern society generally, the values attached to local community have shrunk, the new town is better placed to sustain these values. Despite the limitations of the new town, its demonstration effect remains considerable. Why this is so, will emerge from the broader analysis of the last section.

SOCIAL GOALS AND THE URBAN REGION

Ideologically, we can trace most opinions about the social goals of planning to variations of utilitarian and organic theories. According to the former theory, the goals of planning are firmly anchored in the separate preferences of individuals or families. The planning process is auxiliary to the market (perhaps a very important auxiliary) for the aggregation and maximization of these individual tastes. The planner himself is a technical assistant to processes of bargaining between individual and group interests. According to the latter theory, the goals of planning are related to the social unity and cohesion of a community as expressed through its physical artifacts. The planner has a creative role in expressing and embodying the common aspirations of this community.

This is a very simplified summary of an age-old clash of ideologies. It is obvious that many people hold mixed viewpoints of the two theories. In addition, the utilitarian theory, in particular, has a number of differing subtheories (which cannot be explored here, for reasons of space) of considerable importance to the planner's role. It is also clear that North Americans and, to a lesser extent, Britons and some other Europeans, are much disposed towards the utilitarian or aggregative concept of planning, while Latin and South American countries are more inclined towards organic concepts. Consider the differences between Brasilia and Chicago. The former gives a holistic, monumental interpretation to certain national and communal aspirations, but is somewhat careless about individual tastes. In contrast, many major projects in Chicago (Marina City, for example) represent most sophisticated essays in the maximization of individual satisfactions for those who can afford them, but general city planning dissolves into conflicts among rival interests and goals that must be settled, if at all, by ad hoc compromises.

Even accepting in general a utilitarian position, there remain severe problems over the effective articulation and satisfaction of some types of individual demand. These problems are both sociological and economic. In the former category, the big city hastens the breakdown of the traditional kinship and community bonds of social structure, substituting the specialization and privatization of individual tastes. This process has both liberating and deprivatory effects. It liberates all those who find social bonds oppressive and who are sufficiently independent, vigorous, and, preferably, wealthy to flourish under these conditions. Successful authors or artists, for example, sometimes praise the "village" atmosphere of certain districts of central London; these are villages with their sting drawn, where an individual can remain entirely anonymous or can choose his contacts freely. Yet the same social conditions which delight the young and popular, produce also the miserable lonelyhearts and the isolated old and sick people of our big cities. These latter individuals have lost the supporting bonds of social structure, not thankfully escaped them.

It is true that low-income groups in cities, because of stronger kinship traditions, congested living, and hostile external pressures, generally retain a more vital community structure. These are often areas not of individual but of mass anomie, produced by social and political resentment. The severing of family connections caused by the continuous migration of new families to the suburbs leaves fragmented social systems at both ends, however. Only when living conditions within the inner areas improve and the population becomes more stabilized, can the age and kinship balance be partly restored (Young and Wilmott, 1957).

From an economic standpoint, the efficiency of the urban development market in meeting standardized tastes is offset by its frequent inefficiency in catering to minority or deviant groups, especially if these are poor. The tall man's frustration from inability to buy mass-produced clothes is paralleled by the frequent inability of the aged, sick, and single to purchase the kind of urban environment which suits them. The large, standardized, income-differentiated areas of family housing push urban segregation further than most people want. Unfortunately, planners, through drawing their plans in terms of a stereotyped picture of the average family man, sometimes reinforce instead of combat these tendencies.

Clearly, the physical planner must accept social change and cannot or should not fight it singlehanded. His capacity, however, to propose the physical organization of city or region has appreciable indirect influence upon social behavior, while in another sense physical planning is only one branch of social planning or management. Faced with the exploding, fissiparous metropolis, the planner needs to seek a middle way between physical flux and order, mobility and stability, and the atomistic and communal visions of society. Even if he should follow the demands of a consumer society, he must still try to keep community lifelines open for those who cannot survive successfully within an individualized world. But how are these tasks to be tackled?

THE STRUCTURING OF THE URBAN REGION

The basic choice is no longer between a concentrated or spread city; it is between the unstructured or structured urban region.

Structuring depends upon the creation of sufficient subcenters at appropriate points, which can provide a social and economic focus for a surrounding block of population. There are many theories as to how large these subunits should be and where they are best located. For example, they may be secondary centers within an old urban core (e.g., Croydon in London or Jamaica, Long Island); new towns beyond a metropolitan greenbelt (London, 1944 Plan); larger but closer new towns or satellites (the Paris regional plan); or much more distant new towns (London, 1964 South East Study plan). Finger or sector plans (e.g., Stockholm) locate the new centers along radial lines of communications, divided by green wedges.

A basic social goal is to nucleate a set of economic and social facilities in such a way as to reduce dependence both upon the major city center and upon a highly fragmented pattern of facilities. There are balances of advantages to be struck. As noted earlier, the larger and the more distant the sub-unit, the greater degree of self-containment it will have; whereas the smaller and closer it is, the more it will resemble a traditional suburb. Dispersal along the former lines leads eventually to severance of the new development from easy contact with the parent city and to the creation of a new urban region, although development patterns are now so fluid that clear breaks do not occur. The "new-city" proposals for southeast England are criticized on the grounds that they will have the former effect, severing their residents from access to metropolitan jobs and facilities, while offering no practical prospect of building up an effective alternative complex of opportunities for at least fifty years. Put another way, this amounts to saying that building a new city is different—and far harder and slower—than building a new town. Conversely, one might argue that all recent proposals for New York have been concerned too much with the maintenance of urban concentration, and too little with the creation of dispersed new towns catering to a variety of groups in place of the monotonous spread of low-density private housing across a very large area.

Physical forms hold no miraculous cures for social distempers. The very ordering of urban growth, however, including the promotion of a better balance between homes and workplaces, will give some stimulus to community building. It will provide a framework within which new developments can satisfy a much greater variety of social needs and tastes than does the monolithic suburb. Moreover, the effective management of open space, whether as regional parks or as protected country zones, not only has great recreational and aesthetic value but also aids community identification through the provision of physical boundaries or limits. As noted earlier, the continuing value of the new-town concept is to provide a satisfactory mixture between the self-containment of the community ideal and the advantages of physical mobility and openended opportunities, or, in other words, to realize the potentialities of modern living without sacrificing the older and still necessary values of a structured society. This is why new towns are not period pieces, but, enlarged and adapted, they can form the right kind of framework for an expanding region. Nowhere is this truer than in the United States, where social compression in physical form is frustrating minority ambitions.

SOCIAL MANAGEMENT OF THE URBAN REGION

Quite clearly, our concepts of the social management of the environment are wholly inadequate to the facts and problems of urban growth. The kind of local political system which comprises a giant city government alongside tiny suburban municipalities is one example of this failure. Local planning is an effective conserving force, only in wealthy areas where the citizens join together to protect their amenities and to keep out cheaper forms of development. But for every suburb where the citizens have "arrived" and cherish their environment, there are two suburbs containing spiralists lodging briefly on their upward climb or groups lacking pride or interest in their physical surroundings.

In the big cities, the mechanics of massive development projects and comprehensive regulations leaves little scope for that sensitive micro-planning which alone can save or enhance environmental quality at the very localized level. The subtle tools needed for achieving worthwhile piecemeal changes are difficult to invent. At best, the city government produces tolerable standardized housing on a mass scale, while doing little to maintain the subtler and more diversified features of local districts. These latter easily disappear under the bulldozer.

One attack upon these problems would seem to be a regional-local system of government. The regional level would look beyond city boundaries so as to comprehend the probable catchment area of a city region. It would inevitably be big government, concerned with the strategy of regional development, the location of major sub-centers or new towns, the transportation network, the protection of greenbelts and wedges, and similar functions. Attached to regional government would be public development agencies concerned with the provision of housing and other facilities for poor and minority groups, much of which would be located within new towns and other growth points rather than in the urban ghettos.

In contrast, the local level of government would be concerned with skilful micro-planning, and with the making and enforcement of local building codes, grants for small-scale improvement or redevelopment, local parks and play-spaces, and the like. It would have to draw upon the regional level government for financial and technical assistance, but basically this dependence should be in the nature of aid, not control, since the local plans must be locally determined. It is essential to this concept that each big city be broken down into a number of local management units, since this would be the only way of creating the grass-roots participation and leadership which at present can only be hypothesized (Senior, 1965; Babcock and Bosselman, 1967).

NATIONAL POLICIES FOR URBAN DEVELOPMENT

How far are national policies for urban containment wise and desirable? That is the issue with which this essay began. It is clear that large and growing urban regions are here to stay, and that their improved structuring, to provide a better way of life for all their citizens, is a first consideration. But it is also strongly arguable that *some* city regions, such as New York or Paris or London or Calcutta or Tokyo, are excessively dominant and prone to growth in relation to the second and third ranks of city regions. A policy for securing a better *spread* of total urban development is called for, in terms not only of alleviating the problems of social change and adaptation in the largest cities, but also of spreading the range of social and economic opportunities more broadly throughout each nation.

Further additions to the physical mobility of Londoners, for example, seem likely to yield diminishing satisfactions when compared with the gains of offering greater mobility to the inhabitants of the Manchester or Newcastle urban concentrations. Of course, such mobility is valuable only to the extent that it provides access to more interesting jobs and to more diverse cultural and social facilities. Thus a national policy for urban development must be based primarily upon measures to redistribute the location of employment in such a way that the finger of ambition does not point so strongly to one or a few key cities.

Finally, it may be asked whether the good life of the future must necessarily be pursued for the great majority within an urban regional context, or whether there is scope left for the simpler routines and pleasures of small-town life? It would be dogmatic to assert the former and to overlook the continuing appeal of small-town and rural values to many. These values, however, probably must be pursued within a new context which calls also for deliberate planning and the selection of modest-sized growth points, as a means of overcoming the drawbacks of sparse and scattered population patterns. In the Highlands of Scotland, for example, it seems certain that depopulation and decline will continue indefinitely unless some modest-sized growth point can be created, to act as a focus for the small and scattered settlements, some of which are bound eventually to disappear. Such a growth point, tiny in comparison with a big city, is yet large in relation to existing community systems. Many parts of the world have similar problems. It is only through such policies that a viable way of life, alternative to that of the urbanized region, can in fact be created.

REFERENCES

ABRAMS, CHARLES (1964) Man's Struggle for Shelter in an Urbanizing World. Cambridge: MIT Press.

BABCOCK, R. and F. BOSSELMAN (1967) "Citizen participation: a suggestion for the central city." Law and Contemporary Problems 32 (Spring): 220-228.

HOWARD, EBENEZER (1946) Garden Cities of Tomorrow. London: Faber.

JACOBS, JANE (1961) The Death and Life of Great American Cities. New York: Random House.

New York Regional Plan Association (1967) The Region's Growth. New York: Regional Plan Association.

OSBORN, F. J. (1946) Green Belt Cities. London: Faber.

Royal Commission (1940) Report on the Distribution of the Industrial Population. London: Her Majesty's Stationery Office.

SELF, PETER (1961) Cities in Flood: The Problems of Urban Growth. (2nd ed.) London: Faber.

SENIOR, D. [ed.] (1965) The Regional City. London: Longmans.

THOMPSON, WILBUR R. (1965) "Urban economic growth and development in a national system of cities." Pp. 431-491 in Philip M. Hauser and Leo F. Schnore (eds.) The Study of Urbanization. New York: John Wiley.

Town and Country Planning Association (1968) New Towns Come of Age. London: Town and Country Planning Association.

YOUNG, M. and P. WILMOTT (1957) Family and Kinship in East London. London: Routledge and Kegan Paul.

7

Urban Quality in the Context of the Developing Society

WALTER D. HARRIS

☐ ONE IS ALMOST inevitably tempted to discuss the quality of urban life as if there were a single version of urbanism to be evaluated, appreciated, and criticized. No doubt we can simplify and abstract enough to derive from all cities some highly generalized and theoretical model of "the" city. But the utility of this may be questioned. Even among the most fully industrialized and urbanized societies there are differences in some of the aspects and characteristics of their cities and metropolitan areas; the differences increase when one compares the urban settlements of these societies with those in the less developed countries. Although such variations are rooted in historical processes and conditions of the kind explored in this chapter, they can and should be taken as cautions against assuming that what we are familiar with is inevitable. At the same time, an understanding of the conditions producing variation in the characteristics of urban places should help us to comprehend the

EDITORS' NOTE: *Portions of this chapter are based on the author's forthcoming volume,* The Growth of Latin American Cities, *to be published by the Ohio University Press.*

forces which must be controlled if we seriously intend to reshape any of the existing basic patterns in order to achieve a higher quality of urban life.

The focus here is on the cities of Latin America, with both explicit and implicit comparisons with the urban patterns evident in the United States. Further variation in and on urban themes could be highlighted if the analysis were extended to urbanism and urbanization in Africa, in Southeastern Europe, and in the Middle and Far East. But in the space available it is enough to try to treat a single continent, and even this effort requires a good deal of broad generalization. The observations here are based largely on data about the large cities of Latin America—in the majority of cases the capitals—because most of the reliable information available is limited to these aggregations. It should be remembered in this connection, that the population of the capital city is generally in total disproportion to the rest of the urban settlements of a given country. The range of variation among capitals, however, diminishes the significance of this distortion. For example, we can compare the Nicaraguan capital of Managua, with less than 250,000 inhabitants, to Buenos Aires, with a population well over 6.5 million, and with differences of complexity of the same scale.

SOME COMMON CHARACTERISTICS OF
LATIN AMERICAN CITIES

The spatial arrangement of functions within the Latin American cities is undergoing radical changes, mainly as an effect of the rapid growth of population and the consequent occupation of available land. The resulting network of streets and the patterns of land use are far from being an orderly development; the juxtaposition of dissimilar elements is the rule.

The case of Mexico City is very representative in this respect. Acres that were peripheral ten or twenty years ago, when industries were occupying large tracts of land, have now become engulfed by residential development and peripheral squatter settlements located side by side with good residential districts. The case of Caracas is structurally similar. Since industrialization started at a relatively late stage there, industry still occupies peripheral areas, but they are dispersed in at least five or six major locations around the city. Peripheral squatter settlements or ranchos concentrate more heavily in the areas surrounding the central core, but they are also dispersed along the valleys at regular spacing within the good residential neighborhoods.

In a very general form, the common elements of the structure of Latin American cities can be grouped in the following manner:

(1) Peripheral Settlements of Low-Income Groups

Settlements of this type exist in almost all first- and second rank cities in Latin America, and are proliferating at a fantastic rate. Their most dramatic growth occurred during the 1940's and 1950's and they are continuing to grow in the 1960's. The dwellings are generally first constructed of scrap materials, but in some cases they are improved over time and become permanent elements of the urban structure. Lima provides an eloquent example of this.

Density in peripheral settlements is usually very high, due to the large families and the compact number of individual housing units. The people are generally rural migrants, who either came directly from the country or moved to such areas after a period of residence in the center-city slums. Peripheral settlements are often not degraded districts; on the contrary, their inhabitants present quite energetic features, such as a high degree of organization and desire for self-improvement. Physically, even when sanitary and other environmental conditions are poor, the settlements present several valuable characteristics, such as cohesiveness and interesting clustering forms, as in Lima.

(2) A Mature, Compact Central Area of Representative Character

The form and structure of the central area is heir to the original function of the city in Latin America. With only a few exceptions (for example, São Paulo, the most industrialized city in Latin America), it is possible to define a central core not disrupted by industry around the main open space of the city, the *plaza de armas* or *plaza mayor*. This central area was designed in every case to house the seat of the government and major civic buildings. Within its confines are clearly recognizable sectors such as the main business district, the hotel and theater district, the bank district, and the area of the government palace, town hall, and ministries.

(3) High-Income Residential Areas

These are situated in one sector, toward the outlying areas (Lomas de Chapultepec in Mexico City, Chapinero-Chico in Bogotá, Miraflores-San Antonio in Lima, San Isidro in Buenos Aires, Providencia-El Golf-Las Condes in Santiago). High-income groups have generally located in the direction of the better lands of the city in terms of natural conditions, vegetation, and scenery. At the same time, they have generated new forms of recreation (golf, boating, country club). The physical layout of these sections is remarkably less dense and more homogeneous than any other sector of the city. The characteristic dwellings are one-story houses with front yards, only occasionally disrupted by isolated multi-family buildings. Broad avenues connect this sector (where the number of automobiles per family is the highest in the city) with the central district. In Santiago de Chile, for instance, 50 percent of all the automobiles in the city are concentrated in the high-income sector of Providencia-Las Condes.

(4) Peripheral Industry

In most of the cities in Latin America, industry was a latecomer. For this reason, it was generally located in peripheral areas, discontinuous zones, and all around the developed land along railroad tracks and major routes.

(5) Heavy Dependence on Mass Transit

This dependency makes Latin American cities generally more dense and compact along main routes of mass transit than in North America. It also determines the location of service functions along these lines. However, as mentioned previously, the high-income residential areas have a much lower density, and they reflect the increasing importance of the automobile in Latin America: for example, from 1960 to 1965, there was an increase from one to two million automobiles in Brazil. The impact of the motor vehicle on the cities is aggravated by the deficiency of the road network to handle traffic: traditional streets are narrow, intersections are inappropriate, and parking space is scarce. In short, Latin American cities today face the same traffic problems as the rest of the world.

(6) Commercial Pattern Composed of Scattered Small Shops and Main Shops Centralized in One or Two Centers

Very few planned regional shopping centers exist, and the department stores generally occupy a subordinate position. There seems to be a much higher number of stores than needed, when related to the volume of sales or the purchasing power of the population. One reason is that the small shop is a type of "disguised" unemployment. In cities like Lima and Bogotá the new secondary centers are more important and varied and of better quality than the traditional shopping centers. The latter is characterized by the dominance of a main shopping street of four to seven blocks. Many main streets are enclosed by commercial galleries, a superb example of retail architecture if properly done. For retailing purposes, this device works very well, since it is a sheltered space which permits access from several streets.

(7) Presence on the Urban Landscape of a Natural Element

For one reason or another, cities in Latin America seem to have a more direct relation with the natural landscape than in other parts of the world. The extreme example is Rio de Janeiro, whose form is completely conditioned by the hills, mountains, and the bay. There is almost no city without the presence of a very strong element of the landscape—the Andes in Santiago, La Paz, Quito, Bogotá, Caracas, the river in Buenos Aires and Montevideo, and the desert, the sea, and the Andes in Lima. In almost all these cases, the natural elements have been critical factors in shaping the general form of the city and giving particular interest to different subsections.

This broad review emphasizes two main characteristics of the contemporary Latin American city. First, it is undergoing rapid structural changes, influenced by some "native" factors (the most dramatic being the proliferation of squatter settlements), and some "induced" factors, either from abroad, as an imitation, or as the natural response to the attainment of higher standards of living and technological levels (industrialization of cities, changing commercial patterns, increasing popularity of private automobiles, and improvements in highway construction). Second, in view of these changes, it is extremely difficult to specify historical patterns and from them define what the Latin American city is today. It would perhaps be more appropriate to look at what it is becoming and to consider the integral relationship of rural-urban development policies to national, social, and economic development goals. The external "push-pull" influence on the internal structure of the cities is of increasing importance.

THE BARRIADAS: URBAN PATTERNS AND RURAL POLICY

In Peru the word for an urban squatter settlement is "barriada," in Brazil "favela," in Venezuela "rancho," and in Mexico "colonia proletaria." Each country has its own term, not only in Latin America but in all world regions where the rate of urbanization has overwhelmed planning and building legislation and where cities are growing faster than those in the already industrialized world.

The spread of barriadas is a social phenomenon of our times which could in a broad sense be compared to a similar phenomenon in Europe during the late 1940's. After World War II, with its destruction of many European cities, large numbers of people had to group together under improvised shelters in their search for new homes. In Peru, however, unlike Europe, the barriada problem is seriously affected by the underdevelopment of the labor class and by the huge influx of rural dwellers who continue to migrate to the major urban centers.

In Lima, Arequipa, and other major urban centers, the proportion of resident population each city contributes to its barriadas is insignificant. As statistics show, almost the totality of people in these sectors came originally from the central and southern Andean regions.

From the point of view of modern urban technology, barriadas still constitute "inorganic" portions of the city for they lack the public utilities considered essential to proper urban functioning: water supply, sewers, electricity, paved streets, and sidewalks. The sociologist visualizes the problem as a human transplantation under the most precarious conditions, where health is in constant danger, where children's education is unknown, where economic capital is practically absent, where there are willing hands without the existence of working centers. In short, the pragmatic issue appears to be one of total disproportion between needs and resources.

The barriada problem thus has often been approached as a function solely of the urban place. The solution will probably never be totally complete. But it is clear that the focus of action has to be not only on the urban solution but on the rural as well. The continuous migration of the indigenous population to the coast can certainly not be controlled unless the living conditions in the highlands are raised. It is clear that one problem brings another. The rational answer should be a new approach to each problem with coordinated action involving the government, the church, and every political, social, and economic force of the country, including both national and private capital and on both the rural and urban development levels.

Improving the living conditions in the highlands requires basic rural economic and social development attention. Implementation of the Agrarian Reform Law of 1964 has to be accelerated, but how does such reform affect the barriadas? Peru's economy is primarily based on agriculture; yet the country has less than half an acre of cultivated land per inhabitant. This low percentage, in turn, determines the low income average of the population. The situation is further complicated if one adds the absence of capital and technological instruments to increase productivity of the cultivated lands. In addition, one has to consider that approximately six million people (the majority illiterate) work in agriculture.

An increase in the average income per inhabitant can be achieved in two ways:

(a) through increase in productivity

(b) assuming a certain level of productivity, through an increase in income per man engaged in primary production, in relation to the income of the industrial countries which import part of the production.

Productivity, in turn, is increased by:

(a) adoption of modern techniques

(b) a more efficient distribution of personnel in labor where optimum productivity can be obtained.

Under the new Agrarian Reform Law the native farmer is given a piece of land to work and the necessary instruments. He ultimately becomes the owner of his property by gradually paying his debts to the state with a proportion of his production. But technical training, systems of loans and credit, and cooperatives are a few of the facilities which are still urgently needed on a large scale.

It can be assumed that squatter settlements will continue to exist, since they originate from a number of complex elements. The major factors are related to the agrarian situation, the problem of improvement of rural housing, and, above all, a high rate of population increase. These factors, together with industrial concentration and resulting employment opportunities in urban areas, are the underlying elements which influence the migratory movement of people from rural to urban areas. Legislation simply prohibiting the creation of new squatter settlements and uncontrolled developments in the marginal areas of urban centers cannot be effective unless regional development plans attempt to guide this invasion into predetermined and well-prepared reception areas. Every effort should be made to prevent premature abandonment of productive areas, to delay this migratory movement, and to guide the migration to urban centers where available facilities can absorb the new arrivals. If not appropriately dealt with, the squatter settlements will continue to be a source of instability and decay in the urban centers of Peru.

Whenever a *serrano* leaves the highlands he not only creates a problem in the city where he establishes himself; he also often disrupts the natural economic and social system which exists in his rural area. Partly responsible for the migration is the centralized character of the Peruvian government. To many, in fact, the country of Peru is the metropolitan city called Lima. Political, economic, and social decentralization of the capital is a necessary step for the improvement of the barriada problem, coupled with enlightened dynamic rural development policies and programs. These measures would help decrease the migrational rate and thereby enable the city to provide better economic opportunities, housing, and in general, an improved standard of living in a shorter span of time.

URBAN POVERTY: BARRIADA
VERSUS CORRALON

The typical image which a barriada suggests—particularly to wealthier nations and even the wealthier classes of the poorer nations—is of the worst kind of slum: miserable shanty towns in which the poorest of the world's poor suffer miserable lives. Although this may indeed be the true picture in some areas, this image does not sufficiently translate the total reality of the barriada phenomenon, as the following will suggest.

The barriada population in Lima has grown from an estimated 100,000 in 1958 (then about 10 percent of the total population) to an estimated 400,000 in 1964 (over 20 percent of the total). This rapid growth, however, is merely another reflection of the high rate of urbanization experienced by Lima in the last two decades. The barriadas, in effect, represent the single most important physical expression of the present urbanization process in Latin America: peripheral settlements of low-income groups. But today most barriada populations are not, by Peruvian or even by Lima's standards, extremely poor. This fact is obviously contradictory to the prevailing general opinion on economic conditions in squatter settlements and to the physical appearance of improvised shacks. Generally, the lives that the people lead in their barriadas are a considerable improvement over their former condition, whether in the city slums from which they moved or in the rural towns from which they migrated. A barriada family usually has its own plot of land and at least part of a fairly well-built dwelling which will be completed eventually, even though without many public utilities such as water and sewers.

The frequent and partially erroneous translation of barriada into English as "slum" has resulted in misconceptions. The same Peruvian upper- and middle-class misconception led to paranoic fears of a cinturón de miseria—a misery belt—surrounding the city, ready to cut it off, to invade the residential areas and seize the homes of the well-to-do. From the standpoint of the outside observer, these attitudes and conclusions are comprehensible since a barriada— especially in its earlier stages of development—looks very much like a slum. A valid distinction, however, should be made between the barriadas created before and after 1940. The few dating before 1940 have generally retained their slum characteristics, and a chaotic agglomeration of shacks predominates. In contrast, the majority of those dating from the early 1950's have developed into highly formalized urban geometrical grid constructions. Fortunately, recent close contact with the people and a better understanding of them and their problems have changed official attitudes and policies in Peru; but only a few years ago barriadas were often regarded by technicians and administrators as a slum blight to be eradicated or a cancerous growth to be suppressed.

It is also necessary to differentiate the barriada from the "corralon" or center-city slum, which to the causal observer in Lima might appear to be definitionally synonymous. The construction form of both the barriada and the corralon may appear to be a jumble of cane matting or adobe shacks. However, in spite of the superficial similiarity, there are important differences. Barriadas will almost certainly improve their dwellings either by self-help or government assistance, while the corralons, which constitute the principal reception site for rural in-migrants, will stagnate or decay. The great majority of corralons are situated on small plots near the city center, market areas, or any other source of casual or unskilled labor, while the barriadas are much larger and located at some distance from the city center and along secondary roads. The majority of the corralon inhabitants are appreciably poorer than those living in the barriadas and have lived in the city a much shorter length of time.

The settlement in a corralon is usually a one-stage movement while that of a barriada often occurs in two stages. A population will invade a corralon directly after migrating from a rural town to Lima. Occupation of a barriada, on the other hand, will take place by low-income populations who have resided in the city for some time. Both types, moreover, have different origins and functions as well as different destinies. The typical corralon was previously (twenty to forty years ago) a market gardener's small holding. Then the owner or tenant found it more profitable to sell—or to rent—tiny plots to immigrants or to families evicted from other center-city slums.

The typical barriada has a totally different origin. It is usually started by organized groups of families, some threatened with eviction from center-city slums scheduled for demolition, others unwilling to go on living in corralons and determined to build themselves a better house in more open and healthier surroundings. In most cases, these families are led by a small group interested in the chances of profit through the control of the distribution of land. Hence, they select and invade a suitable area of marginal desert land belonging to the government. The de facto possession of the occupied land, together with the scale of such invasions and the political problems posed by attempting to frustrate them (without providing adequate alternatives), have often guaranteed their success.

The barriada will develop into a typical working- and lower-class suburb, albeit slowly, since it will take the average family about twenty years to complete its house without credit assistance. In contrast, the corralon crystalizes into a typical labyrinthian complex of slum courts which can only deteriorate and eventually be eradicated.

Local geographical and cultural conditions are the bases of the differentiation of these two types of settlement in Lima. Yet in other cities and in quite a few areas of Lima itself, the barriada and the corralon are mixed. Real slums and quite respectable dwellings inevitably exist side by side in most major Latin American cities. But it is clear that if the two situations are confused—that of the immigrant or destitute family seeking temporary shelter and that of the established family seeking better housing conditions—the issue will be erroneously interpreted; hence, unfortunate delays will occur in the search for future solutions.

It should be clear from this review of the barriada phenomenon that the quality of urban life, even in less developed countries, cannot be determined by "urban policies" along—that is, by policies which are intended to affect what goes on within the city but not the demographic and economic patterns of the rest of the society. In the

United States, the problems of the central cities certainly must be seen as the result, not only of the inadequacies of municipal programs and facilities, but also of the massive decline of agriculture, the economic collapse of many smaller industrial towns, the migration of Negroes in search of some semblance of equal opportunity, and the largely uncontrolled exodus of white middle and upper strata into generally unplanned suburbias. Just as Peru and other Latin American nations will be able to deal more effectively with their "urban" problems only as policies are made and implemented for a coordinated rural development, so the "crisis of the cities" in the United States can be met only as some control begins to be exercised over major contributing factors whose spatial location happens to lie outside the city limits.

There is another analogy worth noting: the predisposition in the United States to define slums so heavily in "quality of property" terms leads easily to a disregard for other and sometimes more significant dimensions of the "quality of life." Especially since there is evidence of community development and self-help movements taking hold in American "slums," we need to make differentiations of the kind which enable one to distinguish between barriada and corralon and thus to avoid subjecting both to the same policy.

LATIN AMERICAN URBANIZATION
IN HISTORIC CONTEXT

More will be said later about the impact of these contemporary patterns on the demographic and ecological characteristics of Latin American cities. But it is important to place these and related developments in a historic context. For present patterns must emerge mainly within the constraints and with the degrees of freedom established by preceding generations.

In Latin America, most of the first-rank cities are situated on or near the coast, although Bogotá and Mexico City are notable exceptions. Their general location reflects a port-oriented economy going back to the days of the colonial period. In some cases a coastal city joined the administrative and marketing area and the port (i.e., Lima-Callao, Santiago-San Antonio); in other cases an inner city and the port were located at a considerable distance from each other (i.e., Quito-Guayaquil, São Paulo-Santos). These latter examples can still be recognized today as distinct urban pairs. In the first group, however, as in the case of Lima-Callao, the inner city and port merged to form an integrated urban structure.

A second contrast with North American cities is to be found in the conservation of a population of indigenous origin, left more or less unaffected by the European colonization. One of the most characteristic signs of this phenomenon is the preservation of a series of old indigenous towns that constitute the basic urban systems in Central and Andean America.

The duality between the Indian tradition in Latin America, and the commercial speculations inspired by and oriented toward the exterior can be translated into the coexistence of two urban systems: the first joins the old Indian towns of Central America and the Andes; the second connects the cities of speculation on the sea borders, the product of the colonial period and modern times. These two systems schematically opposed each other in their process of evolution. The older Andean cities had a precolonial knot from which industrial installations and work camps radiated in the course of time. In turn, the cities that were born during the colonial period, generally developing near ports, produced a mixed type of urbanization, associating monumental buildings (products of financial and commercial speculations) with neighborhoods of dockers and immigrants.

Among the first-rank Latin American cities, there is only one located in the high plateaus whose direct foundations go back to pre-Columbian tradition and which today presents the type of growth and urbanism characteristic of the modern city. This is Mexico City, which has grown in population from about 900,000 inhabitants at the end of the World War I to over five million today. All other cities with over one million population are in the coastal regions and on the colonization fronts, such as Buenos Aires (over six million), Rio de Janeiro (over four million), and Sao Paulo (over four million).

In varying degrees, Latin American cities also differ in their internal structure from their European and North American counterparts. This is particularly true in the location of residential areas and other activities where they depart from the general theories of urban form. For the most part, one cannot accept the Burgess theory holding that the structure of the city is a circumference radiating from the center in a plane or surface with an arrangement of land uses changing gradually (usually along radial avenues) according to the intrinsic geometric values of the land toward the periphery. This is not the case in Latin American cities since pronounced topographic discontinuities present real barriers to development. For example, Buenos Aires is on level land limited by the Rio de la Plata. Santiago is located on a valley surrounded by high mountains on every side. Bogota sits on a plateau limited by a high mountain range to the east and the swamplands of the Bogotá River to the west. Rio de Janeiro extends along valleys between numerous mountains and the Atlantic Ocean. It is clear that in every case the topography has had a controlling effect on the form and direction of city growth. The location of cities in the mountain valleys was favored by the more equable climate and the potable water from rivers in the valleys. Those not situated on the ocean were invariably connected by road to the ocean ports, as in the case of São Paulo-Santos.

POPULATION AND TERRAIN: CITY SHAPE AND URBAN HISTORY

The quality of the life experienced by the inhabitants of a city is inevitably affected by its size and shape, for these in turn have an effect on the character and content of the individuals' life experiences. It is more difficult, for example, to segregate the rich from the poor in small, compactly settled cities than in the large, sprawling metropolis. Again, where "natural barriers" such as rivers and railroads break the continuity of residential settlement, ethnic enclaves may remain well insulated from one another and abrasive contact and competition for land can be avoided—until increasingly dense settlements overflow the old boundaries.

We turn, then to a treatment of the size and shape of Latin American cities as abstract, partial indicators of the historic situation within which their inhabitants have worked out their particular versions of urban life.

The shape and size of a city may reflect quite accurately its complexity and stage of development. In general, those of small size conform to a linear or circular form, with little or no functional differentiation between adjacent areas. Large metropolises, on the other hand, acquire more complicated forms (usually fan shaped), with distinct functions between one area and another, particularly with respect to land use. Quite naturally, city shapes evolve over time by growth and transformation from an original circular, semicircular, or linear pattern.

In many cases the phenomenon of growth is explained by the sudden development of areas which were previously considered unsuitable for settlement. The specific reasons for this phenomenon vary from place to place. Thus, in Bogotá large estates were converted for development, and in the southern region of São Paulo artificial lakes were created. Empty desert land around Lima facilitated barriada development while new road construction promoted settlement in peripheral areas of Buenos Aires and Santiago.

The historical development of the shape of Latin American cities reveals a slow progression from the very simple to the more complex. In Buenos Aires, for example, there are at least three main stages in this process. The first runs from its foundation in 1580, until 1870. During this period the city was compact in form, and oriented to the port. The nearby satellite towns were only small agricultural settlements, not integrated into the core urban shape of Buenos Aires. From 1870 to 1940, however, these small towns experienced a rapid growth of their population, and Buenos Aires began to acquire a fragmentary shape. Several factors accounted for this development. The establishment of Buenos Aires as the capital of the Argentine Republic reinforced the importance of the city and attracted heavy immigration. Another factor was the commercial exploitation of the agricultural products of the humid pampa, which brought a network of railroads directed to the port of Buenos Aires. Along the railroad lines new small towns were established and those already existing began to grow rapidly. The third state in the evolution of the shape of the city runs from 1940 to the present, during which time the towns became joined together and absorbed into a single large fan-shaped metropolis.

São Paulo provides a second example of the factors shaping historical growth. It was founded in 1554 as a mission place on a small hill near the confluence of two rivers. The site was chosen because of its facilities for military defense and river communication. The shape of the city did not change from the time of its founding until the second half of the nineteenth century when only very small settlements appeared in its vicinity. The lowlands along the rivers were a main physical barrier to development during this initial period. A second stage may be marked out beginning in the 1870's and running to the 1920's, during which time expansion of the city began to take place due to the increase in the production of coffee and cotton and the establishment of industries. A major influence on the physical shape of the city in this period was the construction of railroad lines along the river valleys toward Rio de Janeiro and the interior of the state and toward the port of Santos. The urban area at this time was composed of fragmentary units divided mainly by the rivers. A third period in the formation of São Paulo started in the 1920's, by which time work on channeling the rivers had begun. The dams and lakes at the southern section of the city were created and a new layout of radial avenues established. Today São Paulo continues to grow along the main avenues, with industrial location taking advantage of new flatland along the rivers.

Generally speaking, the shape of a city is conditioned by: (1) the physiographic characteristics of the site, i.e., mountains (Quito, Bogotá), rivers (Buenos Aires), desert lands (Lima), flat valleys (Caracas, La Paz); (2) the location of external nodes of economic activity in the surrounding regions, i.e., natural resources, commercial centers, and tourist attractions which define the orientation of major roads, channeling and promoting development; (3) the relative magnitude of the different activities that take place in the city. The location of industry, for instance, may largely influence a general shape, since it is a real barrier which segregates surrounding residential development. A railroad line that crosses a city from one end to the other produces a similar effect. High densities may find physical expression in old residential complexes of row houses, quite different in general shape from loose, low-density patterns; (4) the technology of transportation used, as in Buenos Aires, where a mass transit system (especially railroads) defined a finger-like form along its routes. The extensive use of the automobile in Caracas today, as in almost all first- and second-rank cities, encouraged by increased affluence, justified the proliferation of expanded development in the fringes.

The recent urban population explosion, of course, has changed the shape of most Latin American cities but it has not affected all districts in the same way. Generally, the main increase in population density has occurred in the peripheral areas while the central sections have lost population or grown at a smaller rate than the city as a whole. A pilot study of this process for Caracas, Lima, Managua, and La Habana revealed that densities consistently increase along with population until some point (which may be called the "point of maturity") where they start decreasing sharply.

In Buenos Aires, between 1947 and 1957, differential rates of increase in population density occurred in three main areas. One was the central district of the metropolitan city where the rate was lower than that for the whole city. The second was a peripheral area in which the highest land values are found, and the third a transition area with medium values. In Caracas the same pattern of density increase may be observed: density in the center has gone up at a rate lower than that for the entire city, while the peripheral areas have the highest increase. Those with the highest rates within the peripheral areas are located outside the traditional limits of the city. In Lima there is a process of filling the areas between the major centers of the city—Lima, Callao, Miraflores—and a process of growth in the periphery of the metropolitan area. The rate of population density increase exhibits the same pattern as Buenos Aires and Caracas: lower in the old centers and higher on the periphery. In São Paulo, the density pattern presents three concentric rings: the center of the city decreased its population density between 1940 and 1950; the intermediate ring showed an increase lower than the rate for the whole city; and the outer ring experienced a higher rate than the city as a whole.

PERIPHERAL EXPANSION AND INCOME

The increase of population density in the peripheral areas is caused by the centrifugal movement of both low- and high-income groups away from the center. The automobile has allowed the latter to settle in the plush suburbs while the former have moved into the less desirable areas in their search for shelter. The phenomenon is similar to that in the United States with the exception of the low-income group migration to the periphery. Multiple reasons exist for this outward movement, including: (1) the existence of land of little economic value for either commercial or agriculture uses in the outskirts of the city, as in the case of the desert lands surrounding Lima; (2) the presence of large areas of public land on the periphery; (3) the existence of an extremely large housing deficit leading to much overcrowding in the central areas and thus allowing little room for residential increase next to commercial and government areas; and (4) the fact that most of the low-income groups have come from rural areas and prefer the peripheral districts, where at least they have their own shelter and sometimes a small lot to cultivate.

The movement of the low-income groups towards the periphery is most apparent in those cities that have experienced a high rate of population increase and have not had an industrial character. In Caracas and Lima, for instance, a high percentage of the total population lives in squatter settlements (Table 1), while in cities such as Buenos Aires or São Paulo the percentage is much lower. This peripheral expansion of low-income segments of the population fills in areas usually quite different from those occupied by the affluent; and, of course, the low-income communities cannot provide the amenities which the well-to-do can obtain, often as government services.

Peripheral settlements in the majority of first-rank cities today represent between 10 percent to 70 percent of their newly developed areas (see Table 1). In Rio de Janeiro, in little more than a decade (1947-1960), marginal settlements have grown from 17 percent to 38 percent of the total population (Table 2). In Caracas these settlements represent about 38 percent of the total, and in Lima they have brought about an unprecedented structural change in the city. The densities in these areas, moreover, are almost as high as in the center city.

Briefly, one may find three major types of residential developments in the peripheral areas of most first-rank American cities: the squatter settlements, the public housing programs, and the high income suburban developments. Each of these areas is characterized by a different density. The first two have extremely high densities of approximately 400 persons per hectare and the third about 50 persons per hectare.

The residential pattern in the peripheral areas presents a different layout and grain according to the type of development. The older squatter settlements located on hills do not present a regular street pattern. They do not have roads and the housing units tend to follow the natural contour of the terrain in their orientation. Individual property limits are not defined by lots and the development seldom presents an ordered street subdivision pattern. Some such settlements improve with time; their streets become more defined; and they physically resemble low middle-income developments. Contemporary high-income residential areas, on the other hand, tend to follow a loose pattern adapted to the topography and related to automotive transportation. This type of layout presents smooth curves which

TABLE 1

LATIN AMERICA: PERCENTAGE OF POPULATION OF IMPORTANT CITIES RESIDING IN SQUATTER SETTLEMENTS

Country	City	Year	Type of Housing	Population (%)
Brazil	Rio de Janeiro	1961	Favela	38.0
	Recife	1961	Favela	50.0
	Guanabara	1960	Favela	10.2
Peru	Lima	1961	Barriada	21.0
	Arequipa	1961	Barriada	40.0
	Chimbote	1961	Barriada	70.0
Mexico	Mexico City	1952	Colonia Proletaria	24.9
Venezuela	Caracas	1953	Rancho	38.5

SOURCE: U.N., *Economic Study of Latin America,* 1963, pp. 168-169.

TABLE 2

LATIN AMERICA: INCREASE IN THE NUMBER OF SQUATTER SETTLEMENTS

Country	Year	Housing	Occupants	Population (%)
Chile	1952	130,000 (1)	645,000	10.9
	1960	196,000 (1)	1,044,000	14.2
Venezuela	1950	409,000 (2)	2,143,000	45.8
	1961	494,000 (2)	2,488,000	34.6
Honduras	1949	39,000 (3)		
	1961	56,000 (3)		
Brazil				
(Rio de Janeiro)	1947	Favelas	400,000	17.0
	1960	Favelas	900,000	38.0
(E. of Guanabara)	1950	58 Favelas	159,000	7.1
	1960	147 Favelas	337,000	10.2

1. *"Pieza de conventillo," "rancho," hut or "callampa."*
2. *"Rancho."*
3. *Houses of cane-and-straw walls.*

SOURCE: U.N., *Economic Study of Latin America,* 1963, pp. 168-169.

contrast with the traditional gridiron street patterns.

Thus, the urban areas in Latin America are not developing an inner city which belongs to the poor and to racial-ethnic minorities and a peripheral system of suburbias possessed by the more affluent of the majority. Yet the grossest distinctions in the quality of urban life rest here, as elsewhere in the world, on gross differences in income, and these distinctions are made evident by the crudest measures of the most obvious aspects of city life: density of settlement, quality of housing, availability of facilities and amenities, and even the layout of the physical environment.

INDUSTRIALIZATION AND URBAN STRUCTURE

Industry presents a particular case in the development of Latin American settlements. First, it should be remembered that industry has generally had only secondary importance in Latin America. The primary activities (agriculture, mining) still basically dominate the secondary (industry) and tertiary (services) activities in the employment structure. Industry also presents a very slow development. For instance, Peru had the same percentage of total labor force engaged in industry in 1950 and 1960. Second, industry in Latin American countries is usually concentrated in one city (i.e., the capital), but even in such cases, a higher percentage of the city's total labor force is more likely to be engaged in the tertiary rather than the secondary economic sector. In some instances, such as that of Lima, 75 percent of the total industrial labor force of the country is employed within the one city.

Only recently has industry begun to gain importance as a factor in the life of the major cities. Traditionally, the latter have been administrative and not industrial centers. This has had an influence on the location of industrial establishments because they have had to develop in already mature cities—a different pattern from that in more developed countries, where people were originally attracted to the cities by the industries located in them. In the last decades, however, the existing industrial zones of the large cities have expanded while many other such zones have appeared in other different sectors on the periphery. But the advantage of locating industries near the peripheral populations (predominantly comprised of workers) is reduced in every case due to the lack of adequate means of transportation.

The internal structure of a city is a function of its population size and degree of industrialization. In small or preindustrial cities there is relatively little complexity of functions; the main economic and social activities are undispersed and generally take place within one or more localized focal points. However, as the city grows and becomes more industrialized, the activities become more sharply differentiated among themselves so that they tend to locate individually within the city. As a result of this process, different functional foci begin to appear. This point may be illustrated at once by three first-rank Latin American cities: La Paz, a small capital city, of a population between 300,000 and 400,000, with a relatively low degree of industrialization; Lima, with a population around two million and in a process of industrialization; and Buenos Aires, the largest city in Latin America, and highly industrialized.

In La Paz one finds that the main activities of the city are undispersed and take place at the city center, maintaining the traditional pattern of centralization of activities around the Plaza de Armas. In Lima, the general pattern of the internal structure of the city was, until 1940, very similar to La Paz, but since that time a general movement of the upper class away from the center towards the periphery was accompanied by the beginning of the industrialization of the whole city. As a result of this, commercial centers are appearing in the suburbs (Miraflores and Callao) and industrial areas are becoming clearly defined. In contrast to La Paz and Lima, Buenos Aires is the example of a mature city. It presents a hierarchy of centers already established and a spatial differentiation of activities.

THE INNER CITY

The inner city has always been the major retail center of Latin American cities. This phenomenon is due largely to the relatively low purchasing power of the population and the orientation of retail trade toward the pedestrian. The core of the city has the largest threshold and is the point most accessible by means of mass transit from other parts of the urban complex. Thus Buenos Aires and São Paulo have a large movement of daily railroad commuters, and most other first-rank Latin American cities depend on bus or taxi systems or mass transit. The major railroad or bus terminals have a similar location pattern in these cities: they are placed in the transition area of the core, so that they clearly define the central area of the city. The only city in Latin America that has a subway system is Buenos Aires, but this system is mainly a connection between the major railroad stations that carry out the heavy commuter transit. Several points of retail concentration are to be found along major arteries from the central district to the residential areas, or at the centers of smaller towns that have been absorbed by the city. In Buenos Aires, the main suburban commercial centers are located near the railroad stations—at the original centers of formerly independent towns.

In most Latin American cities the major commercial and financial enterprises are still located in the center of the city. The suburbanization of these activities has not taken place as in the United States. More akin to European and Asiatic cities, the chief retail outlets in Latin American metropolises consist of thousands of small shops located in or near the core. Planned regional centers are rare; and when retail stores move out from the central district, they go to the center of the suburbs rather than to the peripheral areas. As the first- and second-rank cities grow, however, the locations of industry, shopping centers, and other activities can be expected to occur simultaneously with new housing developments. In fact, new suburban areas in Bogotá and Miraflores in Lima already indicate evidence of this trend. Here, suburban satellite shopping centers, office buildings, and theaters are the focus of planned residential areas.

The *galeria* is perhaps the most interesting shopping development in the central areas of many South American cities such as Lima, Santiago, São Paulo, Buenos Aires and Rio de Janeiro. As an example of retail architecture, the galeria is comparable to the arcades found in the United States. It usually joins two streets by means of a roofed arcade, broken at several levels and with access from four streets. Small adjacent shops, about twelve to twenty-five feet wide, are arranged at either side of the covered arcade. In São Paulo, for example, a new galeria with five levels on the Avenida São João is in the center of the downtown area. In Lima, Galeria Boza, a smaller scale example, may be found just off Jirón de la Unión, the main shopping street.

Hotels constitute the primary residential buildings in the center of most of the Latin American cities. Their location defines a specific area that is generally near the financial and government centers. Many hotels are located next to the main shopping street, having a close relationship to the major national transport terminals. New hotels are currently being planned for a site in downtown Lima reinforcing the traditional center core location. However, in Guatemala City a new hotel is situated some distance from the central plaza on a main avenue connecting with the airport.

Government activities have traditionally been located in the immediate vicinity of the Plaza de Armas, since the Spanish colonial governments built the viceroyal palace, the city hall, and the cathedral around the main square. In most capital cities the presidential palace has remained adjacent to the central plaza; but with the division of power into the executive, legislative, and judicial branches, government activities are no longer restricted to the area immediately surrounding the main square and have begun to spread out over the central area of the city.

In La Paz, the government activities are concentrated around the Plaza Murillo, where the presidential palace and the congress are located. In Caracas the government offices are concentrated around the Plaza Bolivar—similar to La Paz except that the presidential palace and the ministries have been removed from the plaza to the new Centro Simon Bolivar. This modern center is constituted by twin towers, thirty-story office buildings with the upper floors occupied by the government. The street level and a lower level are devoted to shops, restaurants, and parking facilities, with the super highway Avenida Bolivar passing under the lower shopping level. Lima is more dispersed in respect to government activities, with the congress building located at a distance from the Plaza de Armas.

São Paulo is a peculiar case because it is not a national capital but the capital of the state of the same name. Its importance lies not in its government functions but rather in its particularly highly developed industrial capacity and its financial power. An indication of its increasing importance as an industrial city is the fact that the government buildings, formerly located in the immediate core, have been moved to locations more offset from the center.

Latin American cities have relatively few open spaces in relation to the size of their population. In 1954, for instance, Lima had only approximately 33 square feet of open space per person. (Although the suggested "ideal" figure varies from city to city, some standards have placed it at 250 square feet per person.) This lack is partly due to the continuous housing that makes the center cities very compact, and partly to the Spanish tradition which made the Plaza de Armas virtually the only open space in the city. Thus today the major open spaces are found away from the center of the city, usually in the direction of the best residential districts, the private areas of recreation (such as country clubs or golf courses), and resort places along rivers or the sea. Of these open spaces, the private areas of recreation are generally located in the peripheral areas of the city, but only in the sectors where the high-income groups reside. In Caracas one may find the districts of Altamira, La Castellana, and El Pedregal near the Caracas Country Club; in Lima, the country club is in San Isidro; in Buenos Aires almost all the private areas of recreation are in the northern part of the city, mainly in Palermo, Vicente Lopez, San Isidro, and Tigre.

The main recreational function in the central districts of Latin American cities is represented by the movie houses, theaters, and public parks. In large cities like Buenos Aires and São Paulo, theaters form a special section within the central area; in smaller cities like Lima, they are part of the main shopping street. Public parks are located more in the outskirts of the central area, in the direction of the high-income residential areas. This is the case of the main public parks in Buenos Aires, the Reserva Park in Lima, and the Japones Park in Santiago.

The inner city, then, is in most Latin American urban areas a section which has suffered relatively less deterioration and desertion than is true in many of the larger municipalities in the United States. But its future is far from certain as population growth and movements continue and supporting industrialization lags throughout most of Latin America.

TOWARD A COMPARATIVE PERSPECTIVE

In perspective, one cannot help but sense that the Latin American city is dynamically concerned with growth, while the United States city seems preoccupied with the perils of decay. However, in spite of the different forms and the different influences on them, the cities of both continents are struggling with several basically similar sets of problems, among them: (1) a common battle to handlé population growth; (2) increased automobile ownership, and outmoded transportation systems in both the urban core and the metropolitan region; (3) land uses, transportation, and other activities often diametrically opposed in space and time; (4) extreme difficulties in instantaneous urban creation, as demonstrated in such new-town solutions as Ciudad Guayana in Venezuela, and Reston and Columbia in the United States; (5) the necessity of looking toward the natural environment, whether it be the sea, river, mountains, or plains, in planning development and growth; and (6) the accommodation of low-income migrants entering the city with limited employment opportunities.

These similarities suggest the need for looking beyond one's own country in viewing the form of contemporary cities. In Latin America, for example, the burden of the city's responsibility to the low-income migrants and the resulting mushrooming of self-help housing construction greatly exceeds that of the North American city. Yet the Latin city, perhaps due to the greater magnitude of low-income housing needs, appears to be more realistic in the acceptance of the longer period of time required to evolve improved physical, economic, and social solutions.

8

The Tech-fix
and the City

H. L. NIEBURG

☐ THERE IS A vulgar, self-serving, and optimistic myth that new technology and science offer the best and only solutions to the problems of high-level economic stagnation and rigid social cleavage. As formulated by Alvin N. Weinberg, Director of Oak Ridge National Laboratory, "technologically oriented research" can devise "cheap technological fixes" which will "short cut" the solution of social problems "without having first to solve the infinitely more difficult problem of strongly motivating people . . ." (Weinberg, 1967: 141).

A new expertise is arising with new claims of indispensability, uniqueness, and omnipotence: that of systems analysts, systems engineers, and system managers. The magic word "systems" has come into vogue from wartime operations research and logistical planning; in the postwar period it has become the tool of prime aerospace contractors and of all kinds of experts, both hard and soft. Under the systems rubric, every job undertaken for government becomes essentially a research and development (R & D) job. Government agencies specify requirements that cannot be met by the existing state of the art in science and technology, whether a weapons system or a method for conserving resources, organizing recreational land use, and the like.

The aerospace industries (which now include virtually all large corporations, among them household names like GE, GM, RCA, and AT&T) claim to have proprietary control of most of the systems capability available to serve the nation's interests. As spending for strategic weapons systems and space at last begins to inch downward ($40 billion for aerospace contracts in 1968), and as domestic problems sharpen, the aerospace companies are making this special capability available for the operation of Job Corps camps, management and planning of poverty programs, fighting crime, desalination of water, oceanography, waste management, urban transportation. This development confronts us with a curious paradox. As the international science-technology race at last slackens, as public facilities and domestic needs at last begin to receive official attention, the quasi-public utilities (largely uncontrolled and operated under a code based on a fundamental conflict of interest) are moving into every area, including traditional fields where civil servants of municipal, state, and national agencies or universities, on a nonprofit basis, formerly did the job. Here, too, every problem is defined as R & D and claimed as a special province for the systems capability of large corporations. The paradox lies in the fact that the ideology of these enterprises, faith in the infallibility of science and technology, is directly contrary to the trend in most social philosophy. The tide of opinion among philosophers has come to question the very position now assumed by those with more concrete political and financial stakes in science and technology.

The tech-fixers are the direct descendents of St. Simon, Karl Marx, Frederick W. Taylor, H. G. Wells and others, heirs to the classical doctrine of progress and human betterment through invention, social innovation, and above all, productivity and the machine. At the very moment when this doctrine is crashing down (due to the work of such men as Reinhold Niebuhr, Jacques Ellul, Daniel Bell, and Lewis Mumford), it has achieved unprecedented vogue and public power. Under the pressure of twenty years of permanent diplomacy, space and arms races, limited wars, and nagging political crises, the tech-fixers seem close to a complete monopoly of social purpose, generating a kind of self-fulfilling prophecy of accelerating technological change, ever more complicated systems, more planning and direction from above, in a never-ending spiral. While the establishment clings to tech-fix as a solution to every problem, while conglomerate corporate empires soak up new companies like giant amoebae, and while affluent America buys every new gadget, the poor and the rejected huddle in the big cities which Martin Luther King called "the poorhouses of the welfare state" (Committee on Government Operations, 1966: 2968).

A tendency exists for the new crop of critics to share some of the basic assumptions of tech-fixing. Mumford, Ellul, Galbraith speak of a "technological imperative," technology and science as implacable and impersonal forces. Mumford, for example, writes: "In the most general terms, this basic problem is the control of power, quantification, automatism, aimless dynamism" (Mumford, 1968: 247). There is, of course, no such thing as "blind technology" except as pejorative rhetoric. The problem of controlling technology is more usefully defined as "controlling the technology coalition," that is, limiting the power of those whose political, social, and economic role is legitimized and augmented in the name of technology. In effect, tech-fixing is a special-interest myth of prevailing power groups. More and more public tasks these days, especially "the problem of the city," are defined as "systems problems." A built-in self-serving process is at work by which efforts to solve urgent social crises become instead efforts to advance the state of the arts, to push new hardware development; in sum, to maintain the status quo of social power while attempting to contain, deflect, bribe, or repress the challenge of awakening masses.

Just as every energetic pressure group capitalized on the cold war, the Soviet Sputnik triumph, the Space Gap; and just as educators, scientists, industry, the military, all offered formulas to save the nation, so the explosion of black ghettos, the strikes of teachers and other public employees, the demonstrations of the welfare poor, and the rebellion of the college generation furnish a new wave of special-interest slogans. The massive black army of potential seditionists provides the threat and the promise which every other group wants to exploit for its own interest. The tech-fixers of the prevailing power groups are in command, using the abstractions of technology and systems to exploit the urban tragedy in order to maintain their advantage. They proceed in the name of such abstractions as "the inevitability of machine progress," and "the search for knowledge," all the while reiterating their faith that if only enough money is invested in R & D, a painless solution will emerge which will satisfy everyone without curtailing existing property rights, social inequities, and political jurisdictions.

In terms of political power, there are no such abstract issues, only "Who gets what, when, and how?" and "Who's doing what to whom?" It is a political fact that has defined the crisis of the cities as "a problem." The situation was manageable and purblindly tolerable so long as slum occupants confined their crimes to the ghettos, internalized the disarray of their lives through mental disease, or buffered it with narcotics. But with the arrival of self-consciousness, militancy, and incipient organization, the heat is on. Metropolitan pathology ceases to be an abstract issue to be safely exploited, studied, pacified, but becomes a confrontation, urgent, inescapable, and perilous. Tech-fixing and all other kind of cheap fixes will no longer serve. Even counter-insurgency and police repression become provocative, ineffective, and self-defeating.

THE PATHETIC FALLACY

Faith in economic progress and social justice through innovation and productivity has been collapsing on every hand; but simultaneously it achieves a full panoply of almost unchallengeable power as the mantle of the science-technology establishment. In a kind of self-hypnosis, this coalition and its PR legions (university and government officials, professors, scientists, captains of industry) strive to impose scientific discovery, knowledge for its own sake, space exploration, constantly revised and sophisticated weapons and artifacts of all kinds as the primary purpose of social life. Far from their pragmatic and materialistic self-image, they practice an intensity of moral feeling toward hardware, numbers, computers, abstract schemata and plans that reveals a truly medieval religious spirit. For the "Gauleiters of big business" such journals as *Fortune* are sacramental, McLuhan writes, "conducted as a major religious liturgy celebrating the feats of technological man. . . . a Bayreuth festival in the most megalomaniac style . . . paeans of praise to sheer production" (McLuhan, 1951: 11).

There is ample pragmatic basis for religious intensity and moral duty surrounding science-technology. The American megamachine enables less than ten percent of the work force to produce more than enough food for the entire population. Full utilization of our resources, even without additional technological advancement, would produce a gross national product able to guarantee every family in the United States a standard of living equal to an annual income of more than four thousand dollars. The contribution of technology and organization to productivity is unquestionable. A century ago men and animals did most of the work in industry, with machines supplying about one horse-power for each worker. Today technology harnesses an average of ten or more horse-power per worker, increasing his productivity ten times, making old products better and/or cheaper, and adding an endless wave of new products to make life and work more efficient, if not better.

By any index, the science-technology explosion continues a vertical trajectory. The number of scientists-technologists in the world has doubled every 16 years since the seventeenth century, which means that three-fourths of all who have ever lived are still alive today. The amount of scientific information published in the world every twenty-four hours would fill seven complete twenty-four-volume sets of the *Encyclopedia Britannica,* tallying about 61,320 volumes a year. Were one to divide the volume and number of chemical publications, for instance, into two equal parts, the date of that division would be well within living memory (about 1950). The dividing date which separates equal quantities of metals and minerals extracted from the earth throughout all history is about 1910. Another fact is that about 25 percent of all the human beings who have ever populated the earth are still alive today.

It is the quickening pace of knowledge explosions, the foreshortening of R & D lead-times, the growing complexity and expense of technological prototypes and weapons systems, that accounts for the incipient state of mobilized innovation. The slogan of the developmental engineer ("If it works, it's obsolete") is symptomatic of an almost paranoid fixation. The international science-technology race seems headed toward a reductio ad absurdum in which the rate of innovation itself creates the most severe problems of reliability and overrun costs. Needless complication both technically and administratively leads to escalated confusion and the application of Parkinson's Law. We must run faster and faster barely to manage to stay where we are; build elaborate automated pyramids of systems analysis with systems managers and engineers, a daily saga that rivals for dramatic imagery and noise the Napoleonic campaign in Russia. All seems to be well until an event like the incineration of three Apollo astronauts shockingly disrupts the facade.

Brain factories, future planners, science facilities, laboratories—the single-minded concentration upon innovation has swept the nation in what an editor of *Fortune* calls "the greatest advance in the art of government in nearly a hundred years." Great Society plans have become a commonplace power play by university professors seeking grants and industrialists seeking contracts in a nation which ideologically rejects planning and yet continues to plan indirectly through federal subsidies, grants-in-aid, and procurement contracting. The new think tanks, engaged in simulation and gaming (with or without computers), thinking through and writing up concise reports on all kinds of things that generals, administrators and business men used to do for themselves, are today a multibillion dollar industry. They are not complete without mathematicians, computer programmers, behavioral scientists, economists, and assorted dry-eyed editorial assistants and engineers. These systems analysts and operations researchers constitute the "brain storm" troopers of the futuribles.

The doctrine of tech-fixing holds that anything can be accomplished if enough money is spent to subsidize science and technology. The notion is seriously taken that technology dictates its own surpassing. Subsonic commercial jets make supersonic jets irresistible. The energy of nuclear fission compels response to the siren call of the infinitely greater energy of nuclear fusion. One-billion-electron-volt accelerators make two-billion-electron-volt accelerators unavoidable and pressing. A manned landing on the surface of the moon opens the way and dictates a manned landing on Mars. The tech-fix is the modern "pathetic fallacy" which, unlike that of the romantic poets, fastens upon technology instead of nature, endowing clockwork imitations of life by mechanical dolls with the same fascination and optimistic delight which Wordsworth found in daffodils. The tech-fixers persist in the naive faith that everything can be transformed by dramatic "breakthroughs." This pathetic fallacy is the current myth of legitimacy. Like all historic myths of power, it posits purblind confidence in man the doer, and science as the true and ultimate abolishment of uncertainty. The arrogance of this view, the "can do" delusions of grandeur, combined with great private and public power, is highly dangerous to internal democracy as well as to the realism of national goals.

The myths of tech-fixing have long enjoyed an honored place in the trinity of platitudes (along with motherhood and the flag). The invocation of onward-and-upward knowledge and science is the common pap of perorations. Typical are President Kennedy's declamation: "Let us explore the stars, conquer the desert, eradicate disease, tap the ocean's depth. . . "; and President Johnson's assertion: "Scientific research is the key with which we unlock the door to the future. . . . The task we have set for ourselves is to wrest from nature

THE TECH-FIX AND THE CITY [217]

the intellectual treasures with which we will build the world of tomorrow." Like most noble sentiments, these avoid clues to hard policy choices and priorities. Which comes first, the moon or the slums, the unexplored or the unemployed, security or solvency? The range of choice is endless and conflict lies hidden in any attempt to order the priorities. This is the heart and soul of public policy and imposes political deficits almost more numerous than assets. The art of statecraft is to order priorities and its key virtue is courage in developing standards of judgment and accepting the political costs of applying them.

Extreme claims are common fare in the public prints, too frequently purveyed by experts with high credentials. We are told of anti-aggression pills designed, not to rob man of ambition, but only to control the "pathological desire to fight"; cryogenic freezing of the old and the diseased, to be thawed decades or centuries hence when science has achieved cures for both infection and aging; of the elimination of all labor through automatic factories, and the conversion of the household into an euphoric womb-like tank, chore-free and stress-free, liberating all of man's energy for art, philosophy, self-examination, enriched personal relationships, fun and games, relaxation and atrophy; of the elimination of communicable disease and annoying insects through DNA genetic engineering; predetermination of the sex and physical endowment of our progeny. The genetic engineers will design supermen to live in societies devised to optimize some system of values which presumably such an existence will generate without strife. The president of the American Chemical Society, Dr. Charles C. Price, has proposed that the synthesis of life should be a national goal on the order of the moon landing, a proposal which elicited from the New York Times the editorial comment that the most "unfortunate result . . . would be the mounting of a new international competition, a race that would produce the same tension and needless duplication that the space race has produced."

Herman Kahn's Hudson Institute, a nonprofit research center spun off the Rand Corporation, recently issued a staff-report listing the one hundred "very probable" technical innovations expected by the end of the century: a method to let people decide before they sleep what they want to dream about; artificial moons to provide "cheap" nighttime illumination for any area of the earth; slave robots to bear the same relation to the owner (for such things as letting out the cat, retrieving a shoe from under a bed, and weeding the garden) as does the automobile in providing transportation, recreation, and status display; life expectancy beyond one hundred years; supplementing one's thinking by connecting a brain directly to a computer; foods which give all of the satisfaction of eating without adding weight; and weather manufactured or modified at will (whose will?) (Kahn and Wiener, 1967).

A fundamental and persistent dictum of tech-fixing is unquestioning faith that science and technology in all forms, both pure and applied, regardless of direction, constitutes the underlying secret of economic growth and hence "always" pays for itself in the long run by productivity advance. If automation shrinks a payroll, cuts market purchasing power, and fails to lower prices, the fault must lie in the perverse unwillingness of the obsolete production-line worker to become an engineer or computer programmer, or in a backward educational system, or in the social disorientation of school dropouts. If space and defense research fails to spark new products and open new markets, the blame is laid upon the inertia of private initiative in the civilian market or lack of effective information systems to implement "technology transfer" (see Lesher and Howich, 1968).

The fad of public investment in research and development for computerized classrooms, programmed instruction, talking typewriters, and similar innovations is representative of many broad new undertakings of the antipoverty programs, which, it is argued, constitute a fresh approach to problems long resistant to conventional efforts. In fact, the amounts spent on technologizing far exceed the funds previously available to support the dedicated human beings who sometimes succeeded in establishing contact and breaking the cultural isolation of a few needy and evasive poor. Such breakthroughs may come as by-products of tech-fixing, but the sudden influx of new funding for such programs is not aimed so much to help "the disadvantaged" as to combat the "poverty" of giant aerospace and electronic firms whose other government contract business is lagging and whose R & D teams are considered an invaluable resource for future government needs. In spite of the poverty programs, the tide of emigration and relief payments among the rural poor grows twice as fast as the gross national product, and our major cities and universities have become permanent festivals of confrontation, hardening the vicious circles of alienation.

A leading tech-fixer, General James McCormack, once Air Force R & D administrator, now head of the Communications Satellite

THE TECH-FIX AND THE CITY

Corporation, told an audience of educators that the way out of our difficulties is more innovation: "Technology makes it possible and not enormously expensive to make cities of light, . . . rather than cities of terror. . . . Turn to your own fields—the humanities, science, the arts—and determine how to use the fruits of technology." Only by this means can "the quantum jump" (i.e., major breakthrough that transforms the whole situation) which, in his words, "so astonishes society that we don't know what to do with it," provide "more opportunity for the things that raise the spirit and uplift the soul" (New York Times, Sept. 16, 1966: 2).

The tech-fix syndrome leads to gold-plating (that is, unnecessary sophistication or mechanization) of every aspect of our lives, uncritically, wastefully, and wilfully widening the caste and class cleavages between consumer groups. This result is seen in the haste of educators to take General McCormack's advice. With the aid of ample federal grants and contracts, the top institutions of higher learning have built multi-million dollar computer facilities which they have hardly learned to use. At MIT it is common to talk about teaching methods "BC" and "AC"—before and after computers. The complex at Harvard has maximized "access" by mounting fifty keyboards around the campus' five major computers, with twenty-five additional consoles located in student dormitories. Similarly, all the leading institutions (such as Michigan, Cal Tech, Dartmouth, Carnegie Tech, Texas A&M) are adopting automatic data processing to courses far afield from engineering and mathematics. Music and art students compose. Social scientists invent games and build models that lead inexorably to the conclusion already contained in the programming scenarios. Then the conclusions are treated as empirical results which they are not. Too often, the mantle of data enables the "experimenter" or program designer to don the impersonal authority of the machine; the impenetrable and interminable data conceal all his hunches, values, and preconceptions in a toga of anonymity. Tons of meaningless data are accumulated. "When you don't know what you're doing, you count something!"

There is a tendency for computer fetishists to equate "information" with knowledge and wisdom, and to assume that the business of teaching is mainly, if not entirely, information transfer. Robert Tschirgi, Dean of Planning of the University of California has this to say: "Is it any less comprehensible to imagine a generation with nostalgic memories of one's old computer-mentor than to have cherished remembrances of ivy-covered walls?" (Ridgeway, 1968:

54). One evaluation of a machine-taught sociology course concludes: "All the computer did was to print up some basic definitions in an introductory course, which one could get just as well from reading a book;" a minute portion of the course was on the computer and more time was taken for programming than for students (Ridgeway, 1968: 55).

The B. F. Skinner form of mechanistic behaviorism is very appealing to computer scientists. They confidently expect that eventually machines will duplicate the human brain, learning from its own experience and discovering a principle of creativity and selection beyond mere malfunction or accident. This is a naive expectation. So long as man designs, builds, and programs an instrument, no degree of sophistication or infinite regress of self-programming programs will endow it with ultimate adaptability in completely unprecedented, unanticipated and untransferrable problem-solving. Man has the unique ability to test for significance in nonlinear, imperfectly comprehended patterns which on some levels are nonsymbolic, intentional, and capable of adaptation and initiative (Adler, 1967; Bertalanffy, 1967).

The tech-fixers have kind of a flat-earth view of man. Confidence in the machine is blithe and unsullied. Emmanuel G. Mesthene maintains: "Computers and associated intellectual tools can . . . make our public decisions more informed, efficient, and rational, and less subject to lethargy, partisanship, and ignorance." (Mesthene, 1968). Ithiel de Sola Pool expects that the computerized, and therefore more powerful, behavioral sciences will generate behavioral engineers as middlemen between knowledge and its social applications (Pool, 1968). "The behavioral sciences" will quantify and resolve the indeterminate variables of human nature, thus encompassing the whole equation of both physical and human environments for computerized systems analysis. The social scientists have been reluctant to clarify their own limitations, and the hardware types are infinitely believing. The National Research Council has recommended setting up a National Institute for Advanced Research and Public Policy which would enable "the decisions and actions taken by the President, the Congress and the Executive Departments and Agencies" to be based on the knowledge and methods of the behavioral sciences (*New York Times,* September 3, 1968: 16).

Mumford suggests that computerized projections might be useful in calling attention to undesirable trends whose recognition should induce avoidance. Instead of this, he grieves, the systems analysts treat the readout of the data as instructions to continue repeating the old errors (Mumford, 1968: 179). To reduce the human factors to those the computer can handle is to eliminate the better half of life. The depersonalized manipulation of persons as though they are things is, said Martin Luther King, "as much responsible for the perpetuation of grief and misery in our cities as is the absence of wealth and national resources" (Committee on Government Operations, 1966: 2977). Yet it is exactly this kind of treatment and perspective that is embodied in the computerized systems approach to urban problems. Werner Z. Hirsch, an economist, typified the syndrome in his statement before the Congressional Joint Economic Committee: "Information systems and cost-benefit analyses provide the decision-maker with the kinds of vital information he needs for making wise decisions. Careful planning today can provide a large variety of beautiful and yet efficient cities for tomorrow" (Joint Economic Committee, 1967: 8-9). If cities are to be saved, the most vital information to which decision-makers must respond concerns the action capabilities and values of new groups that no longer will be passive. One does not require and cannot adapt computers to process the impatient demands of the new organizations of the poor.

THE CITY AS ABSTRACTION

In its approach to urban problems, the tech-fix mentality grinds out predictable schemes, concentrating on physical materials, architecture, spacing, automobiles and highways, computerized analyses and controls. In doing so, it ignores the troublesome human dimension: systematizing abstractions rather than living groups, individuals, and neighborhoods; converting face-to-face services into hardware and "black boxes"; inventing new ways by which government can increase subsidies for maintaining and containing slums; offering more and more research as a surrogate for action programs; emphasizing manipulation and centralized management by the old Establishment rather than participation and self-determination by emerging minorities. When a problem presses hard, it is cheaper to enrich the research establishment with moderate funds, as if some magic solution will be found to make massive expenditures unnecessary and leave vested interests untouched. Technology and science become an alibi, a political delaying tactic, an escape from reality, and a tool of self-interest, ignorance, short-sightedness, and procrastination (Nieburg, 1966).

Like the largest part of urban renewal, the New Towns movement and model city plans are swiftly corrupted by special interests and tech-fixers of all varieties, who share the common fallacy of emphasizing tangible physical components (readily manipulated) while ignoring the overall social context which might make the concept work. The president of the Franklin Institute, Athelstan Spilhaus, suggests that the only solution for cities is to dismantle and disperse them utterly, and rebuild from scratch, not with initiative and control by the people who will live there, but by "the leadership, imagination, and enthusiasm of scientists, industrialists, and educators. . . . We must utilize the most advanced methods of construction, transportation, communications, waste removal, and city management" (Spilhaus, 1968). This approach typifies what Mumford calls the city as a "disposable container," an aspect of the throw-away society created by modern technology.

Edmund N. Bacon, director of the Philadelphia Planning Commission, has done much to make segments of the historic inner city green and open works of art. However, the enormous funds expended to achieve an artful setting for Independence Hall do not resolve the separatism and decline of nearby Negro slums. The new squares and malls, in fact, have been a problem for the police to keep safe for tourists, suburbanites, and white-collar workers. Everywhere in the rest of the city has appeared the spreading scrofula of open parking lots, tinny motels, gas stations, and similar insightly uses, a phenomenon rampant from coast to coast and neither contained nor redirected in Philadelphia. In his recent book, Bacon tends to equate city planning with architectural design, failing to grapple with the dynamism of a complex and increasingly restless social environment (Bacon, 1967).

The city is a complex living organism, of which its technological and physical environment is as much a response as it is a condition. In a sense, the physical growth of a city responds to the interests and desires of those who exercise political, social, and economic power. The problem of the city is the emergence of a new capability for social bargaining by previously submerged groups. Attempts to isolate and treat accessible physical forms may pursue apparent needs, but these are peripheral, tend to serve the prevailing power groups, leave the basic social equations (which create the impetus and incentive for change) to fester and squirm beneath the surface, ultimately to break through and deface the facade.

One of the most earnest and interesting New Town projects has been Reston, Virginia, on a 7,400-acre tract of rolling countryside, eighteen miles northwest of Washington. But Reston is not going well, in spite of Secretary of Interior Udall's effort to make it a "cross-section of America" by locating a major federal facility there, and by HUD's low-interest loan for the construction of 200 units of low-income housing. So far the cheapest houses have sold at the mid-twenty thousands and rents for three-bedroom apartments start well over $200 a month. Dedicating the new federal facility, Udall said that "in this land of equal opportunity, no town, despite 'the brilliance of its design and the insight of its planning,' can claim to be truly American if it is an enclave of the well-to-do or the private preserve of any single ethnic or racial group," (quoted in *Science,* November 10, 1967: 752). Reston's idealistic founder, Robert E. Simon, Jr., unable to carry the heavy debts of lagging development was eventually forced out by the Gulf Oil Corporation which has a $15 million investment in the project. Gulf is more interested in promoting sales and minimizing risks than in "a cross-section of America."

While a major model cities demonstration program remains stalemated by Vietnam war costs, a few large corporations in the United States have entered the area of slum renovation and New Town development. These ventures have often been good business for the company while further reducing the housing available to the poor. For example, the United States Gypsum Company has been tearing out the inside of slum apartments and rebuilding them as fully equipped modern apartments (with three coats of epoxy paint sprayed in the hallways to make the walls impervious to almost anything). First choice for re-occupancy goes to the previous occupants; however, rents are usually more than doubled and beyond their means. Virtually all of the work (except special security guards) is performed by lily-white labor unions (*U.S. News and World Report,* September 2, 1968: 58-60). Present model city plans give lip-service to opening construction jobs to ghetto residents, but enable trade unions to classify blacks as "special trainees" not eligible for union membership.

Two large aluminum companies, Reynolds and Alcoa, are involved in big urban renewal projects and now have multimillion dollar realty holdings. Both want to use reserve capital for diversification and are conducting research and development in the renovation uses of aluminum. Similarly, General Electric has acquired experience in small community development through investment and control of "all-electric" suburban projects in San Francisco and West Haven, Connecticut. GE now has a major role in developing a New Town between Baltimore and Washington which is called Columbia. In connection with the latter, the company is interested in the possible use of small electric cars as exclusive vehicles inside the city. Through its subsidiary, General Learning Corporation, GE hopes that municipal government, schools, hospitals, and industry will share the time of centralized computers for bookkeeping, while a GE nuclear power plant will heat and air-condition the entire city. This is tech-fixing with a vengeance and hardly seems calculated to solve the most pressing problems of urban America. The Department of Transportation has recently recommended a federal financing role for "Dial-a-Buses" and electronically guided personal vehicles for commuters of the future with anticipated costs over the next decade of about $1 billion.

Recent years have witnessed a fad based on the phenomenon known as "the Disneyland effect." According to this development, the anomie of urban life is largely due to the absence of elaborate, imaginative, and expensive fun parks. Such areas will provide convenient access to a world of ersatz fear and romance. Like Disneyland East now being built from scratch on forty-three square miles of swampy land in central Florida, they are in the process of opening near every major center of population in the country. Such entertainment centers will, it is alleged, be a powerful regenerative force in bringing prosperity to the surrounding area and, with adequate security guards, providing a substitute for city parks now too dangerous for walking or sitting. In effect, the Disneyland phenomenon is a corruption of the New Town concept.

In other approaches the romance and fear are all too real. The Cornell Aeronautical Laboratory (Buffalo) is working on a variety of nonlethal incapacitating (mace-like) aerosol gases for use in limited wars and domestic riots. The Center for Research in Social System (CRESS), affiliated with Washington's American University, contributed a 1966 research report titled: "Combating Subversively Manipulated Civil Disturbances" which suggests intelligence infiltration by police of potentially subversive groups, the handling of crowds by electric shocks sent through streams of water, the use of intolerable noise and eye-burning light, and the distribution to police of tranquilizer cartridges that can be fired from conventional ordnance. The Institute of Defense Analysis has come up with other forms of social therapy, including itching powders, sticky blobs to glue rioters together, and foam generators which "lead to psychological distress through loss of contact with the environment" (Ridgeway, 1968: 148). Clark Abt through his Abt Associates Inc., of Boston adapted his highly successful counter-insurgency game (Agile-Coin) to depict urban counter-insurgencies (Urb-Coin) and found that children in the Boston slums were enthusiastic about it. Based upon the success of counter-insurgency in Vietnam, the prospect does not inspire confidence. J. Sterling Livingston of the Harvard Business School, one of the big wheeler-dealers in government contract business of all sorts, is also very big in developing games to play with and on black people (Ridgeway, 1968: 70-75).

Ralph E. Lapp, former nuclear physicist, states the technological imperative as a natural law. Every invention or discovery takes on an independent imperative and *will* be used. The corollary is that each invention creates and furthers its own demand (Lapp, 1965: 67). The computer becomes a common necessity and one can no longer function without it. The computer industry grows from nothing to $6 billion a year in less than a decade, more than half of it in sales to the federal government and its contractors. Many examples, the telephone, the wiretap, the transistor radio, must be made faster and smaller, if not cheaper.

Lapp's Law is merely a principle of advocacy and a popular creed. There are many potentials in the present state of the art that are not undergoing development and subsidy, such as the Skinner box, air-conditioned domed cities, large settlements at the North and South Poles, and excavation of the Rocky Mountains. The files of the U.S. Patent Office contain vast numbers of ingenious inventions that somehow never were commercially exploited. Obviously the tech-fixation is highly selective and consequently behooves a more careful analysis. Some process of social choice and priorities is at work. Technological breakthroughs in themselves are meaningless. On the other hand, as Joseph Schmookler and others have pointed out (Schmookler, 1966), when society recognizes a problem and begins to devote attention and resources to its solution, scientific and technological change results, invention and development occurs, reexamination of relevant existing knowledge for its transfer to new uses is brought to bear. Dramatic technological breakthroughs which have important implications in other areas of social priority may or may not occur. In any case, one way or another, new knowledge is acquired and new artifacts created to solve the problem and possibly to bring into existence the potential for reducing the difficulty and cost of solving other problems which otherwise might be neglected. Our society has a highly flexible multipurpose knowledge base amenable to development at virtually all points, subject only to the incentives of the political and economic process.

Technological innovation derives its importance not from technical attributes alone but from the way it fits into its total environment. As Schmookler notes: "For this reason an important invention in the United States is likely to be an unimportant one in India, and an important invention in the twentieth century is likely to be an unimportant one in the nineteenth—and the converse of these propositions is true" (Schmookler, 1966: 64-65). This is the major reason why nuclear research reactors in the Congo or Thailand do little to bring about economic development and political stability. It is the reason why the magnificent panoply of science and technology of the advanced nations does not guarantee either diplomatic influence, a surefire formula for a new paternalism, or tranquility and justice throughout our own land. This is also the reason why the vast damage inflicted upon the physical plant of Europe by World War II could be completely repaired in less than a score of years; while aid programs limp and lag in the maze of the underdeveloped world.

A dreariness is found in the proliferation of reports by presidential panels and the National Academy of Sciences. Most of them have the same structure: the problem is difficult and not enough is known; more research and therefore more students and fellowships are needed; and more government money is required. There is a self-serving tech-fix pattern here, one which many interest groups support as a substitute for social reform and action programs. Science often becomes another way of evading solutions by subjecting issues to more study, new reports, further research, and ultimately inaction or tokenism. Scientists too often and too willingly degrade their craft by serving the purposes of social evasion and irresponsibility in the name of science for its own sake.

THE REVERSAL PRINCIPLE OF TECHNOLOGY

There are internal self-limiting factors in technological change which require our attention. We may already be in the throes not only of a technology explosion but also of a sort of negative feed-back which, although dangerous, tends toward the preservation of older values and protects the incorrigibility of human nature. A trend seems to accompany the tech-fixing syndrome, a kind of human defense mechanism against efficiency, speedup, and convenience, which reverses and undermines both positive and negative effects of innovation.

The reversal principle is most readily apparent in military tech-nology: every new offensive weapon begets a defensive system or forces a competitive system into being—in either case, erasing incremental advantages and sometimes rendering the whole system obsolete. The common response of the tech-fixers to this process is to accelerate the innovation process as though to outdistance the reversal principle in some decisive way.

This reversal principle is embodied in the nuclear stalemate. Great powers are required to have arsenals of versatile missilery and nuclear bombs, generating the science-technology race, but at the same time rendering each additional increment both more expensive and less relevant to diplomacy and military requirements. As weapons become more destructive and sophisticated, actual conflict retreats into the historic paths of earlier and more primitive military tactics and hardware. In the late 1960's, we see the aircraft manufacturers reaching back to pre-World War II and even World War I aircraft types for limited warfare in Southeast Asian jungles or for military needs throughout the underdeveloped world. Modern well-supplied soldiers in the field find it impossible to cope with poison bamboo spikes and tree limbs used for booby traps.

A curious reversal of the historic trend of industrialization is occurring which adds enormous expense. So rapid is the pace of innovation that it is impossible to freeze design of new space and defense technologies. The result is that aerospace corporations abandon mass production techniques in favor of custom-built, hand-crafted products. Scientists and engineers replace production-line workers. More and more each product is a one-time operation to be drastically altered when the next block is constructed. Missiles and aircraft contain thousands of electronic components and subsystems and require coordinated efforts by thousands of separate companies; and all of their parts undergo redesign and change both before and after assembly. Ten years of engineering design may go into a system which will be obsolete before it is completed. The reversal from mass production to handicraft is a result of a change from quantity to quality, partly due to the symbolic diplomatic nature of the present arms race. Each further step is more demanding than the one before and no limits can be formulated. Keeping up with the Joneses, Veblen's concept of the driving force in a well-heeled economy, is replaced by keeping ahead of the Russians in all forms of military and civilian technology which may impress the natives of Katanga or Kwansi provinces. The principle of mass production is to make things cheaper by turning out exact models of a frozen design on a fixed-investment production line. The science-technology race negates such cost accountancy and defies economic logic. Thus it tends to run down and requires new tech-fixes and additional public resources continually. And the entropy builds and builds.

The reversal effect in warfare is well known and has been extensively analyzed. But there is a tendency to regard it as highly exceptional; sort of an accidental side issue to the strategic technological stalemate between Russia and the United States. There is persuasive evidence that far from being exceptional, the reversal effect is universal. By the same mechanism that creates the military hardware race, diplomacy among the nations finds its own level and searches for advantages in all the less dangerous occasions for conflict. The urge toward classical imperialism engaged in by the great Western powers for over 300 years came from the stalemate of the balance of power in the cockpit of Europe. Tensions and conflicts moved into the remote and primitive reaches of the world where the Great Powers competitively sought to colonize tribal regions through the most primitive forms of violence, trade, and persuasion. In this sense, the trend toward present day interventionism in the task of policing the process of political change among the new nations is driven by the same dynamic which under other conditions came to be known as Western Imperialism.

The reversal principle can be generalized beyond diplomacy and warfare. It can be seen in the tendency toward the breakdown of large automated systems like the electric power distribution grid of the nation. During the last ten years most of the switching mechanisms (which balance surges and peaks of power among regional systems) have been automated. This automation improves efficiency and reliability, but it also amplified small failures to become major power blackouts in vast sections of the United States. Not only are the effects magnified, but causes of failure are concealed, and unprecedented time is required to trace the path of malfunction. Next-generation computer surveillance and control systems are designed and superimposed upon the flawed and chastized older generation of machines, and the problem is augmented. The ultimate controller must be human, but his attention and skill tends to be degraded by the awesome speed and imperturbability of the machine. He trusts them because they "cannot fail" and his own adaptive capabilities atrophy. Unfortunately, the automatic systems are incapable of adaptive behavior in unprogrammed situations which, when they come (and eventually they must!) will doubtlessly bring augmented blackouts and disaster.

In the same way, the vigilance of electronic early warning systems are degraded by the necessity for failsafe devices, without which the world would face a nuclear catastrophe precipitated by radar echoes from the moon, or migration of geese. If the failsafe system is too safe, it will fail to react properly in the situation it was designed to serve. This results in several paradoxes: need to degrade the failsafe in order to avoid undermining the deterrent value of the system, and the accompanying need to reassure our enemies that only by mutually degrading the alertness of our deterrents can either of us gain safety from accidental or preventive attack.

The same reversal principle operates on the magnificent new computerized systems that at this moment are keeping track of all the ships at sea, all the planes in the sky, all the satellites, boosters, and space junk in orbit around the earth; and that are processing all the fire, theft, casualty, and life insurance policies issued by major companies, routing long distance telephone calls, setting newspaper type, making sausages, navigating ships and planes, mixing cakes and cement, preparing weather forecasts, directing city traffic, diagnosing illnesses, and cashing most of the checks in the nation. If we were to introduce a failsafe system in all these activities in order to prevent the augmented disaster of a breakdown, we would find ourselves returning to smaller, multiple, and segmented systems with human intelligence once again operating at many interfaces. But this is not the direction we have chosen. So highly articulated and complex has the technology of advanced civilization become that, in the words of Hubert Humphrey, "I was in New York City yesterday. They had a little snow which wouldn't be enough where I live to even entice a child to take a sled out and yet it snarled up the traffic, the power lines came down, the telephone system was in trouble" (Committee on Science and Astronautics, 1966: 6).

There is danger in "overspecialization" because man's unique survival reflex has been flexibility—his ability to adapt to change. A society organized as a rigid megamachine—wedded to a blind, high-speed, narrowly specialized process of technological change—may thereby forfeit some of its options for the future. Any society whose existence is overly dependent upon one or two major staples suffers instability. The tendency to treat science and technology as separate and apart from the familiar processes of public skepticism imposes a fixed charge on the psychic life of the entire community tending to dominate the flavor of its culture. Some curious paradoxes threaten us: the most productive economy can stagnate at a high level; the most democratic society can see political opportunities rejected in favor of violence and protest; the most prosperous society can create anger in the poor and terrible anxieties in the affluent young. Technological inversion tends, in short, to reenact in a totally new environment the dilemmas of a banana republic. This is reversion never dreamed of by the classical optimists of technology.

In every area where new technology challenges human needs, man finds ways of degrading and circumventing progress. Another example can be seen in private communications. Snooping in all its varieties has become a normal part of government investigations, private legal actions, and industrial espionage. The technology of eavesdropping and wiretapping has outpaced all the means for control and detection. A solution cannot be found through legislation since the very ease and low cost of snooping makes it a simple matter to evade enforcement. Under such conditions, new laws are bound to be futile. So long as the technology for the invasion of private communications continues to improve its efficiency, decrease in cost, become easier to install and operate than to detect, and so long as there is a market for information available by this means, enforcement will remain spotty and ineffective at best.

While these developments are taking place, however, the private man finds ways of overcoming the technological advantages of the snooper. The new world of professional eavesdroppers and wiretappers, countereavesdroppers and wiretappers, double-agent eavesdroppers and wiretappers in government, private industry, and private lives force him to go underground, to adopt evasion techniques in private communications as a normal way of life. Evasion and nonuse of the conveniences becomes for some a regular pattern which defeats bugging and counterbugging. The social norms of privacy survive at the high cost of degrading, discarding, or limiting the full potential of radio, television, satellites, and transistors. The reversal principle generates a defensive cultural adjustment which, as a way of preserving the coin of communication intact, in turn degrades the technology itself. As in professional spying, confidential private or business communications return to the primitive conditions of whispered words in crowded airports or noisy bathrooms, the antithesis of technological convenience and progress.

Other dimensions of the reversal principle have received comment. Much has been written about the impact on ecology and the negative environmental toll which mounts throughout the world. The first fruits of industrialization are spread abroad where they dramatically cut infant mortality and then generate an explosive population increase. The latter, in turn, widens, rather than narrows, the development gap between the advanced and the backward areas of the world, sharpening all the political antagonisms that arise from the fact of malnutrition. In low-calorie countries, population since World War II has increased faster than food production. As a result, most of these areas now suffer a lower per capita food intake than they did at a lower level of technology prior to World War II. Similarly, as growing technological change sweeps over the Western World, conservation of air and water purity becomes a central issue of man's future. The squandering and pollution of environmental resources is certain to bring a larger government role apart from all the other factors of social integration and centralization.

The reversal principle is inseparable from scientific and technological change. As an ideological search for infinite security against war, want, and weariness, tech-fixing must face the fact that the search itself may be futile and unavailing. The reversal effect is expensive and exasperating, although a healthy and dependable human reaction to find ways of subverting the megamachine to more humane if primitive realities. Science and technology constitute the basis of the modern industrial state. Yet the key problem of advanced economies is to continuously juggle and balance the contradiction between the rate of change and the economics of change. The input-output cost must be judged in terms of the overall human condition and social policy. Change in itself is unavoidable and necessary; but too rapid or uncritical change passes a point of diminishing returns and becomes too costly to sustain.

Vietnam is the archetype of reversion, the testing ground for highly sophisticated technology and massive striking power versus elemental and primitive forces of political organization and loyalties. Typical of the devices employed are nylon body armor for U.S. troops (and for riot squads at home); "people sniffers," a chemical and electronic instrument designed to smell the body odor of concealed enemy troops; complex fragmentation weaponry in infinite numbers launched against enemy personnel from air and ground platforms of virtually unrestricted range and mobility; night-seeing infrared radar; and defoliation chemicals.

American strategists explain that Vietnam is a testing ground where "international Communism" seeks to demonstrate the efficacy of "wars of national liberation" as a means for communizing all the underdeveloped regions of the world. It might also be said, from the Vietnamese point of view, that the war was a proving ground to demonstrate that the infinite might and power of the United States could not impose a doctrine of "welfare imperialism." It cannot be said that the doctrine of dampening "wars of national liberation" had been demonstrated; indeed, the contrary is the case. As inconclusive as is the Vietnam military situation, the insurgents have already made their point. More than anything else, the war shows that warpower is of little use; that other kinds of reality are far more powerful, however primitive; and that pure force must fail.

In the *Myth of the Machine,* Lewis Mumford points out that highly complex technology and integrated societies appeared in history prior to the development of hardware and artificial energy sources. A megamachine can be built with soft human and animal power through rigid social organization, as in ancient Egyptian and Assyrian civilizations. The problems of the latter were not different from those of the modern megamachine of Western civilization. In Mumford's words:

> With mordant symbolism, the ultimate products of the megamachine in ancient Egypt were colossal tombs, inhabited by mummified corpses; while later in Assyria as repeatedly in every other expanding empire, the chief testimony to its technical efficiency was a waste of destroyed villages and cities, and poisoned souls. . . . (Mumford, 1967: 12).

The megamachine, he asserts, tends toward dehumanized power-centered culture which "monotonously soils" the pages of history "from the rape of Sumer to the blasting of Warsaw and Rotterdam, Tokyo, and Hiroshima." He asks: Is this association of power and productivity with violence and destruction purely accidental?

The power and magnificance of hardware technology is inordinate in the simultaneity of its reach and power. Mankind enjoys new comforts and benefits, but pays a price. Through better nutrition and sanitation children grow taller and straighter than their parents but, while more of them live past infancy, they may ultimately have trouble controlling their weight and many will die of heart attacks earlier than did their grandparents. The scale of good is indeed augmented and evil become massive and monumental. Technology may mark the difference between the random extermination of American Indians compared with the systematic destruction of European Jewry. The difference between Andrew Jackson and Adolph Hitler may be entirely based upon advances in technology and social integration.

There is a kind of grotesque reversal in the statement of Johnnie Scott of Watts (before a Congressional committee) describing the chill of sitting at night in his room in a brand new housing project, watching dog packs and old people in the alley fighting over garbage cans. There is something of a reversal in the fact that cities are tending to become places for the very rich and the very poor, a situation that one finds also in the most unstable countries of Latin America. There is a kind of reversal that along with new programs to renew the city comes the thunderous crescendo of violence and extremist militancy. Jacques Elull provides many examples of the fact that "every technical advance is matched by a negative reverse side" (Ellul, 1967: 106-107). There is a lag, he suggests, before "the monster is apparently tamed." But, he adds, the most disastrous consequences come after apparent adaptation has been reached, and these are of long term effect.

Aldous Huxley reiterates the theme in *Brave New World Revisited:* "These amazing and admirable advances have had to be paid for. Indeed, like last year's washing machines, they are still being paid for and each installment is higher than the last" (Huxley, 1958: 22). McLuhan says that proliferating technology stimulates anti- or counter-environments: "Today technologies and their consequent environments succeed each other so rapidly that one environment makes us aware of the next. Technologies begin to perform the function of art and make us aware of the psychic and social consequences of technology" (McLuhan, 1966: ix).

German physicist C. F. Von Weizsacker writes: "The modern world is a tree in which many birds build their nests, and it is an all-consuming fire.'" It is blind to its own ambivalence. Weizsacker likens the inevitability of ambivalence and the half-truths through which man's mental life fluctuates to Jesus' parable of the wheat and the tares: "I have never seen a clearer description of modern times than this growing cornfield in which the tares unavoidably grow up alongside the wheat" (Weizsacker, 1964: 179). Present in human relations and in each human being are great natural forces comparable to hurricanes, earthquakes, and sudden natural disasters. The megamachine sustains a magnetic plasma of all the inherent powers of human relations; but this is a volatile fusion not yet fully harnessed and never fully congruent with the requirements of a highly integrated society. In its modern form, the megamachine may suffer the all-consuming fire which destroys all the benefits which should preserve it.

THE CRISIS OF REPERSONALIZATION

The most baffling aspect of the reversal principle has been the dropout of vast numbers of people of all social classes and age groups, especially those who have been the beneficiaries of the comforts, conveniences, and wealth generated by an expanding technology. The mounting contradictions of the internal American turmoil do not fit the classical categories and mock the neat ideologies that have confidently marked the forces of change for over a century. Mumford, always the incisive diagnostician, sees its as "a pathology that is directly proportionate to the overgrowth" of the metropolis, "its purposeless materialism, its congestion, and its insensate disorder . . . a sinister state" that manifests itself in the enormous sums spent on narcotics, sedatives, stimulants, hypnotics and tranquilizers, not by the hippie generation, but primarily by the great mass of middle-class adults, an adjustment to the vacuous desperation and meaningless discipline of their daily lives (Mumford, 1968: 194-195).

The golden optimism of the early philosophers of technology has not stood up to the great disappointments that have accompanied America's rise to wealth and power. We have had disappointments in peacemaking after wars, in conducting diplomacy, in maintaining economic stability, in putting too much trust upon a faddist succession of arts and sciences, psychoanalysis, social reform, professional experts of all varieties. The terrors of nuclear annihilation have not ended the tribulations of diplomacy, the inexhaustible power of atomic energy has not transmogrified the poverty, desperation, and hunger of excluded minorities in our midst or of eager, struggling, and hostile peoples abroad.

Not only is political ideology dead, but scientific ideology is dying—not only in its mystical aspects but also in its practical and necessary methods. The great initial success of science and technology built, in all cultures, a faith in an objective and pragmatic real world, shattering the transcendental images by which prescientific man sought to grapple with the conditions of his life. Today we see strong signs that even the best parts of this new pragmatism are beginning to crumble: "God is dead" in many senses.

The proponents of tech-fix talk much about the importance of the individual and the wealth of options that technology and science offer him. Yet, we have surrounded him with pollution, radiation, megalopolis, intimations of daily disaster, rigidifying pockets of social disorganization, unfamiliar human jungles, all of which appear to be of such impersonal power and scale as to be beyond individual and social control. As Donald N. Michael asks: "How does a man see himself in relation to his espoused ideal of individual autonomy when he also sees man-made circumstances as awesome and implacable as acts of God framing his destiny?" (Michael, 1966: 134). The optimists still preach a completely technologized society that will make possible a rationalized world in which there is no injustice; but there is a creeping anxiety that such a world is subject to cataleptic seizures and, like molecules of a gas or of a crystalline solid, where all are equal, none are free: "One molecule, one vote." In our growing sophistication, we come to understand that it is the imperfect justice of all human relations, past, present, and future, that generates the dynamics of politics and gives individual freedom its meaning. Justice for all is a lovely creed, but is not to be had except as the rigors of political bargaining give it provisional status and degree for those who prevail in the shifting compromises of the bargaining process.

Classical Liberalism, the visions of Karl Marx, and the doctrines of democratic socialism have scant relevance to the problems of modern technological societies. They are "fuddy-duddy" and are rapidly going the way of the Anarchists, the Syndicalists, the Wobblies, the Technocrats, and other such groups based upon tech-fix optimism and faith in productivity. Their heirs are contemporary tech-fixers, the Pentagon Cold-Warriors, the captains of major corporate empires, and the naive scientists, all of whom find themselves increasingly besieged by recalcitrant reality. The modern megamachine dissolves the old divisions of public-private, caste-class, rich-poor—and new disjunctions threaten. In the advanced nations the rich get richer and the poor get richer; but the poor get richer more slowly and begin to view other disparities as more important than those of simple survival. The rejection of rationalism in favor of subjectivism challenges our confidence in the homiletic scenarios of linear growth that constitute much of what passes for the history of Western civilization. Both Liberals and Conservatives are stuck with increasingly irrelevant and meaningless slogans.

We are faced with problems so unfamiliar that social scientists must begin anew, discarding old concepts in the face of an ambiguous and uncodified reality right under our noses. It is no longer possible to run the modern industrial megamachine, with all its massive power complexes, on the basis of the eighteenth century profit system, the Protestant Ethic, or the reformist and revolutionary creeds of the nineteenth century.

The tech-fixation is a kind of culture lag. It cannot long endure. In the late 1960's one can detect a change in the discussion of public affairs in America. The illusion that we can do anything is yielding to widespread doubt. A growing sense of futility surrounds attempts to deal with the nation's problems. This is an extreme recoil from faith that science and technology can solve all problems conclusively. Perhaps it is in part the reductio ad absurdum of the science-technology race. Perhaps it is the lingering, "sorry-about-that" war in Vietnam. Perhaps it is the spontaneous combustion of the urban ghettos, or nostalgia for the loss of John F. Kennedy, our own loss of innocence and the abiding problems of the Great Society, the War on Poverty, the receding vision of a compassionate and vigorous nation living in decency and order. There is a sense of weariness without solemnity, and our children do not take us seriously.

The impersonality of the technological society is at once cause and effect. Its effective exploitation is supported by the Protestant Ethic, which subordinates all things to efficiency. Technology promotes social distance between the various skill groups of the megamachine who are as much its cogs and wheels as the hardware kind. It buffers and insulates men from the human reactions and consequences of their decisions and actions. It promotes patterns of behavior which focus attention on technology rather than human relations. Disguised as an irrepressible natural force, technology becomes both an escape mechanism and a reality that requires further escape.

Hiding behind palace guards of computers, automated production lines, management consultants, systems engineers and managers, the tech-fix becomes the preferable means of dealing with all human problems. It maintains faceless social distance and minimizes the possibility of human confrontations, thereby sparing the self-righteousness of the powerful. The soldier activating electronically controlled missiles has no basis for empathy with his target; the industrial manager prefers to automate rather than deal with union grievance committees; the state legislatures delegate the problems of traffic control and waste disposal to systems engineering firms rather than deal face-to-face with the requirements of county boards, mayors, aldermen, and townspeople. The megamachine thereby imposes heavy burdens on human empathy, and this reinforces the technological imperative. Facelessness and human avoidance become the means of personal stability and a safeguard of such privatism as survives in family and personal life. There is a peculiar congruence between the impact of this syndrome on the strategically placed decision-makers and on those who are merely the tools and victims, especially among the college students, who are preparing themselves for future roles in the managing elite, or among the toilers, who are possibly prolonging their freedom at the expense of their parents or at the cost of accepting a permanent place outside the established culture.

The Newtonian God is dead in all the pervasive cultural embodiments of our scientific ethos. At the very time when the megamachine is embracing our lives as the building of pyramids did the slaves of Pharoah, the minds of many of our brightest youth reject it and seek solace and individuality in expanded self-consciousness and positive alienation. In the words of Daniel Bell, "people no longer can have a sense of linearity, a beginning, middle, and end, foreground and background" (Bell, 1964: 58). The simultaneity of electronics, and the collapse of distance and physical terrain are tending to construct a cultural mode from ingrown multiplicity of sensation and experience. Marshall McLuhan sees in this a return to tribalism in the midst of advanced technology. He finds this positive and interesting and considers it the primary message conveyed by the mass media. On the other hand, Bell sees it as a threat, a growing distrust of explicit language, culture forms, and social roles: "The whole breakdown of the rational cosmology is imminent and will ultimately create the most serious problems for society as a whole because of an alienation of the modes of perception about the world (Bell, 1964: 59). The withdrawal of the young may be a passing fad, but it will also leave a trace upon the larger culture. The most highly articulated and centralized society ever known may be in the process of creating a new generation which has lost the "zap" necessary for continued functioning of the megamachine.

Contemporary subjectivism (of which the collegiate hippies and their successors are an expression) has at least not chosen the role of the Luddites who destroyed eighteenth century English factories. Technology and automation are not (except by older labor leaders) apprehended as a personal threat that should be destroyed. Things are far too serious today and the megamachine too vast and insuperable. Rather the new wave is fleeing to personalism, transforming alienation into a new kind of culture. "The organization man" is inverted, the music, art, psychedelic movements occur within the context of the very environment which causes the alienation. This is a search for new adaptive patterns of life which may be better suited to an economy capable of sustaining great diversity (made possible by technology), and a response to the tribalistic simultaneity brought about by the instantaneous, physical, visual, and auditory sensations of the world village (made possible by technology). The need for a more satisfying reality denounces objectivism and finds a transcendentalism of self in deepening awareness of internal sensation and self-consciousness, the result of the womb-like technologized environment itself. Absorption in oceanic sensation and self-sensuality may embody a survival reflex by the human organism which, locked in the technological megamachine, prefers to quest for the infantile illusion of omnipotence.

McLuhan has said, "We have put our central nervous systems outside us in electrical technology" (McLuhan, 1966: 60). The new subcultures build a numbness barrier, renewing the inward look, perhaps an action of some relevance and significance, the middle-class equivalent of the Black conversion from killing each other to killing policemen. This society with all its "suffocating abundance of machine-made goods and gadgets, has resulted in a dismally contracted life, lived for the most part confined to a car or a television set," so empty of "first-hand experience," so lacking in tangible goods and spiritual values that "it might as well be lived in a space capsule, travelling from nowhere to nowhere at supersonic speeds" (Mumford, 1968: 223-224). The subjectivism of so many of today's youth is a search for first-hand experience and a rejection not only of soap, but also of soap opera.

If nothing else, the self-searchers make a contribution by demonstrating that MAN is not yet ready to become, what many tech-fixers would make him, an acronym for Meaningless Archaic Nonentity.

TECHNOLOGY AND THE CONVENTIONAL WISDOM

Beneath the protestations of tech-fixing is an accelerating wave of corporate conglomeration based upon new technology. A handful of huge industrial organizations are becoming quasi-public utilities dominating all areas of production and distribution, creating a self-sustaining, self-serving, self-justifying and self-perpetuating industrial oligarchy which claims efficiency and progress as its necessity. This process coincides with too many trends which together generate alienation and confrontation. Centralized and concentrated institutions give rise to a volatile and dangerous brand of political pluralism, but apparently the only kind that can challenge a computerized military-industrial society. Technology and science today are a special interest myth of these power groups.

There is a convergence of politics and economics which is a response to, and a reflection of, important changes in our national problems and institutions. The old guarantees of pluralism based upon private property have all but passed away, revealing underlying political power relationships in a manner which intensifies new constituency formation and interest-group confrontation, converting virtually all traditional areas of private economic bargaining and contract into political activity. The political process, always close to the surface anyway, breaks through and increasingly asserts its predominance. This breakthrough is induced by war and diplomacy, by the multiplication of instruments for public and private planning and intervention in the economy, by the massive role of government, the chronic tendency of the economy to lapse into high-level stagnation, the intensified concentration of private economic power, and with it, control over the patterns of consumption, the allocation of resources, and the impact of accelerated technological change.

The rise of the giant conglomerate corporation is a fact. In the name of science, technology, and private enterprise, the giants control their environment and immunize themselves from the discipline of all external controls, especially those of the market place. Through separation of ownership from management, they are emancipated from stockholders. By reinvestment of profits they escape the capital market and the financier. They insulate themselves from consumer sovereignty by brainwashing, and they dominate both suppliers and customers by concentrated market power. By government contract, pressure-group activity, personnel interchange, and similar means, they share power with the state. Whatever the corporation cannot do for itself, it does through government: through policies and programs, to maintain full employment; through subsidies to research and development, and support to education, to supply the necessary scientific and technical skills. The industrial giants come to perform society's planning functions while preserving the myths of an anti-planning culture.

The role of science and technology has made the conventional wisdom of democratic politics more pertinent and essential than ever. No elixir of success is to be found in science and technology or in the skills of any group of experts that can relieve the citizen of his responsibility to know, to participate, to criticize, and, if necessary, to demonstrate. Mathematical formulae, nuclear reactions, electronic circuitry, none has rendered obsolete the immemorial usages of man. The human factor and values continue not only to outweigh man's technical achievements, but also, through politics, to share the direction of discovery, invention, and change.

The problem is not how to control science and technology. Some are controlling them already. The problem is to recognize which interest groups are exerting preponderant influence and for what purposes—in order that we may seek the time-honored correctives of pluralism, namely visible public accounting and countervailing power. If there is, as Admiral Rickover frequently asserts, an antithesis between blind technology and individual liberty, it is an antithesis between coalitions of narrow group interests able to allocate national resources toward ends not shared by other large groups. Our theme, therefore, is the need to assimilate the gothic mysteries of science and technology to ordinary political analysis, common-sense political judgment, and plain English. Obviously, the nation cannot deny itself the aid of augmented science and technology in facing the serious problems of the day. But neither can it blindly accept all those claims made in the name of science and technology as inexorable natural forces. Scientific and technical change are far from unstoppable and automatic, but are rather the result of, and responsive to, public policy. The interested public can gain access and predict consequences in this, at least as well as in any, area of policy choice; and all areas today are complicated, highly specialized, and jargonized.

The social scientist carries an obligation to facilitate this access so essential to the resilience and safety of a democratic society. When a technologist informs Congress it is now "technically feasible" to accomplish some task, this information no longer carries the implication that the task therefore should be funded to carry out the mandate of technological progress. The hard questions fly thick and fast: Technically feasible relative to what? What are all the options for dealing with the problem? The standards of judgment are the old familiar ones: How much good will it do, for whom, at what cost? What are the alternatives? How much political power do those who favor or oppose have? How does action on this problem relate to the priorities of competing problems? Faith in science is no longer an irresistible talisman for persuading public policy to endorse any and every proposal. More and more it becomes clear that there are no "pure" scientific problems—only human problems and political choices in allocating values. Every policy act has many and diverse motives behind it and the practical politicians who mediate group conflict are frequently more interested in, "Who favors what?" and, "What will they do about it?" than in what is favored.

The social scientist must assert the death of the technological imperative. Just as in physics the God of Newtonian Order is dead, so the myth of "scientific" certainty in public policy deserves political exile. For many years the scientist and the computer intellectuals have moved freely through the corridors of the Pentagon, the State Department, and the Congress like Jesuits through the courts of Madrid and Vienna 300 years ago. But increasingly, the Congressmen have discovered these new priests divided on all points of every issue, as were the lawyers and economists before them.

The present disjunction means that the tech-fixers have failed and that the conventional wisdom of a diverse society full of political change is asserting its primacy. Man is rehumanizing and repersonalizing his social and physical environments, rejecting authoritarian centralization by the tech-fixers and proving that they cannot manipulate all of the people all of the time. The disintegration we experience can lead to a regrouping of elements and social purposes toward pluralistic goals. The social hygiene offered by the tech-fixers treats the symptoms of our problems as though they were the enemy. In reality, the symptoms are friends: where there are symptoms, there is conflict, and conflict indicates that the forces of life which strive for expression and reintegration on a human scale are still in the ring.

Reintegration has not yet come and may take long hard years. However, if the tech-fixers of the power economy seek to contain it by extreme measures they will equally fail, although at an unnecessarily high price in terms of what might have been and what still might be. A society so close to disintegration as our own must choose between escalated violence or a creative effort at reconstruction and reconciliation. It would be rash to predict the outcome— both trends track one another indecisively.

What must come is a system of values and institutions which will replace economic initiative and private property as guarantors of political independence and pluralism. As economic pluralism disappears, only political pluralism safeguarded by new institutions of representation can make the exercise of power both responsive and limited. A heightened and more representative infrastructure of interest groups is necessary at all levels of society in private and public institutions, from local school boards to national regulatory agencies, including as many emerging new constituencies as are capable of demanding inclusion on the basis of such self-consciousness and organization as has already been tempered by the heat of confrontation. The contemporary politicization of the economic system is creating new constituencies whose ultimate effect upon public policy must be to revise and reform the status quo. Those who have too long held a monopoly of social purpose cannot escape the challenge of the cities, and the inevitable redefinition of the city's problems by the ghettoized residents themselves.

REFERENCES

ADLER, MORTIMER J. (1967) The Difference of Man and the Difference It Makes. New York: Holt, Rinehart, and Winston.

BACON, EDMUND N. (1967) Design of Cities. New York: Viking Press.

BELL, DANIEL (1964) "The post industrial society." Eli Ginzburg (ed.) Technology and Social Change. New York: Columbia Univ. Press.

BERTALANFFY, LUDWIG VON (1967) Robots, Men and Minds. New York: Braziller.

Committee on Government Operations, Subcommittee on Executive Reorganization, U.S. Congress, Senate (1966) Hearings, Federal Role in Urban Affairs. Washington, D.C.

Committee on Science and Astronautics, U.S. Congress, House (1966) Proceedings, Sixth Panel on Science and Technology. Washington, D.C.

HUXLEY, ALDOUS (1958) Brave New World Revisited. New York: Harper and Row.

Joint Economic Committee, Subcommittee on Urban Affairs, U.S. Congress, House and Senate (1967) Hearings, Urban America: Goals and Problems. Washington, D.C.

KAHN, HERMAN and ANTHONY J. WIENER (1967) The Year 2,000. A Framework for Speculation on the Next 33 Years. New York: Macmillan.

ELULL, JACQUES (1967) The Technological Society. New York: Vintage Books.

LAPP, RALPH E. (1965) The New Priesthood. New York: Harper and Row.

LESHER, RICHARD L. and GEORGE J. HOWICH (1966) Assessing Technology Transfer. Washington, D.C.: U.S. Government Printing Office.

McLUHAN, MARSHALL (1966) Understanding Media: The Extensions of Man. New York: McGraw-Hill.

———(1951) The Mechanical Bride. New York: Vanguard Press.

MESTHENE, EMMANUEL G. (1968) "How technology will shape the future." Science (July): 161.

MICHAEL, DONALD N. (1966) "Some speculations on the social impact of technology." Dean Morse and Aaron W. Warner (eds.) Technological Innovation and Society. New York: Columbia Univ. Press.

MUMFORD, LEWIS (1968) The Urban Prospect. New York: Harcourt, Brace and World.

———(1966) The Myth of the Machine. New York: Harcourt, Brace and World.

POOL, ITHIEL DE SOLA (1968) "Behavioral technology." Foreign Policy Association, Toward the Year 2018. New York: Cowles Education.

RIDGEWAY, JAMES (1968) The Closed Corporation. New York: Random House.

SCHMOOKLER, JACOB (1966) Invention and Economic Growth. Cambridge: Harvard Univ. Press.

SPILHAUS, ATHELSTAN (1968) "The experimental city." Science (February): 159.

WEINBERG, ALVIN N. (1967) Reflections on Big Science. Cambridge: MIT Press.

WEIZSACKER, C. F. VON (1964) The Relevance of Science. New York: Harper and Row.

Part III

OF PROBLEMS, POLITICS, AND PLANNERS

The American Scene

Introduction

THE preceding two sections constitute an effort to set our concern for the quality of urban life generally, and in the United States in particular, within a broadly historical frame of reference and to see the urban environment in terms of some of the main dimensions of its structure and culture. The seven chapters which comprise this segment of the book constitute more specialized probes into the character of the urban milieu: more specialized both in the subject around which each contribution is organized and in the focus on the United States. Breadth in each case is necessarily sacrificed for penetration, but there is throughout a continuous awareness that each author's "piece of the action" connects in diverse and critical ways with other parts of that matrix of social systems and institutions which, taken together, shape the immediate world in which the vast majority of this nation's citizens now live.

This section provides a bridge of realities connecting the preceding critique and commentary with the concern for reform in the concluding section. Proposals for reform must in the end be grounded in the actualities of human conduct; innovations may aim at some general reconstruction of society but they must in fact deal with recognized problems in some field of action such as education, mental health, or urban leadership and management, and with the predominant institutionalized patterns which characterize specific urban systems—schools, mass media, cultural enterprises, and government itself. At the same time, the interactions between distinctive urban systems and institutions may significantly enhance or impede efforts within each to achieve needed changes.

Something should be said about that which the reader will not find here. Space limitations required some strategy of omissions, and judgment in this regard was based largely, but not entirely, on the extent to which a subject has already received attention. Thus, there is no chapter devoted mainly to the city's architecture, its zoning, or its financing. An immense amount has been written about the "proper" shape, scale, layout, and surface appearance of urban buildings. However, almost none of the conflicting conclusions about what is desirable and what is reprehensible is grounded in empirical research on how people actually respond to their material surroundings. The field is drenched in ideological doctrine, and most urban physical structures are compromises between the philosophies of architects and the personal preferences of those who can command the allocation of resources, whether for profit or glory. At the same time, most city-dwellers display behavior that often seems surprisingly inelastic (within rather extreme limits) in relation to the physical packaging which contains it and provides its visual background, whether in central-city public housing or in suburban developments with curving streets bearing picturesque names.

Where the literature about the face and form of the city constitutes a field fraught with speculative argument, the voluminous treatment of zoning, traffic, and fiscal support tends to be preoccupied with descriptive research producing little or no real theory and surfeited with technical minutiae. Most of this represents an engineering approach in the narrowest sense, ignoring such underlying determinants as the metropolitan power politics of private enterprise and the class-based private pleasure principle which dominates the allocation of economic resources. The consequence has been chronic deprivation of the public sector and the failure to escalate our "war on poverty" above the level of a continuing skirmish (see volume II of this *Annual Review* series).

The contributions to this section do not ignore either physical form or economic wherewithal, but the focus is on the experience of living in an American city and on the characteristics and conduct of those who have some power, by virtue of their locations in institutionalized offices and positions, to have more say than most of their fellow citizens about the character of the environment within which this experience will occur. The first three chapters center more on the life of the adult urban dweller: on the psychological dimension of his existence, on the imagery and expressive opportunity provided for him by the "performing arts," and on the information about, and the symbolic organization of, the city developed (some might say "perpetrated") by the mass media. The next two chapters deal with the school systems within which the city's children and youth spend a substantial and still increasing part of their waking hours. The first of this pair is concerned with the quality of the education actually provided, especially through problem-besieged inner-city schools, while the next takes up the processes by which the operational values and resource allocations of school systems are determined. This discussion leads into Chapter 14 which treats more generally of the quality of leadership in urban institutions and the recruitment of various types of people into positions of influence. The final contribution in this section deals with the problem of "urban management" in more abstract terms, conceptualizing and illustrating the difficulties and possiblities inherent in the fact that life in our cities must be more planned and administered than we often care to admit.

Levy and Visotsky open Chapter 9 with a review of conditions that make urban life at best unpleasant and at worst utterly miserable for far too many city-dwellers. In the face of so many possibilities for deprivation, most of which are not likely to be quickly eliminated, they call for much more attention to possibilities for enhancing the generative and supportive elements in this often forbidding environment, especially for combating the consequences of both isolation and enforced contact and for strengthening personal identity and participation in the face of alienating factors within both work and leisure. To help achieve a psychologically more viable community, they argue, those in the field of mental health must get beyond clinical treatment and make a major contribution to the solution of environmental problems.

John Suess also is concerned about the expressive dimension of life in the city, which has traditionally been the place of performers and audiences. Without any pejorative connotations, he divides his area of interest into the "serious arts, folk and traditional arts, and commercial arts," all of which coexist within today's metropolis, influencing each other and the milieu of most urban dwellers. The middle and upper strata have always been the patrons and patronizers of the serious artist, though there have also been forms of counter-exploitation; but professional artists (as against proliferating cadres of amateurs) have not enjoyed much increase in economic support in our society, except from the universities. Nevertheless, the technology of the urban-industrial order is being increasingly explored, as in "mixed media" shows utilizing film, electronics, and other components. And the "urbane" artist is more and more relating to, rather than retreating from, the problems and potentialities of the modern metropolis, whose denizens both need and resist that expressive recognition of their situation which only the serious artist can provide.

On quite a different dimension the American city with its diverse immigrant subcultures has been the recipient of a whole anthology of folk and traditional performing arts and, as Suess points out, even suburbia is generating its own forms at meetings and parties. But ethnic traditions have slowly deteriorated and waned under the impact of mass media and assimilation, except in the segregated Latin and black ghettos. Moreover, both serious and folk performing art is regularly co-opted and corrupted by commercial presentations oriented essentially to profits and exemplified by "wallpaper music . . . to shop by, to eat by, to sex by, to relax by. . . ." Suess concludes that both substantial economic support for the performing arts and aesthetic education for the general public are essential if the city of mass media and mass marketing is also to be the city of cultural diversity and authentic individuality.

The intimate relationship between mass media and mass marketing is explored by Gene Burd in Chapter 11, in which he further underscores the power of the media to shape the imagery and expectations that city-dwellers have of their environment. "The manipulators of the mass media are like urban middlemen who market the metropolis," a fact which gives political import to much of what they say and portray. Commerce and history have made much of the American mass media's vested interest in the central business district, where their principal offices and properties are located, thus tending to bias their approach to many urban problems. Control of the media's content also constitutes a major power base from which those who own and manage the main press, radio, and television facilities of the metropolis can enter into other community decision-making roles. Not only vested interests, according to Burd, but also the newsman's conventions about what constitutes "news" conduce toward a superficial as well as conservative treatment of the city's problems and inadequate criticism of its planners. Still there are hopeful potentialities in the development of the underground press, the black press and radio, and the community-oriented suburban press and radio—all contributing diversity and alternative views to those propagated by the centralized metropolitan mass media. However, we have yet to find the means for effective citizen and internal criticism of the urban mass media, for modernization of schools of journalism, and for reorganization of reportage itself around meaningful, issue-oriented rubrics that would help illuminate the central issues confronting and confounding American urban citizenry.

The next two chapters, dealing mainly with public education in urban America, continue the unresolved dialectic between hopeful potentialities already evident as a kind of minority report on the American urban scene and predominant patterns reflecting prevailing values and a powerful politics of vested interests, both class and institution based. Although acknowledging the need for major improvements in the education of affluent Americans, Havighurst and Levine consider the most pressing issue confronting contemporary urban school systems to be the failure to provide a large proportion of children from low-income familites with an education at least good enough to open real opportunities to them in a "knowledge-transacting society." Achievement of this kind of "product" is likely to require a complex "mix" of changes including desegregation and destratification of student bodies and new approaches to instruction that incorporate but also go far beyond what is now known as "compensatory education." This means, in turn, not only substantially more money for the schools, but also ". . . imagination, innovation, and careful evaluation, three characteristics which have been all too rare in American education."

Havighurst and Levine, while sympathizing with many of the complaints of "anti-institutionalist" critics of the schools, argue that most of the more revolutionary changes that have been proposed probably would not bring much improvement to education within inner-city schools and might well compound existing problems. More hope, they believe, lies in the necessarily trial-and-error experiments with decentralization now underway in most big cities, in a variety of efforts to improve school administration on all levels, and in improvement of instruction and extension of educational opportunity to younger children. In any case, they conclude, only when much has been done to deal with the admittedly short-run problem of education for the disadvantaged in our society will we be able to give effective attention to what education ought to be, so that, given an economy of abundance, we can "teach ourselves to live an abundant life."

If Havighurst and Levine may be thought of as discussing what could and should be done about the quality of urban education in the fairly immediate future, Decker and Masotti provide in the next chapter a counterpoint analysis which seeks to explain why present patterns obtain and what the probabilities are that they will change.

Using a systemic model, they ask hard questions in what may strike some readers as a hard-hearted manner: what are the inputs of the system and how does it operate to combine them so that any given output results? Where schools are concerned, active elements include administrators, teachers, pupils, parents, and others who intentionally or inadvertently do much to determine what the mix of behavior and resources will be. It is their respective values and relative power to actualize these values within school systems, which will determine what criteria for "quality of education" actually obtain and how well these criteria are implemented in the everyday life of the schools. Decker and Masotti examine each of the component sets of actors within the typical American school system and find meager evidence that rapid and extensive change is likely, for those with power show little predisposition for an "educational revolution." Readers who dislike such a conclusion must either find sources of new behavior among schoolmen, teachers, pupils, parents, and voters that the authors of Chapter 13 have overlooked, or dispute the weights they give to various trends and tendencies now observable, or reject the whole systemic model of analysis.

Confronted by pressing problems whose solution often seems blocked by the inertia of social systems and the stickiness of institutionalized patterns of conduct, many Americans fall back on the traditional cry for new or better "leaders." While such pleas are usually naive and simplistic, they are not without some merit since those we identify as leaders do much to establish operating values and wield power in various urban systems. In Chapter 14, Charles Adrian asserts that this leadership today is more knowledgeable than in the past and increasingly willing to tackle the formidable problems that confront our cities; but it is also, in part, "inexperienced, naive, angry, uncooperative, uncertain . . . frustrated . . . and . . . sometimes too strife-ridden to make decisions." He provides a colorful taxonomy of contemporary urban leaders, briefly sketching the characteristics and history of each type: professionals, advertisers, hobbyists, status-seekers, ideologues, and members of ethnic groups. Within this constellation the most significant change now underway is the emergence of "black power." Adrian concludes a short history of this latter development with the expectation that it is now, and for the rest of the century will be, a leadership source likely to be high in drive, intelligence, and skill.

The principal images of local government which are held differentially by different types of leaders emphasize, alternatively, "booster, amenities, caretaker, and brokerage functions." Professionals in municipal administrative bureaucracies have tended to be associated with establishments oriented to amenities and brokerage functions and to accept most of the "givens" of the existing regime. Those interested in professional positions with more leadership leverage have, until recently, tended to move into federal and some state bureaucracies; but today a noticeable number of younger individuals who see themselves as "change agents" have begun to seek entry into local administrative positions, creating additional new potential for urban leadership—but so far only potential.

In a "conceptual essay" well laced with empirical material, Nathan Grundstein asks how we might best move from the existing inadequate management of our urban environment, whose actors Adrian has described, to one more capable of deliberately selecting and actually achieving "desired environmental states," that is, conditions which diminish constraints on the attainment of our goals. Such management can be achieved only through organizations whose own characteristics permit and facilitate increasing lead-time in dealing with undesirable environmental states, responding quickly to environmental changes, determining fields of action and choices within them, and cumulating managerial knowledge and skills. Such organizations are rare among urban systems in the United States today. What we require, according to Grundstein, is not so much further improvement in the technology of urban management as a strategy which will take full advantage of the still emerging alliance between urban planning and management, which originally developed independently from one another in the early 1900's, and of the growing science base which each now enjoys.

Well-developed "managerial urban planning" should be able to operate as a "dispassionate resource" for the contending institutions and systems which comprise the urban milieu, facilitating choices by the provision of "an adequate diagnostic and prognostic base" for decisions, a base which deals with both the whole metropolitan system and its component subsystems with more than opportunistic criteria for proposing priorities and alternatives. But this kind of knowledge must be joined with a sophisticated and well-informed understanding of the actual "dispersion of power and of the conflicts of personal and social interests" for the generation of effective "politico-managerial" decision-making. Both increased federal intervention in urban affairs and the decentralization movements now underway have, according to Grundstein, created opportunities to restructure urban management so that the "politico-managerials," "urban general managers," and "program operationalists" could fulfill their respective functions in more effective and coordinated ways. Once again, however, there is no assurance that this potential will be actualized.

And that is the theme which runs throughout this section. In every sphere of concern and action our urban society proliferates both new problems and exciting possibilities. The latter are continuously being seized and explored by cadres of innovators; but they are to date minorities, often small ones, while the majority of those in power continue the very patterns of conduct and relationship which have produced and exacerbated the problems. Yet we must sustain hope for what is at least potential in this situation for improvement in the human condition during the era of urban man, lest despair induce paralysis. This, however, is a subject developed and dealt with more fully in the concluding segment of this volume.

—W. B. Jr. and H. J. S.

9

The Quality of
Urban Life

An Analysis from the
Perspective of Mental Health

LEO LEVY

and HAROLD M. VISOTSKY

☐ THE QUALITY OF urban life and the presence of massive unsolved problems in our large urban areas are issues which directly concern the mental health professional. It may be well at the outset to list and catalogue some of the prominent and critical problems generated by large concentrations of persons in limited geographical space. The first of the following two groups comprises those problems, most or all of which could be effectively dealt with by the elimination of poverty. The second consists of those not directly related to poverty but quite general for all social classes.

I. Poverty-related Problems

   (a) substandard housing

   (b) unemployment

   (c) presence of filth, trash, garbage

   (d) vermin, roaches, other pests

EDITORS' NOTE: *This chapter is a revision of a manuscript which originally appeared in* Urban America: Goals and Problems. *Materials compiled and prepared for the Subcommittee on Urban Affairs of the Joint Economic Committee, Congress of the United States (Washington: U.S. Government Printing Office, 1967), pp. 100-112.*

    (e) health problems

        1. lower life expectancy
        2. high infant and maternal death rates
        3. poor access to health care facilities
        4. high prevalence of illness (including mental illness
            and mental retardation)

    (f) family instability

    (g) unavailability of social services

    (h) inferior recreational facilities

    (i) harassment & discrimination by official agencies (police, courts)

    (j) poor educational facilities

    (k) large numbers of poorly educated and untrained persons

    (l) lack of privacy (high dwelling-unit density)

    (m) general sense of powerlessness (feeling of inferiority)

    (n) lack of money (deprivation of commodities)

    (o) high birth rates (many unwanted children)

II.   Nonpoverty-related Problems

    (a) air and water pollution

    (b) traffic congestion

    (c) high area population densities

    (d) inadequate public transportation

    (e) unsafe streets (exposure to crime and violence)

    (f) generally unaesthetic surroundings

    (g) poor control over urban growth and development

    (h) unsatisfactory patterns of socialization (isolation, anomie)

    (i) high noise level

    (j) general absence of large natural preserves

    (k) inadequate sense of identity (place in social structure unclear)

The above compendium, which might appropriately be characterized as "a house of horrors," is set forth simply to denote the immense scope and complexity of the problems created by our large urban areas. Obviously, behavioral scientists have limited contributions to make in dealing with many of these issues. Some of their contributions, however, have direct relevance to mental health and most of them relate indirectly to issues of individual social competence. The distinction between poverty-related and nonpoverty-related problems is also of some significance in that a good deal of data has accumulated pointing to a positive relationship between general poverty (low socioeconomic status) and the prevalence of severe mental disorder.

From the point of view of the behavioral sciences generally and mental health in particular, we have selected certain of these problems for discussion in this paper. These cover family instability, high dwelling unit and area population density, isolation, anomie, inadequate sense of identity, poor control over urban growth and development, inadequate recreation, and unemployment. These and other problems created by the megalopolis are the professional concern of many groups including architects, engineers, city planners, public health personnel, transportation and highway planners, mental health practitioners, and behavioral scientists. Our purpose in this chapter is to indicate how mental health professionals and behavioral scientists relate to problems of urban life. We limit ourselves to the issue of the psychological effects of urban environments on individuals, particularly with regard to their mental health, and the failure to achieve and maintain it (mental illness).

ROLE OF MENTAL HEALTH PROFESSIONALS

To illustrate how we conceive our role, we cite two examples, one pertaining to urban renewal, the other to migration. An area of a large metropolitan city is slated for urban redevelopment. Sites selected for such a project are generally chosen on the basis of the substandard condition of buildings in the area, and the fact that the neighborhood is blighted and needs attention. As mental health professionals, we are prepared to aid in the process of determining the psychological costs and benefits which might accrue from such a project. The zoning of the redeveloped area, the kinds of buildings which will be placed there, and the varieties of functions associated with it, are seen as issues best addressed collaboratively by behavioral scientists, architects, engineers, and city planners. The effects of the demolition of people's homes, and the relocation of these individuals, coupled with the general impact of such a project on an existing neighborhood or community structure, are mental health concerns. An example of such concern is Fried's interesting and carefully executed research on the reactions of people dislocated by urban renewal from the west side of Boston (Fried, 1963). This study furnished exhaustive data demonstrating that a substantial number of these people suffered severe reactive depressions as a result of their dislocation, and that, in fact, a viable community had been destroyed. Urban renewal efforts have frequently ignored cultural and ethnic variations in reconstructing neighborhoods and have proceeded by applying a bland middle-class American standard. This approach has had two immediate effects. The first is to create neighborhoods into which the persons dislocated cannot or will not return. The second is to take one of the vital and interesting qualities of our large cities—cultural and ethnic variation—and destroy it.

A second example has to do with problems raised by migration. We are a highly mobile society in which about one-fifth of the population changes residence each year. Migration poses many problems—rehousing, reemployment, change of school for children, and a host of other difficulties—which bear on the mental health of the individuals concerned. A rather extensive literature of provocative research has developed in this area indicating that geographical (as well as social) mobility produces a uniquely high-risk group for mental disorder (Kantor, 1965). For some, migration offers an opportunity, but for many others it poses a crisis and as such may interfere with effective social role performance.

The general framework in which mental health professionals operate when approaching the complex problems associated with life in communities and cities calls for an analysis of the various stress-inducing circumstances which exist in such settings and the counter-active supportive forces. This dyad, stress-support, poses two factors which constitute a system and which therefore must be examined as related and not separate. In other words, one can never state in absolute terms that stress has a specific debilitating effect on an individual. The effects it will have on him depend on the supportive mechanisms on which he can rely during any given stressful period. A man with a viable intact marriage, a satisfying home life, and stable employment, can withstand higher levels of environmental stress than an individual who is isolated. Well-documented findings pertaining to the prevalence of severe mental illness show that patients in mental hospitals are more likely to be single and unemployed, or marginally employed, prior to hospitalization. Widowed and divorced persons stand a higher risk of becoming severely mentally ill than married persons, although the risk for these two groups is proportionately less than for the never-married individual.

Stress means one thing to a person living alone without the support of a family, and quite another to a person who is supported by a spouse and possibly by parents and children. It has been demonstrated in a variety of experiments that humans and other animals respond entirely differently to stress and frustration when they are alone than when they are in the presence of other friendly persons. These may be parents, siblings, or friends. A wealth of scientific literature has been accumulated in the area under the general title of "Social inhibition and facilitation of behavior" (Liddell, 1964; Harlow and Zimmermann, 1959; Davitz and Mason, 1955; Masserman, 1943). Auto insurance companies and credit bureaus know, on the basis of actuarial studies, that married persons are better risks. Mental health professionals know that married persons are also lower risks with regard to developing a debilitating mental illness. Marriage and family is a primary supportive institution which counteracts stress.

Our concern with family life does not end here. We observe with interest the various possible surrogate familial arrangements possible when a primary family unit does not exist. Foster family placement is not only beneficial to luckless children but also to persons suffering from chronic schizophrenia who can exist outside of a state mental hospital within such a contrived familial context. Experiments with group living arrangements such as half-way houses for ex-hospital patients and group experiences for narcotics addicts and others have proved successful. Social clubs, taverns, and even street gangs have been useful in providing surrogate familial experience for persons otherwise unattached.

The decline of the three-generation family is a related concern. We ask ourselves what are the gains and losses to all parties involved when children leave their parental home relatively early in life, marry and set up separate domiciles, often quite distant from the parents. They then have children who, in the space of a generation, repeat the same cycle. We are interested in the effects of this cultural pattern on the young mother and father, on their immature offspring, and on the aging grandparents. The grandparent-child relationship has always been a special one, and in our view, a constructive one for both parties. We look with some concern at the mounting numbers of psychiatric casualties among the aged who increasingly collect in our state mental hospitals and nursing homes. Many of them, in fact, are there only because no familial context is available to them in which they could be maintained.

Stepping from this level of concern, we may ask similar questions with regard to the breakup of the small city. The quality of urban life in the megalopolis differs in distinctive ways from that of the small American city. In the latter, the individual has always been under greater social pressure to conform to community standards because he is more visible in his behavior. The large city, on the other hand, is often described as a place in which one loses his identity. He becomes an anonymous face in a vast throng. He is less an integrated participant in the social and political fabric of community life. For deviants, this is an ideal and perhaps absolutely necessary condition for survival. The freedom to deviate from social standards is important and is encouraged in every enlightened society as an essential aspect of individuality, as long as the form of departure is not destructive of and inhibiting to the freedom of others. A democratic society should not attempt to regulate human behavior too stringently. Yet, as has been observed many times, laws attempting to regulate human conduct in our society are so numerous that it is difficult for a normal person to avoid breaking some one or another municipal ordinance or state statute each day. That our criminal courts and jails are not completely deluged is testimony to the fact that many laws regulating behavior are unenforceable (and unnecessary). One of the purported advantages of living in a large city is that one is free to experiment with new social roles (Cook, 1963) and to indulge oneself in behavior which, in a small city, might bring immediate reproval and punitive social sanctions from the community at large.

As all of us are aware in this age of bigness, the small city is in decline. In 1800, 95 percent of the country's then five million citizens resided in rural areas. By 1960, 70 percent lived in urban communities of 25,000 or over, and 63 percent in those of 50,000 or over. By the year 2000, the urban-rural differential which existed in 1800 will have been effectively reversed, with about 90 percent of the population residing in large urban areas. The trend, moreover, is unmistakably toward the megalopolis—the vast multimillion person aggregation concentrated in limited geographical space. By the end of the present century, the ten "supermetropolitan" areas in the United States are projected to contain one-third of the nation's total population. These ten megalopoli will house 107 million people, or an average of 10.7 million each. As mental health workers, we see it as one of our tasks to introduce into the large urban areas some of the positive elements of small communal life, helping in this way to protect the individual against the damaging effects of isolation and anomie, while at the same time retaining the advantages which such areas offer.

ISOLATION

Human beings do not tolerate loneliness. Freud once defined anxiety as "the feeling of being alone in a strange place," and Harry Stack Sullivan is reported to have remarked that loneliness is worse than anxiety. Perhaps one of the most cruel punishments ever devised by man is the concept of solitary confinement. In our penal institutions this treatment has generally been reserved for the most recalcitrant, belligerent, and dangerous prisoners. There is a long series of well-executed studies in the area of sensory deprivation, our scientific analogue of solitary confinement (Myers, 1964; Schultz, 1965). In experiments of this nature, the individual is placed out of contact not only with other persons, but also with all visual stimulation (i.e., blindfolded) and all auditory stimulation. In many instances he is even prevented from experiencing much tactile stimulation by the restriction of his limbs in such a way that they do not touch. Under these circumstances, individuals characteristically suffer a great deal of personal discomfort and also frequently exhibit behavior quite similar to that of schizophrenic patients in that they begin to hallucinate. The need for sensory stimulation in humans is so pervasive that if they are totally deprived of it, mechanisms begin to work within the organism to compensate for its lack. In this way, sensory stimulation is "bootlegged," so to speak, in the central nervous system.

These experiments, among other things, demonstrate that people cannot bear to be alone and that they cannot in any sense achieve mental health in isolation from other individuals. A wealth of literary works, clinical observations by psychiatrists, and our own experience, illustrate repeatedly that even peculiar and perverse relationships are often supportive experiences. A recent excellent contribution to this literature was Edward Albee's play, "Who's Afraid of Virginia Woolf?" Despite the friction, antagonism, and sadism which is inherent in the relationship described in the play, the author leads one to the conclusion that this relationship is basically supportive and necessary to the maintenance of social functioning for both parties.

ENFORCED CONTACT

A consideration of the issue of isolation inevitably leads to a consideration of the opposite; enforced and excessively close contact with others. It is by now common knowledge that a serious over-population problem exists in parts of the world today and threatens to become a problem of cataclysmic proportions in the years ahead. In 1825, the world contained one billion persons. By 1930, 105 years later, this number had doubled; and by 1960, 30 years later it had tripled. Projections for 1977 indicate a population of four billion and for 1995, six billion. Living in large urban areas, moreover, is a relatively recent phenomenon. Cities over one million persons were unknown prior to the nineteenth century. Concomitant with the move into large urban areas, not only population but also population density is increasing. In Chicago, for example, average densities are 16,000-17,000 persons per square mile. In some crowded areas of our large cities, they run as high as 1,000 or more persons per square block.

The problems which accrue from high population densities in our inner cities have been discussed in a large body of literature emanating mainly from the disciplines of sociology and social psychology. Tentative findings tend to associate a number of social ills with overcrowding; delinquency and proneness to racial rioting are two examples. Some interesting findings from the field of animal ecology have as yet been unrelated to human affairs. There are studies indicating the existence of self-limiting mechanisms which determine maximum herd size in certain animal species. When this size is surpassed, animals mysteriously, and to outward appearances inexplicably, die off, thus balancing the group at a certain number. This phenomenon occurs in spite of an adequate food supply. Other studies indicate the emergence of pathological behavioral traits and social restructuring under the impact of unusually high density and confinement in certain animals. In the Norway rat, for example, overcrowding leads to aberrant maternal behavior which results in high infant mortality and the stabilization of herd size even in the presence of abundant food and water supplies. (Calhoun, 1962)

No support exists for the claim that there is an optimum size for the human group, nor can it be stated with confidence that there is any particular space requirement for an individual. Living arrangements tend to be highly culturally relative, and what may be considered a high degree of privacy in one culture may, in another, be viewed as intolerable exposure. Although living arrangements are quite varied cross-culturally and even within a culture, we believe certain principles obtain in all settings. Some arrangement for solitude and voluntary isolation, as well as opportunity for interaction, is always provided. In western cultures, the tendency is to provide each family member a room generally designated as a bedroom, and to create other rooms specifically designed for interaction (living rooms, dining rooms, kitchens, etc.). Although this practice may be cited as a middle-class value, lower-class persons protest strongly the sharing of private rooms and the use of public rooms for bedrooms. It would appear that even beyond culturally relative (learned) values in this regard, there is a biologically determined distancing mechanism in people. Where *physical* distancing becomes impossible, as it does in prisons, concentration camps, army barracks, and slum apartments, people seem to make use of *psychological* distancing mechanisms. Enforced physical contact often leads to the maintenance of "emotional distance." If people cannot escape into their private rooms, they will escape into their private thoughts. It is possible that such emotional distancing in crowded city areas may contribute to indifference to the suffering of others and unwillingness to "get involved" with, for example, neighbors or people who may be attacked on the street by hoodlums.

It should be noted in passing that a stable and significant correlation has been observed between the prevalence of poverty and severe mental disorder. That this correlation exists has been conclusively demonstrated. Why it exists is still a matter of debate. Part of the explanation may lie in the fact that poor people who reside in cities live in overcrowded conditions and are thereby subject to many stresses which may exacerbate symptoms of severe mental disorder. The fact that Negroes show higher rates of severe mental disorder may point to a similar mechanism (i.e., Negroes are generally poor; the poor live under conditions of high dwelling-unit density).

RECREATION AND WORK

Let us now turn to a different area of human endeavor, namely, recreation and work. We classify both of these categories together because we believe differences between them to be quite superficial. One generally gets paid for work but not for leisure and, further, these two classes of activities tend to be dissimilar. We also expect that in leisure time people will pursue radically different activities than at work. Beyond these distinctions, however, the differences are superficial. For example, popular conceptions of recreation and leisure regard these activities as entailing little physical exertion. A moment's reflection enables us to dismiss this notion as myth. For some persons, recreation consists of intense physical exercise which far exceeds the exertion they normally put forth on their jobs. Conversely, a recreational activity may call for the highest level of mental exertion while a job may call for little such effort.

People generally require about eight hours a day for sleep or total inactivity. They require perhaps four additional hours for performing such biologic functions as eating and eliminating. This leaves approximately twelve hours a day during which individuals require varying forms of activity. Lacking this activity, they become bored, restless, and disturbed. Up to very recently this matter was dispatched quite readily via the twelve-hour work day, six days per week. Today for the bulk of our population, the weekly work period turns out to be six to eight hours a day for four or five days. Thus, the employed population is left with considerable amounts of time to be occupied in other than work activities. The problem is more substantial for the unemployed—large numbers of women, older persons, unskilled, and Negroes.

The use of leisure time can be a problem for many individuals. One of the ready-made societal solutions for it—a solution we consider ill advised—is the institution of passive-receptive activities such as watching television, going to the movies, and spectator sports. The problem with such passive recreational activities is a subtle and interesting one which has, in our view, mainly to do with the concept of personal identity. There is a strong need in man to assert his individuality, to leave his distinctive mark on things, to participate actively in the life process. This central striving or need may be accentuated or suppressed by any given culture, but we believe it is a fundamental biologic characteristic which is encountered universally. People who compulsively watch television and who indulge generally in passive-receptive leisure time activities frequently express disgust with themselves afterwards. Some justify their behavior by saying they have nothing else to do while others appear unable to verbalize beyond expressing the feeling of dissatisfaction itself. This feeling appears to arise principally from the sense of nonparticipation.

There is a drive towards creative expression in all men. It can be inhibited by both work and leisure activities which forbid self-expression. With the passing of the era of the craft, and the introduction of the age of mass production, a period in history began in which work activity became progressively more depersonalized, automated, and dehumanized. Today, most persons gainfully employed cannot point with pride to a product which is uniquely theirs. It is rather difficult for one to feel that a wire fixed in place on a television set on an assembly line or a series of adjustments made on the chassis of a car as it goes past is in some way distinctively *his* contribution to society. A painting, a woven tapestry, a piece of pottery, a hand-forged iron gate, a patient cured of disease, are accomplishments men point to with pride and which contribute to their positive sense of identity. We are faced with a problem peculiar to the latter half of the twentieth century: a majority of Americans today experience no sense of identity in work and little or none in leisure. We must attend to these problems of work and leisure and attempt to achieve better solutions in these areas for people if we intend to maintain and improve their mental helath. The problems are accentuated today by so-called automation and forced early retirement in the face of a lengthened life span. Constructive, participatory, identity-confirming activity for all persons is of utmost importance. Advanced study and self-expressive modes of work and leisure activity are vehicles to this end.

Mental health is not an easy term to define. If we were required to select its cardinal aspect, we would choose the concept of identity. A mentally healthy individual has a sense of participation in the life process. He has a sense of his individual worth. He has a sense of dignity, of knowing who he is; and he does not have to contend continually with the problem of justifying his existence to himself. The hallmarks of identity tend to be rooted in the performance of a limited number of social roles: work, familial identification as a parent, child, or spouse, and constructive participation in communal life. With regard to the latter, many individuals in our society, lacking a significant role and significant power to make decisions for their own community, suffer from role ambiguity as citizens of a community. In spite of the many exhortations to be motivated as a participating citizen, this role is not clearly defined nor are opportunities provided for many individuals to participate in communal life.

THE VIABLE COMMUNITY

To achieve adequate social role competence and participation, one must exist in a viable community. This means, in concrete operational terms, that the concept of the small city must *somehow* be recreated within the large city. It means that there must emerge a new concept of living relevant to the last half of the twentieth century. The evolutionary form for this emergent product seems to have been the neighborhood within the large city. Unfortunately, neighborhood structure and social organization has too often been little understood and disregarded by physical planners. In their genuine enthusiasm to rebuild the slum, the city planner has often interfered with neighborhood and communal life by imposing solutions on a community to which it was hostile and which it had no part in formulating. Not all planners are insensitive to this dimension. In particular, C. A. Doxiadis has developed a conception of large urban areas for which he coined the term "dynopolis" (Doxiadis, 1963). This concept of urban growth and evolution allows for the small city within the large city. It calls for the building and enhancement of neighborhoods which in certain respects resemble small towns. It belts the small communities with large traffic arteries which promote automobile transportation to other sections of the city. At the same time it makes entry into the center of small communities difficult for the automobile. The communities are self-sustaining in certain respects. They have their primary and secondary schools, and their small library and basic amenities such as small shops and recreational areas. On the other hand each community is linked with other communities and with a downtown area which contains the major cultural assets of the city, such as the opera house, the symphony hall, the art museum, and the university. It is also linked with the industrial complex which furnishes jobs and revenue for the city as a whole.

The city, in the final analysis, must reflect the biological and social needs of its inhabitants. It must provide contrasting experiences, smallness and bigness, work and play, solitude and company, activity and repose, intellectual and emotional stimulation, noise and quiet, tension and relaxation. The course of a man's life follows these dimensions. With the immense technology available to us, the vast resources at our disposal, and the accumulation of centuries of knowledge, we should be able to make our cities into places which are a joy to behold and a pleasure to experience. Much so-called mental illness and antisocial behavior should abate when these solutions to environmental problems are forthcoming. Mental health professionals and behavioral scientists have a role to play in their solution. Perhaps this paper can serve as a general guideline for our participation.

REFERENCES

CALHOUN, JOHN B. (1962) "Population density and social pathology." Scientific American 206: 139-148.

COOK, DONALD (1963) "Cultural innovation and disaster in the American city." Pp. 87-93 in L. Duhl (ed.) The Urban Condition. New York: Basic Books.

DAVITZ, J. R. and D. J. MASON (1955) "Socially facilitated reduction of a fear response in rats." Journal of Comparative and Physiological Psychology 48: 149-151.

DOXIADIS, C. A. (1963) Architecture in Transition. London: Hutchison.

FRIED, MARC (1963) "Grieving for a lost home." Chap. 12 in L. Duhl (ed.) The Urban Condition. New York: Basic Books.

HARLOW, H. F. and R. R. ZIMMERMAN (1959) "Affectional responses in the infant monkey." Science 130: 421-432.

KANTOR, N. B. [ed.] (1965) Mobility and Mental Health. Springfield, Ill.: C. C. Thomas.

LIDDELL, H. S. (1964) "The challenge of Pavlovian conditioning: an experimental neurosis in animals." Pp. 127-148 in J. Wolpe et al. (eds.) The Conditioning Therapies. New York: Holt, Rinehart & Winston.

MASSERMAN, J. H. (1943) Behavior and Neurosis. Chicago: Univ. of Chicago Press.

SCHULTZ, D. P. (1965) Sensory Restriction: Effects on Behavior. New York: New York City Academic Press.

10

The Performing Arts
and the
Urban Environment

Some Observations on Relationships

JOHN G. SUESS

☐ THE PERFORMING ARTS have always been considered mainly
within the domain of the city in Western civilization. It is the urban
environment that has provided the audiences and the basis for
financial support. And it is primarily the urban centers that have
given the patronage—public, private, individual, and institutional—
which inflames the creativity of the performing artist and furnishes
a market for his work.

It is within the city limits that we will observe today's performing
arts in their native American habitat, including their changes of
function and relevancy and the changes imposed by mass
communications media. No definitive treatment of this subject is
intended here. Studies of such a nature are virtually nonexistent and
are only beginning to be conceived and executed. There are,
however, two recent analyses which are helpful in the area of the
serious performing arts, in that they define some of the major
problems. One is *The Performing Arts; Problems and Prospects, A
Rockefeller Panel Report on the Future of Theater, Dance, and
Music in America* (New York: McGraw-Hill, 1965); and the other is
The Performing Arts—The Economic Dilemma by William J. Baumol

and William G. Bowen (New York: The Twentieth Century Fund, 1966). Both of these studies deal primarily with the economic problems of maintaining existing performance institutions.

For our purposes we shall consider the performing arts to include music, theater, and dance, and also their manifestations in television, films, and radio. In addition, we will briefly consider the rage of the sixties, mixed media, which incorporates various combinations and permutations of these performing arts.

THE PROBLEM OF DEFINITION

Not only are we aware of the varieties of art media which constitute what is commonly termed the performing arts, but we are also cognizant that each of the specific media is in itself plural in theory and manifestation. Each of the arts may be both urban in geographical context and urbane in sophistication. Thus the symphony orchestra and the jazz combo are both urban products but of different musics.

Immediately upon mentioning multiple musics, theaters, dances, and the like, we realize that one of the "sacred cows" of tradition is being threatened, and rightfully so. Differentiation based on traditional associations of "good" music as opposed to folk music need not, and in our case cannot, have any meaning. The sophisticated theater has no implication of superiority over less sophisticated expression and the luxury of the ballet dancer thumbing his nose at the soft-shoe expert is anachronistic. We say this despite the realization that many of these absurd and artificial differentiations will exist today, not only in the responding audiences, but also among professional performers. Each avenue of the performing arts, whether "living theater" or Broadway musical, electronic music or the blues, has its rightful and undisputed place.

There are, nevertheless, meaningful differentiations which can be made between the various types or levels within the performing or "time" arts. The discussion here will be limited to those based mainly on function and social context. Consequently, the designations *serious arts, folk and traditional arts,* and *commercial arts* will be used. These differentiations are to be considered only as terms of convenience without any deep psychic images or implications attached to them. Nothing is implied by the term "serious arts" to indicate that the other two categories should not be considered as serious manifestations with different functions. The difficulty of finding terms that carefully reflect all the proper nuances has led many to distraction. We should also keep in mind that these terms are not mutually exclusive. Attribution of specific works to these categories will be based on emphasis rather than exclusion. There are serious art works that are commercially successful, witness the plays of Tennessee Williams and Arthur Miller, and folk arts that have made an impact on the serious arts. This especially is true in the realm of avant garde jazz performers and groups.

For our purpose, we will consider the designation of serious arts to refer to those artistic manifestations created mainly for aesthetic communication on a relatively intellectual level without a specific nonaesthetic function necessarily in mind. Since this is only a working definition, we will not present any lengthy defense of its use except to point out that it includes such performance media as the symphonic orchestra and band, the traditional musical chamber groups, opera, avant garde and the classic jazz soloists and groups, the repertory theaters, much Broadway theater, ballet, modern dance groups, avant garde films, and some special creations for radio and television. These references are deliberately kept general in order to allow the reader his privilege of flexibility in fitting in his particular exceptions.

By folk and traditional arts, we mean those artistic manifestations that tend to emphasize their folk origin or traditions; they need not stress sophistication since they are intended to function mainly as entertainment and self or group expression. This category incorporates some of the most fertile sources of the urban performing arts through religious, ethnic, and racial groups. The arts it encompasses are usually based upon some type of oral or nonwritten tradition which is handed down from generation to generation as part of a particular heritage. Here the sociology of the performing arts has enriched the palette of urban color and experience.

Lastly, we refer to the commercial arts where the function is primarily to create successful musical goods that will bring a fine profit. This type of artistic manifestation is frequently directed toward a specific function such as a television or radio show, or for a specific market such as the products of "Tin Pan Alley." The aim here is to get the highest ratings on the discotheque market or have the longest run in the film industry. Profit and entertainment are the chief criteria but seldom the exclusive ones.

In all these categories, the performing arts are equally concerned with the solo performer or creator as well as the group performance and creation. Moreoever, the emphasis may also change between the creator's intent and the performer's rendition. Thus if a play has been taken out of its original context and recreated into a film script, there may be little relationship between the original art work and its manifestation on the screen.

THE SERIOUS ARTS

Traditionally, in Western culture, the serious performing arts have been closely linked to urban centers. As such, they have virtually always reflected the taste, or at least the desired image of the taste, of the middle and upper classes in the socioeconomic ladder. This relatively small audience of intellectuals, pseudo-intellectuals, and dilettantes has had the economic power to provide the necessary patronage for supporting creative and performing artists as well as performance costs. It has always been fashionable and prestigeous for this audience to be connected on intimate terms with the "bohemian" or "hippie" artists and become sponsors of creation. In many ways this traditional view continues in the more sophisticated guise of foundation and educational grants. The "arty" arts have been and still are the meat of the urban intelligensia and the panacea of pseudo-intellectuals, the fashion and fad followers.

THE ECONOMIC DILEMMA

The economic dilemma of the urban-centered performing arts has been made clear by the two studies previously referred to. That by Baumol and Bowen, in particular, stresses this aspect:

> It is a thesis of this study that the root of the cost pressures which beset the arts is the nature of their technology. For the economy as a whole, productivity (output per man-hour) has risen at a remarkably steady rate of roughly 2½ percent per year over the last half-century, and there is every reason to expect that the discovery of new knowledge and the invention of new techniques of production and capital accumulation will yield comparable increases in production per man per hour in the

future. But the technology of *live* performance leaves little room for labor-saving innovations, since the end product is the labor of the performer. While increases in money wages in an industry such as auto manufacturing are offset, either partly or in full, by increases in productivity, the corresponding increases in salaries in the arts are directly translated into higher costs. The more successful such industries are in keeping up the rate of increase in their productive efficiency, the more will the cost of the living arts rise relative to costs in general [p. 390].

Thus the difficulties of the performing arts are inevitable given the technology of live performance and the facts of economic growth. With the technology of live performance there is no effective way to increase productivity of a string quartet or a symphony orchestra in concert hall performances. In addition, as productivity increases elsewhere in the economy, wages also rise; and it is only natural that performers as well as creators will be discouraged from professional careers in the arts unless their compensation keeps up to some extent with the general level of wages and salaries. The same economic approach may well be applied to the area of education if one considers the technology of "live" instruction.

There appear to be three major conditions required for a healthy and stimulating environment for the performing arts. These are professional excellence in the work of the creative and performance artists, a well-educated audience, and financially healthy performance organizations.

The Rockefeller Panel study of performance organizations—professional opera, symphonic orchestras, commercial theaters, and dance groups—reveals a number of interesting facts and problems in this regard. First, the recent cultural explosion has been reflected largely in the expansion of amateur rather than professional groups. Second, problems common to all of the performance arts include the general economic disadvantage of the performer in the serious arts when compared to the general wage level of the society, the second-class training generally received by professionals, the lack of adequate facilities, the lack of development and stability in sponsoring organizations, the attitude of crisis financing (waiting until the emergency strikes), and the unfortunate neglect in planning and research for resources. The Panel's report, like the Baumol and Bower study, emphasizes the need for governmental assistance for the performing arts. In doing so it sums up the relationship between private and public support in the following words:

In summary, the panel concludes that while private support should remain dominant, the federal government—together with state and local governments—should give strong support to the arts, including the performance arts, by appropriate recognition of their importance, by direct or indirect encouragement, and by financial cooperation [p. 148].

Besides the economic predicament, other difficult relationships frequently exist between the performing and creative artist and performance organizations. The professional artist in the serious arts is traditionally associated with a performance organization which usually operates at a deficit. The patrons of these organizations frequently become the taste-makers and the result is the generally sterile standard repertory with little room for innovation. This statement, however, is least accurate in the realm of repertory theater.

Closely allied to this traditional type of oppression by the directors of performance institutions is the dichotomy of the "arty" colony of creative artists, be they composers, playwrights or choreographers. Many of the same sponsors that provide the inhibitory taste-making machinery in the performance institutions are the very ones who encourage "originality" among the creative artists they wish to support. Almost all urban environments usually have an "arty" colony of creative artists aggregating, not only for reasons of artistic stimulation, but also because of economic necessity and loss of communication with an understanding audience. The creative artist has been led to seek new outlets within his own artistic community or in the university because of the hypocritical tendency to encourage originality at virtually any cost in the true "Madison Avenue" sense of "gimmickry" and the frequent lack of an opportunity to perform experimental works through the framework of traditional performance organizations.

Largely because of these restrictions in the larger communal setting, the university has become an important haven for much artistic innovation as well as for the creative artist himself. This frequently means that the natural habitat for creative performance has now left the larger urban environment for the university campus, where traditions need not be an obstacle to innovation. Most of the experimental theater, experimental choreography, and experimental music now has its home in the university, urban and nonurban. The university has provided a secure and at least adequately paid home for the creative and performing artist and has made available to him a vast array of performance media as well as enthusiastic student support for excellence in innovation and performance. Such a combination of events, facilities, and enthusiasm has not occurred since the patronage of nobility in the eighteenth century, with composers, choreographers, and playwrights flocking to the university communities. The result is that in most cases the performance arts are thriving on the nation's campuses. Those situated in urban centers where the creator and performer have direct contact also with the professional world seem to be particularly attractive to the creator-performer. It is important to note in this latter connection, however, that the need for direct physical contact with the urban environment has been considerably lessened by the advent of the mass communications media and advanced electronic technology.

The mass media of television, radio, film, and the long-playing record have brought the potential of mass dissemination to the serious performing arts. But how have these media been utilized to provide an impact upon the serious performing arts? The answer, unfortunately, is "relatively little and utterly insufficient." Outside of the FM radio station, the long-playing records, and educational television (NET), there has been a woeful inadequacy in the use of these media for the serious performing arts. The function of such media for the dissemination of all levels of the arts is potentially awe-inspiring, but the "vast wastelands" still exist in an unbalanced emphasis on the commercial arts. It is undeniable that the mass media have in some cases disseminated the serious performing arts from the urban environment to a larger and diverse audience, but the profit motive in particular has again inhibited such progress to a large degree.

THE IMPACT OF TECHNOLOGY

More significant, however, is the impact of the urban complex of industrial and electronic technology. Not only has it opened new vistas to the traditional performing arts, but it has also provided the means for the discovery and invention of new kinds of performing and time arts. It is basically within the last decade that film has become a new medium for serious expression; film-making schools have sprung up like mining towns in a gold rush. The new underground films now provide a new outlet for artistic expression. The new film art has also tended to focus on special social problems, such as the urban revolution, in the most vivid and sophisticated manner. The capability of conquering relatively difficult technical feats in film-making by new processes and the new stop-action and animation techniques have brought this new art to life. In addition, film footage is also capable of making social criticism more lifelike and "real." Experimental films and new film techniques have become commonplace and many are closely linked to the urban environment through subject matter and criticism.

Another impact of urban technology has been in the area of electronically generated and manipulated sound for musical purposes. This development takes the form of electronically amplified instruments, electronically generated sound put on tape, or taped sounds manipulated through mechanical or electronic means. Through such means the composer and performer have enlarged the repertory of aural experience to virtually infinite possibilities. It is now possible to create and manipulate never-before-heard sounds and the listener must learn how to cope with them. With the advent of the tape recorder, the performer has the opportunity of hearing himself and testing his performance capability before the live event occurs. The choreographer has the opportunity for greater control over the music he uses, the actor hears himself as others do, and the musician listens to his own performance.

Urban technology has also had a considerable impact on the development of a variety of new visually conceived types of performing arts. The new kinetic sculpture with its mechanically or electronically controlled program has opened an entirely new dimension of the visual arts. Real time and performance media have become an integral part of the visual arts: the importance of light, white or colored, has become an art medium by itself rather than simply a part of staging theatrical productions. The "light shows" which permeate avant garde art exhibitions reflect the sophisticated urban technology through the use of changing light schemes as structural determinants of space, and the created patterns may be controlled by mechanical or electrical time devices. In these ways the urban environment has been responsible for positive changes in the performing and time arts.

One of the most recent artistic manifestations of the serious performing arts in urban centers is the mixed-media performance, either by carefully calculated organization or by chance "happening." Here the various performing and time arts are generally combined or rather respond to one another as if they revealed the urban proximity of man reacting to man. The performing arts thus become arts of involvement by the performers who reveal the close relationship between art and life, especially the sensitive awareness of urban proximity and the multiplicity of stimulae. Again it is the urban environment which seems to provide the means and the spark for such interrelationships.

IMPACT OF THE URBAN ENVIRONMENT

The question of the impact of the urban environment upon the serious and creative performing artist is an area of strong feelings, but little studied. Nevertheless, there are certain phenomena which reveal such an impact. To say that the artist is a reflection of urban life is virtually meaningless since it but repeats the truism that the artist is a product of his time. What is more interesting is to review briefly the serious performing arts in view of their awareness of their urban environment and the reactions they manifest. This is best seen in some of the new directions in the performing arts.

On the one hand there is the reflection of urban man and his twentieth-century scientific accuracy and technology through which all the aesthetic variables can be controlled in the artistic performance carefully organized through man's reason. Such is the case in the deterministic art as exemplified by Ernst Krenek and Milton Babbitt in music and Arthur Miller on the stage. The orderly procession of artistic events along a logical direction, even if these events include carefully planned irrational happenings, remains a very strong tendency in many new artistic avenues. The major point is one of control by the creator and performer; in such an environment improvisation does not exist.

On the other hand, the decade of the 1960's has revealed another reaction to the modern urban environment, a reaction totally in sympathy with the present social revolution and reflecting the complementary aesthetic of this social change. Some of the primary considerations in this aesthetic direction are those of chance, human feeling, lack of predictability, and an attempt to make art and life identical. The importance of individual, immediate, and impulsive "artistic" reactions, no matter what the stimulus, emphasizes the importance of human involvement. Such is the meaning of the mixed-media "happening" where everyone has a chance to do their "thing" based on reactions to the "thing" of the other person; such is the action of the "living theater" with its audience involvement. The desire is for a breath of fresh air amid the urban skyscrapers, crowds, numbers, and general loss of identity. It reflects urban man's cry for significance in a society where the computer is becoming the unconcerned keeper of his bloodless identity.

It is in this vein that John Cage writes for "performances" without any traditional instructions. Aleatoric music, dance, and text, as well as chance film clips or programs for kinetic sculpture, are presented to a traditional audience with the all-too-frequent result of loss of communication. As with social revolution, the creators and performers in much of the avant garde serious performing arts are seeking new aesthetic solutions to old problems.

The logical extreme in this direction is the loss of the identity of the creator and the total anonymity of the performers and audience. In this situation, art and life become identical, and the consequence is a folk art with the function of involving and entertaining all who "dig it." The question is really whether or not there will be enough people to "dig it" to make it folk art. Will there be an audience of sufficient numbers to receive the aesthetic communication which the new arts wish to offer?

The Rockefeller Panel Report reflects the concern for an educated audience when it says:

> The panel is motivated by the conviction that the arts are not for the privileged few but for the many, that their place is not on the periphery of society but at its center, that they are not just a form of recreation but are of central importance to our well-being and happiness. In the Panel's view, this status will not be widely achieved unless artistic excellence is the constant goal of every artist and every arts organization, and mediocrity is recognized as the ever-present enemy of true progress in the development of the arts [pp. 11-12].

This point of view makes it clear that both the professional performers and the audiences need top-caliber education and experiences in the serious performing arts. As the report states, "the effective exposure of young people to the arts is as much a civic responsibility as programs in health and welfare" (p. 192). Again, the education field is the most obvious recipient of this pointer.

It is depressing to view the urban educational system with respect to aesthetic education in the primary and secondary schools as well as in most universities. This is the view of a number of studies on education of various performance arts in the school systems

THE PERFORMING ARTS [279]

sponsored by the Federal Department of Health, Education and Welfare (e.g., *Seminar on Music Education* at Yale University, 1963). If the challenge to understand the new performing arts is not soon met, the distinct possibility exists that only a few will receive the benefits of aesthetic exposure which the Rockefeller Report cites. The loss of contact between creator-performer and audience is particularly tragic in the urban areas where the vast majority of performances of the serious arts occurs. This problem is not limited to the primary and secondary schools, although it is here that the greatest number of future audience members may be reached. It also occurs on college and university campuses, where only a small number of students have direct contact with courses in the performing arts. Our educational institutions have yet really to come to grips with this problem at any more than a token level.

The serious performing arts, as indicated earlier, are basically urban in location and urbane in sophistication. The problems associated with performance organizations, artist and audience education, are largely urban in scope and quantity. The opportunities to meet these challenges also appear to depend mainly upon urban initiative and leadership. It is, thus, the urban areas that must become the leaders in aesthetic education, opportunities, and artistic encouragement in the serious performing arts.

FOLK AND TRADITIONAL
PERFORMING ARTS

It is, perhaps, in the area of the folk and traditional performing arts that the rich cultural diversity of the urban environment makes itself most clearly understood. This diversity encompasses all the distinctions—ethnic, religious, racial, and social—which frequently have differentiated urban neighborhoods, or blocks, or even apartment buildings. The mutual strength of the groups provided, and in some cases still provide, most American cities with a wide pallet of experience for their inhabitants. Frequently taking the form of group activities and entertainments such as vocal and instrumental musical groups, local theatrical groups, and dance groups, these performing arts, based largely on oral tradition, reveal a variety of local customs and traditions associated with their special heritages.

For hundreds of years the immigrant populations have sought political, economic, religious, and social freedom in the United

States, but never at the cost of ignoring their own culture, customs, and traditions. The huge American melting pot has received yearly inputs of fresh artistic manifestations from the major cultures of the world and the immigrants themselves certainly continued much of their original folk and traditional performance arts fiercely and religiously. One such block was Irish, a second was Italian, and a third was Chinese.

Other means of urban cultural, and sometimes physical, separation have been based on religious affiliation, race, and socioeconomic class. Churches and synagogues are frequently the center of specific populations and their facilities the center of particular activities in local traditions in the performing arts. Jewish neighborhoods are nationally famous as an example of an ethnic-religious group that has fostered its own theaters, musical performance groups, and ethnic dance groups, as well as magnificent delicatessens. Other ethnic-religious groups of the Greek or Russian Orthodox faith have also developed neighborhoods that emphasize their particular performance arts.

In a similar vein the recent "march to the suburbs" by the middle-class American can be considered a type of cultural focus in the performing arts. "Suburbia," which must still be considered part of the urban environment in that it circles and feeds on the city in an economic sense, appears to be in a process of establishing "folk arts" all of its own. The common bond of supermarkets, PTA groups, distance from the urban center, and generally high-level education, is encouraging rituals and ceremonies with functional performing arts. These include the many amateur performances of commercially oriented music, theater, and dance related to neighborhood parties, the entertainment for social clubs, the conservative orientation of country club entertainment and social dancing, and the cocktail party ceremonies with background music. The adaptation of the commercial performing arts and their products to a community and its internal "folk" function differs from the true folk and traditional performing arts in that heritage and oral traditions are nonexistent; instead, the traditions are being created.

THE ETHNIC INFLUENCE

When we consider the American urban environment in the first half of the twentieth century, we envisage multiple neighborhoods with rather distinct ethnic emphases and with particular cultural manifestations in the performing arts going on parallel to each other. These folk and traditional arts were therefore continuing in their functional vein in accordance with the customs and traditions of each group. While all of these neighborhoods were generally aware of each other, they appear to have generally had only a small impact on each other in the performing arts. This diversity, with several significant exceptions, seems to have largely disappeared.

Much of the original differentiation based upon ethnic and religious grounds appears to have disappeared since World War II largely as a result of the physical disintegration of tightly interwoven neighborhoods, changes in the socioeconomic conditions of the immigrant population, breakdown of the significance of ethnic and religious heritage among the second and third generation offspring of the original settlers, and the impact of mass-media encouragement of common American experiences and traditions. Much of this does indeed seem to be true in the realm of the performing arts.

The breakdown of the tightly interwoven ethnic neighborhoods, or ghettos as some sociologists would have it, seems to be closely linked to the rise in the level of the socioeconomic conditions of the immigrant populations. These foreign-born Americans have fought and worked hard to become integrated into American society and many have moved to "suburbia" or "exurbia" as they won their struggle over economic hardships. Unfortunately, the very family that has made such a move for better schools, more fresh air, and assimilation with middle-class America has too often consciously lost its link with its own heritage. A major result then is an unfortunate loss of identity with the commercial performing arts serving as a replacement for their authentic folk heritage.

This loss of ethnic and religious identity is true particularly of European immigrant neighborhoods. The German, Italian, Scandinavian, Slavic, and other ethnic organizations appear to maintain little more than a token allegiance to the folk and traditional performing arts. Their second or third generation American-born leaders are usually more interested in their organizations' social activities than in the authenticity of their dances, music, and theatrical productions. Their language itself frequently becomes as watered down with Americanisms as their performing arts.

The ethnic and religious groups seldom have the tight solidarity they once displayed, while the American commercial arts, provided so freely by the mass media, continue to dilute the individualities that were once fiercely guarded. This is not to say that the youth of foreign extraction does not respect or even attempt to continue the traditional performance arts, only that they no longer have the vital functional force which they once possessed.

In addition, the ethnic folk and traditional performance arts have been assimilated into the commercial arts of the mass media. Polkas are simply another part of the musical merry-go-round on "ethnic" radio and television programs. This trend is also witnessed in the consistent closing of folk theaters and foreign language film houses, except for those showing the art films. It is a sad fact that the younger generations seem to have insufficient interest or desire to maintain the authentic folk and traditional performing arts on any consistent basis. Only the "old folks" still encourage the ethnic distinctions while the younger generations strive to conform to the norm of the commercial performing arts.

It is interesting that the college campus in particular has recently been the scene of a revival of ethnic performing arts, especially folk and traditional dancing, song literature, and even attempts at recreating folk idioms in music. Some special community centers have also encouraged this, but these manifestations are primarily for the entertainment of all, rather than the functioning of a community.

THE GHETTOS

The major exceptions to this trend are the Spanish-speaking peoples and black ghetto residents of the American city. Each reflects a different phenomenon in the performance arts. The former, generally immigrants from Puerto Rico and other islands of the Caribbean, seem to be carrying on the basic cycle of immigrant tradition based upon the socioeconomic necessity of living in tightly interwoven neighborhoods which resemble the ghetto conditions of yesteryear. They maintain their independent language and culture, and therefore their functional folk and traditional performing arts. But the mass media and educational systems are working hard to "integrate" these people and soon they, too, will lose their cultural identity. What is so unbelievable is that many city administrations deem it not only advisable but necessary to discourage the indigenous folk and traditional arts of these groups in order to make it easier to assimilate them into the conformity-minded American economy and society. To be different is still a dangerous attribute in American life.

The black ghetto reveals a totally different situation, as black intellectuals and leaders have pointed out so well. We need only to read the significant literature written by LeRoi Jones, Stokely Carmichael, Rap Brown, Dr. Martin Luther King, Jr., and many more to understand their point of view. Arriving in America not by choice, but by force, the Afro-American slave culture maintained its folk and traditional performing arts to a high degree of authenticity before the Emancipation Proclamation. Since then, despite the political, social, and economic suppression of black American culture, the Afro-American performing arts have had an enormous impact upon the development of indigenous American folk and traditional performance arts. Jazz, dancing, and gospel singing, in particular, have become significant enough to achieve the status of serious performance arts in many cases.

Let us make two points, without going into an extensive historical discussion. First, although many of the performance arts of black America began with the southern plantation and other rural locations, it is in the urban environment that they started to gain artistic and commercial distinction. Secondly, although jazz, dancing, and gospel singing have their roots in folk and traditional performance arts, they long ago entered the arena of serious aesthetic expression through such magnificent performers as Louis Armstrong, Mahalia Jackson, John Coltrane, and many others.

The primary emphasis in the black American folk and traditional performing arts is in music and dance. Granted that the original African oral traditions were the basis of their performance arts, there is little question that the Afro-American forms of these traditions have spawned new combinations and permutations with the white American performance arts. The consequence is an indigenous folk and traditional performance art virtually independent in its realization. All the recent writing on the history of jazz focus on the amalgamation of urban white music with black performance practices. It is the individual interpretation that makes a piece of music sound like jazz or that makes a black gospel singer sound as he or she does. It is, in other words, the performance practices that particularly tend to distinguish the black performance arts.

Unfortunately, very little of the authentic black folk and traditional performance arts are practiced even in the black ghettos of our urban centers, and this is a major concern of black intellectuals and leaders. Authentic black jazz, dance, and vocal music such as gospel singing and "blues" were seldom economically successful in themselves despite their innate popular appeal. Folk and traditional performing arts seldom produce large profits since they are for special groups in American society. The black performer, however, was given the choice of watching his white counterpart utilize jazz techniques for commercial gain or of adapting his art to a wider market to make it a commercially saleable item, for it is the commercially successful performer who reaps money and fame. Fortunately, there are some jazz and blues giants who still maintain their identity in the face of the commercial onslaught, but they are few. Too often, the black musician or dancer turns to the market place—for which he can hardly be blamed—and the result is diluted jazz and "jazzed up" arrangements for dance bands.

On the other hand, the stars that guide the new directions in jazz, such avant garde musicians as Thelonious Monk and John Coltrane, soon lose contact with the commercially attuned public. These stars of jazz festivals have evolved jazz into a serious performing art of the first caliber and are supported by a small minority of devotees.

It is primarily a reaction to commercialization that has led the younger generation of black leaders to call for a return to the more authentic souces of Afro-American culture and thereby to find their true cultural identity. Thus it is that the performing artists in the black ghettos are turning more to themselves for the true folk arts and rejecting the very mass media that has had such a large hand in popularizing, as well as prostituting and diluting, their art. This is the new breed of dedicated folk artists in the black community who have no place to go for other than technical instructions in their art. This is also the new breed which may provide some fascinating new directions in the folk and traditional performing arts of black America.

There are, of course, many other manifestations of folk and traditional performing arts in the rural areas of America, but we are not really concerned with the phenomena. We have become aware of country and western folk traditions not because they are urban or are the results of urban environment, but because they have become commercially popular through the mass media and the efforts of Madison Avenue. In addition, the mass migration of "country and western" peoples to the urban centers has made much of this music a popular urban manifestation.

THE COMMERCIAL FORM OF
THE PERFORMING ARTS

In discussing the commercial form of the performing arts we are concerned with the direct application of urban industrial technology within the competitive enterprise system to create an attractive product for mass consumption and for high profit. The "industry," as it is often called, is primarily interested in sales appeal to the largest market. The result is naturally the application of high-powered advertisement for a product with the greatest common denominator. Thus we find the most successful form of the performing arts, in economic terms, aimed at the entertainment value. It has come to the point where Madison Avenue frequently is telling the public what its taste should be in order to create a large demand for the product the seller wishes to market.

The popular performing arts were at one time—before they became big and profitable business—closely linked to our category of folk and traditional arts. This urban manifestation of the performing arts industry arose mainly out of the arrival of mass media and probably first through radio, the Hollywood motion picture industry, and the recording industry. With the object of reaching the largest possible market, common standards and conformity of taste were analyzed and developed through the channels we now call Madison Avenue.

In order to build such a market, individuality and aesthetic interests are sacrificed for instant universal appeal, a must in making a profit. Some of the major results in the music industry, for example, are country and western music, commercial jazz, commercial folk music, and comercially "piped" music. In the theater some of the major results are the Broadway entertainment theater, radio and television theater, and the Hollywood films. Unlike music and theater, the dance arts have adapted their individuality to those forms of music and theater where they generally provide a pleasant divertissement. All phases of the folk and traditional performing arts, as well as the serious arts, have been corrupted in some manner for making the "fast buck."

"Tin Pan Alley" has become notorious for its almost scientific examination to determine the ingredients of instant universal appeal, to which is added a little built-in obsolescence. Then the tunes are ground out in mass production for mass consumption. Fortunately, there are always exceptions to the rules. The marketing formula of massive advertisement and distribution leads to the recording industry's best-seller lists, which measure the economic success of the instant universal appeal. We need not go into the incredibly sophisticated psychological machinery which is employed by the music industry and, especially, the advertising industry to witness its effect around us.

Jazz, which is still one of this country's greatest source of original Afro-American music, has become successful through its commercialization into ballroom dance music and adaptation to the standard commercial repertory. As mentioned before, it is rare to find performers capable of improvising in an authentic black American manner, or an arrangement written in a musical style that is not in a commercially acceptable idiom.

Music for radio, television, and Hollywood production fits into the same common commercial standard as the show tunes for Broadway musicals. Only once in a great while does a *West Side Story* come forth to provide a refreshing change from the same old mill. In mass-media usage, music seems to have three primary functions: to provide the proper background or mood music to a script, to provide the proper accompanying arrangement for the "torch singers" for songs high on the hit parade, and to provide the large sound to accompany large dance production numbers. Illustrations of these functions are available every day on the radio, television, or in motion pictures. The television cartoon shows are frightening in their prostitution of musical experience for dramatic effect, while the subliminal learning of musical commercial standards of sameness is stifling the aural growth of future audiences.

With the advent of "piped" music and mood music, the commercial music industry has hit an all-time high in profits and an all-time low in taste. Music to shop by, to eat by, to sex by, to relax by, to stimulate by, and so forth has become commonplace all over the country. This kind of "wallpaper music" is nihilistic to the aural sense in that the listener now accepts music in the same manner as he becomes accustomed to automobiles driving by.

Commercial formulas are equally applicable to the success of a Broadway show, but there are other variables that can bring economic disaster—such as critics. The Broadway theater is seldom the place where any experimental or repertory theater occurs, since these are not necessarily sure profit-makers or intended as big business. Such productions have had to find homes in off-Broadway theaters and universities where audiences are more receptive to innovations. Broadway theater is a hallmark of conservatism and audiences can generally predict the character of the play and its performance.

It is radio and television theater, however, that reveals the dramatic taste equivalent to that found in popular music. The constant soap operas, horse operas, comedy series, game shows, and the like which we experience on radio and television are but a few examples of the diet offered to the public as popular commercial performing arts. The question here is not one of categorically condemning the present shows in these media, but of bemoaning the sheer quantity of their production. A soap opera need not be chided because it is a soap opera, for the quality varies as in any artistic medium. The problem is one of quantity, for even six excellent soap operas shown consecutively would reach the point of supersaturation. What is to be deplored, in short, is the lack of imagination and variety that characterizes the media.

In dance, the commercial tradition of providing a choreographic diversion also has its history in Hollywood film spectaculars as well as vaudeville and burlesque. The flexibility of being able to adapt to the need at hand is based on the entertainment value of the dance arts and the opportunity of displaying attractive girls rather than any intrinsic aesthetic function. A glance at today's television and motion picture production numbers is sufficient to illustrate the point.

There are, nevertheless, a number of new performing arts phenomena, especially by and for the younger generation, which began as popular arts. Such is the case of early "rock and roll" groups and some folk singers and groups. The quick eye of the industry, however, soon picked up these opportunities and cashed in on their originality. Aware of the innate appeal of these relatively new products, the industry immediately adapted its machinery to utilize and make big business out of them. Perhaps an outstanding illustration of these effects is the meteoric rise of the Beatles to international fame and riches. This group invented and disseminated a fresh new approach to commercial popular music and soon began to be imitated throughout the Western world. After years of the same diet, they became saturated with their own style and began to experiment. It is interesting to note that when the Beatles decided to make a much more serious turn away from the market, their record sales diminished significantly.

The popular "folk singers" and groups fall much into similar situations. The traditional folk singers like Josh White and Pete Seeger were not on the best-seller list for many years. As the social revolution began to be supported by the students in the 1960's, it was soon realized that this appealing and subtle folk music could be an excellent conveyor for provocative lyrics. The new "folk singers" and "folk groups" quickly appeared to utilize this performance art for their own purposes. The commercial interests immediately grasped the profit potential, and the ubiquitous folk singer and folk group are now at the height of fame, fashion, and financial success. The relationship between authentic, anonymous folk songs as a cultural heritage and the carefully composed commercial folk song now have meticulously contrived similarities.

It is significant, nevertheless, that the folk and traditional performance arts have become a major resource for means of communicating and reflecting many of the urban problems. The circle of folk art to popular art and back to folk art is almost complete, only now it is commercialized.

As mentioned earlier, folk dancing is becoming a popular art and the proliferation of dance studios, community centers, university groups, and social organization which feature folk dancing reflects the growing interest by amateurs in this performance art. Presently, however, the commercialization of popular folk-dancing groups has yet to be done successfully.

The ability of the commercial arts to adapt, absorb, and manipulate virtually all phases of the performance arts for large profits serves to indicate the efficiency of such business enterprises. Only as a result of the incredible urban technology could efficient mass production methods be coupled with the mass media and the most sophisticated advertising campaigns to serve and create public taste.

SOME SPECULATIONS ON THE
PERFORMING ARTS

There is a clear and intimate relationship between the urban center and the performing arts on all levels of sophistication. The cities have been the traditional center for their creation and performance. Recently, numerous new problems have arisen to threaten the very existence of many of these arts.

In the serious performing arts, the problems center around the economic dilemma, associated with the proper financial support of creator, performer, and performance organization, and the educational dilemma of providing first-rate professional training and audience education. The 1965 Rockefeller Panel Report provides a convincing array of suggestions based upon the division of financial support between individuals, private enterprise, and public support. We should like to emphasize the great need and convincing arguments presented for reasonable federal government financial support for the creators, performers, and performance organizations.

What is required is a competitive salary scale for the professional creators and performers to encourage them to remain in the performing arts and not to look elsewhere for a decent and secure living. Along with this, the performance organizations must have adequate facilities and economic support in order to achieve the highest form of excellence desired of them.

Finally, the problem of professional and audience education is just beginning to be acknowledged by our professional schools, universities, and primary and secondary school systems. The enormous cost of providing top-quality professional performers is a very important hurdle to be overcome if mediocrity is to be avoided. Professional institutions, as much as the performing organizations themselves, should be supported by the same economic sources. The problem of aesthetic education is one that until now has been generally left in the hands of higher education. The numerous recent reports on the lack of aesthetic education in the public school systems is alarming and a new awareness of its importance is absolutely necessary. This means that professional performers, educators, and communities must work together to reform the school curricula so that it includes some form of aesthetic education from kindergarten through high school. Enough reports have been accumulated, especially by the Department of Health, Education, and Welfare, to document the needs and recommend specific and direct action.

The immediacy of the need for aesthetic education is especially pointed when the relation between the performing arts and increasing time for leisure is considered. Sebastian De Grazia's work, *On Time, Work, and Leisure* (New York: the Twentieth Century Fund, 1961) clearly indicates the ever growing need for establishing closer ties between the arts and leisure. Aesthetic education is certainly the first step, and a large one, in the right direction, for it is this which leads to the next rung, that of amateur involvement in the performing arts. This involvement in turn means greater interest and broader support for all phases of the performing arts. The latter, it should be reiterated, are not for the priviledged few, but for all to experience meaningfully and become involved in.

The folk and traditional performing arts are on the verge of losing their identity and individuality in an urban culture emphasizing sameness and conformity. For logical reasons, the groups that originally supported these diverse and colorful performing arts are no longer doing so with the same vigor. These particular arts are thus in danger of becoming extinct and there seems to be little that can be done to save them. Even if they are renovated at this time, it would be mainly for entertainment purposes or for academic reasons. On the other hand, the folk and traditional performing arts of the Spanish and black ghettos do still have functional identification, and it is the responsibility of the entire nation to make sure that these arts are maintained. Cultural variety is the food that helps to encourage respect for diversity of dress, opinion, belief, race, and heritage.

Of primary concern is the attempt by the black American to find his Afro-American identity. The performing arts should be a major avenue in this quest: private and public funds must be found to provide facilities, develop instruction programs, and support creative as well as performance training and programs. Since black American culture has provided the nation with some of its finest indigenous art forms in the performance arts, everything must be done to nourish these creative endeavors before the mass media commercialize them more.

The commercial performing arts are thriving in America as a result of the use of urban technology applied to the performance product. This enormous industry is providing jobs for the labor market, large accounts for the advertising firms, and a mass-produced product comparable to that of the Detroit automobile industry. It is the close coordination between the powerful commercial interests in these performing arts and the mass communication media that is largely responsible for the vast nationwide market of consumers of mass-produced performing art products.

We should be quick to emphasize that this is not necessarily the fault of the commercial performing arts industry. This industry in a capitalistic society is predicated upon the profit motive, as it should be, and it naturally provides those products which have the greatest market appeal. If there is any complaint about the taste revealed by the popular performance arts, it should be directed at those areas which are socially responsible for the development or nondevelopment of the taste level, namely the educational media. What is needed is not a cry of corruption as much as an overall perspective on the encouragement of aesthetic education. We have already pointed out the sad state of neglect of such education in our school systems. What we have not mentioned is the equally sad state of affairs that exists in teacher certification in this regard.

Aesthetic education must be encouraged at all educational levels, and all private, public, and community institutions must become aware of the seriousness of this problem in order to provide the most effective measures for solving it. Television, radio, and motion pictures could conceivably become as strong a force in developing aesthetic education as they now are in entertainment and amusement. The most imaginative minds must be employed to help man recognize his aesthetic needs and fulfill them in a variety of ways rather than leave him to be simply part of one mass market.

The urban university has a major role to play in making all phases of aesthetic education available to the public. Such things as community theaters, musical groups, and dance groups could easily be linked to the urban university, but only if the professional instructors first shed their blinders and begin to mingle with the public. Only in such a manner can the urban university provide a relevant link to the performing arts on the one hand, and the audience on the other. Such a function could indeed help significantly in making the university an integral part of the community environment. It could also contribute inmeasurably to the enhancement of the quality of urban life.

11

The Mass Media in Urban Society

GENE BURD

☐ CITIES AND COMMUNICATIONS bear a close relationship to each other. Throughout history the range of the latter has affected the size of urban concentrations and conditioned both the pervading image of the city and the conceptions of life within it. These in turn have helped to shape man's notions of urban citizenship and the quality of urban environment.

Plato conceived the ideal size of a city to be such that all the citizens could be addressed by a single voice at any given time. Many urban communities have, in fact, been limited in size by the number of people who could respond promptly to messages coming from within their boundaries. Mesopotamian cities, for example, had an assembly drum; those in the medieval period used a bell in a church tower; the sound of the Bow Bells once defined the city limits of London; and in World War II, English church bells were rung to signal a German landing when normal communication by telephone and telegraph was disrupted (Mumford, 1961).

"Cities were evolved primarily for the facilitation of human communication" (Meier, 1962: 13). They have served as the seat of security and surveillance, the repository and custodian of knowledge, the locus of power, and the disseminator of information—religious, economic, and political—in short, as the control and communications center of the total society (Greer, 1962).

The city is not only the center of a communications network but also, in Mumford's words, a "special receptacle for storing and transmitting messages."

> This metropolitan world, then, is a world where flesh and blood are less real than paper and ink and celluloid . . . The swish and crackle of paper is the underlying sound of the metropolis. . . All the major activities of the metropolis are directly connected with paper and its plastic substitutes. . . . the journals, the ledgers, the card-catalogues, the deeds, the contracts, the mortgages, the briefs. . . the advertisements, the magazines, the newspapers [Mumford, 1961: 547].

While the traditional city has been linked with the print media and writing, the future city is developing as an "information megalopolis" (McLuhan, 1964: 99). Modern transportation and newer means of electronic communication reaching into outer space have indeed transformed the world into a veritable city.

Urban values have been transmitted across geographical boundaries via the mass media, so that the city has become as much a state of mind and style of life as a geographic necessity. Even psychological and physical security are tied to communications rather than to central place: satellites warn of hurricanes, transistor radios explain a power blockout in megalopolis, and television screens caution citizens on tornadoes and the "Ten Most Wanted" fugitives. Modern communication has not only made environmental surveillance feasible on a wide scale, but has done much to make the total society urban in nature. The city is the habitat of the mass media with its high order of symbolic behavior, which in turn reflects leisure and affluence, a large audience available for mass merchandising techniques, and the market for the consumption of the mass culture of quantification.

MEDIA FUNCTIONS IN A MASS SOCIETY

It has been suggested that "the structural trends of modern society and the manipulative character of the communication technique came to a point of coincidence in the mass society, which is largely a metropolitan society" (Mills, 1956: 314). It is in this urban setting that the mass media enforce the norms of the community, dispose of threats to those values, transmit the social heritage to the next generation, and help maintain consensus in a highly specialized and changing environment.

In a very real sense, the city is what people think it is. "The city we know personally—the city of the mind—largely determines the world in which we have our life's experience and through which we strive to gain many of our daily satisfactions" (Carr, 1967: 199). As Anselm Strauss has noted, "The city as a whole is inaccessible to the imagination unless it can be reduced and simplified. The streets, the people, the buildings and the changing scenes do not come already labeled. They require explanation and interpretation" (Strauss, 1961: 12). Characterization of the city and the life within it is indispensable for organizing the mass of impressions and experiences to which every inhabitant is exposed and which he must collate and assess, not only for his peace of mind but also for carrying on his daily affairs (Wohl and Strauss, 1958: 523).

Thus, each citizen of the metropolis, unable to experience first-hand most of his environment, must rely on the mass media to inform him about it with symbols that represent the large part to which he may not be directly exposed. This, more than the direct shaping of specific opinions, constitutes the power of the mass media. The words and pictures give the citizen his only summary image of the city: the camera shots from high buildings, the picture postcards, the aerial photos of the metropolis, the city model, map or diagram, the landmarks and landscapes, and the legible nodes and districts of the community. These become a kind of civic stereotype, civic icons, and abbreviated "civic shorthand" which help the citizen identify his urban environment.

The manipulators of the mass media are like urban middlemen who market the metropolis. The media translate the urban experience and become in effect the urban reality for many people. Indeed, the words, pictures, print and celluloid are often as permanent as the civic cement of the steel skyline, and more so, when they are preserved in the historical society photo files and in the memory of earlier residents. The city's songs, scenes and mottoes are often perpetuated in radio station identification, in television screen scenes, on magazine covers, and on newspaper mastheads. (Legend has it that the city of Cleveland's name was due in part to a newspaper's not having a sufficient supply of the letter "a" to properly spell the name of the community's namesake, Moses Cleaveland.)

The news media convey the tradition of civic ceremony and link the past to the future. Local sports teams with name and uniform grace the sports pages devoted to civic identity. Unique and distinct local products, from autos to beer, appear in advertisements and on billboards and neon signs. The individual community often becomes inseparable from its most dominant symbol (Burd, 1968d). The images of the city in the mass media help shape its physical nature; they are the dream preceding the reality of bricks and concrete. They affect what the city visitor and outsider see, and they have an impact on convention and industrial site selection, insurance rates, and bond and investment transactions. They do virtually everything from affecting the politician's message to influencing the attraction or distraction of artistic talent.

The image of the city conveyed in the media also is significant for collective action and inaction. "A vivid, integrated physical setting, capable of producing a sharp image, plays a social role as well. It can furnish the raw material for the symbols and collective memories of group communication" (Lynch, 1960: 4). In addition, the mass media act to legitimize and confer status on public issues, persons, organizations, and social movements, and perform a "narcoticizing dysfunction" whereby media content channels the energies of individuals from active participation to a vicarious and often passive role" (Lazarsfeld and Merton, 1948). In less sophisticated language, V. M. Newton (1961: 306), managing editor of the Tampa, Florida *Tribune* said ". . . a newspaper with a soul must be an integral part of the community. It must share all the fortunes, the sadness and joys, and all the adventures, big and little, of its readers. It must share them as a part of understanding the great human family."

But it is all too obvious that the newspapers and the other media, whatever their other functions—from selling goods and services, through collating the intriguing trivia of everyday urban life (births, deaths, divorces, auto accidents, petty felonies), to creating and manipulating symbols of civic identity—have failed to facilitate that kind of understanding and motivation among urban readers and listeners and watchers which would lead to action to meet the problems of the metropolis. "The crisis of the city is thus, in the beginning at least, an intellectual crisis. . . . The action crisis in the metropolis cannot be disengaged from the intellectual crisis for the very definition of a metropolitan problem is dependent upon one's picture of the city and the kind of life it should contain. . ." (Greer, 1962: 21).

Why have the mass media, whose functions would seem to constitute so powerful a force for motivating and facilitating the renovation and reconstruction of the urban social order, apparently failed in large part to realize that potential? Why have they so often appeared to so many social critics as mainly supporters of a status quo of deteriorating quality rather than as partisans of innovations which might halt the trends of decay and initiate some new and needed kinds of enrichment of urban life?

Serious dilemmas arise in any discussion of mass media and metropolitan problems. The economic need of the media to sell coincides with the democratic notion of reaching large mass audiences, but this marriage of necessity and invention creates other dilemmas. First, to reach large numbers of people with specialized complex news content may mean a dilution of its substance and quality. Thus, mere entertainment may result from an honest attempt to educate (Burd, 1968h).

Secondly, the record of citizen participation in a democracy is not flattering. In megalopolis "locals" find comfortable isolation, and only a few, highly dispersed metropolitan citizens (often media personnel) identify with the city as a whole. They take part in the decision-making, but they usually live, buy, and vote outside the central city where they work, and they are often linked only by the metropolitan news media, which is a kind of rudimentary metropolitan government, but without power to govern. If the media allow themselves to become urban actors, they destroy their self-image as objective observers. If they merely report all viewpoints disinterestedly, continuous "democratic debate" may delay or prevent any action.

Third, wide reader and citizen attention may be attracted through "entertainment" and "human interest" and news of conflict, but they may not propel the reader, viewer, or listener into actual involvement; instead they may pacify or narcotize him, providing him with a defense against any civic commitment. Furthermore, news arising out of cities via controversy puts decision-makers in a crisis context, and also creates an adverse civic image. In the 1920's, for example, the city of Chicago sued the *Tribune* for affecting the city's credit rating because of stories on corruption; and in 1967, Milwaukee city officials considered suing the press because of adverse publicity on civil rights demonstrations.

Thus, further exploration of the treatment of major issues by the mass media requires seeing how their action has been conditioned by the institutional characteristics of at least the locally based mass media as private business corporations, by the interests and actions of their owners and managers as members of the local decision-making elites of most communities, and by some of the perceptual and reportorial habits of their staff personnel, habits which have become the American tradition of "news."

THE MASS MEDIA AS BUSINESS
ENTERPRISES

Though written communications are as old as cities, printed mass media has existed only during the last few centuries of the eight thousand years of urban history. From their earliest origins written records were related to market transactions, and the earliest cities developed a special role as communications centers. In American cities, the press was to follow the business civilization which created urban regions. The newspaper, as Weisberger points out, was part of this growth, responding to it and nurturing it. Being a form of social communication, it went wherever society was going, to the very limits of settlement. It served as a handmaid of commerce, and as the latter increased, the newspaper page grew in size and in the frequency of publication (Weisberger, 1961).

Growing cities meant more news, more circulation, more consumers, more advertising. The larger the audience, the greater the capacity of the newspapers to attract advertisers. The shorter the headlines and the larger their type, the quicker commuters and busy city residents could read them. The newspaper circulation area, the trade area and the city's image coincided in the days before the full impact of telephone, television, teletype, and radio began to offset the need for urban clustering. The press provided a territorial viewpoint otherwise lacking.

To be successful, the newspaper's influence and prosperity had to parallel the growth of the city in which it was published (Grubb, 1940). The American city was a creation of business and the market, and the press a creation of them all. News was just one more highly perishable product which had to be marketed quickly. "As private business, all branches of the press are subject to the economic logic of being successful in accordance with the principles as well as the pressures controlling private enterprise" (Wolseley and Mott, 1958: 2). Even the pattern of a newspaper's organization is conditioned by the nature of the locale it serves, including sociocultural, economic, and political pressures (Wolseley and Mott, 1958: 37). Small wonder then that "freedom of the press" and "free enterprise" found each other's company compatible, and that the notion of newspapers as "the watchdog" of government operations (budgets and taxes) in cities which were "bigger and better" (profits and more profits) prevailed. Frank Thayer expressed it in his advice on *Newspaper Management* (1934: 410):

> The success of a newspaper depends, in large measure, upon the success of the town in which it is located. . . . If the newspaper management understands the commercial texture of a community, it can see what commercial and industrial policies would make for a better city. . . . The newspaper publisher must ask what the industrial leaders outside the town wish to find in the town or city in which they are to invest their capital. . . . One difference between a "dead" town and an enterprising town [is] a town with increasing population, growing bank deposits and substantial programs of municipal development. . . .

In America the definition of urban problems and the proposed solutions also are related to the locale of the news media, especially the daily metropolitan press (Burd, 1968c). Press facilities are usually located in or near the central business district, close to news "beats"; and near, also, to media real estate investment and the highest points of land value in the metropolis, and to the heart of the advertising and circulation base. This central location contains the convergence of telephone, telegraph, postal and other nonmedia communication, the means for transporting newsprint, and usually the highest elevations in the metropolis, where television antenna and radio signal points may be placed atop tall buildings. Indeed, news media structures often form part of the civic skyline: the Time-Life Building and Times Square in New York, and the Playboy building and Tribune Tower in Chicago, the latter built in 1920 as the "gateway" to Michigan Avenue.

Not only the media's image of the city emanates from the central business district, but also its prevailing notion of news:

> If it [the central business district] is the center of communication, it is,
> by the same token, also the news center, in both the journalistic and the
> market sense of what news is. The relation between news and speculation
> has, historically, been not only close, but causal. The rise of the great
> insurance firm of Lloyd's of London cannot be understood aside from
> the relation which risk-bearing bears to the availability of reliable
> news. . . . The central business district as the locus of the market is then
> the place where both news and credit are created and concentrated and
> from which both are distributed [Johnson, 1957: 253].

Thus, the mass media's concept of the urban "public interest" and
the quality of urban life is usually subsumed under the philosophy
of: what's good for business is good for the central business district;
what's good for the central business is good for the city; what's good
for the city is good for the press and, indeed, for the metropolitan
area it serves, and so on. But in their support of metropolitan
government, the metropolitan media, despite their territorial view,
have failed to convince voters, who generally reject it. Ironically, the
media have most often presented to them an image of the city which
does not fit a metropolitan approach. The media managers, however,
are apparently better able to "see the necessity" of such an
approach, which might prevent further decentralization from the
central city's economic, advertising and political center of control,
for "to complete the process of metropolitan monopoly . . . one
further step is necessary; the effective monopoly of advertising,
news, publicity, periodical literature, and above all, the new channels
of mass communication, radio and television" (Mumford, 1961:
537).

Political fragmentation prevents full communications integration
in the metropolis, as "editorial writers and publishers of central city
newspapers . . . manage . . . to be true to both halves of their schizoid
selves by touting the virtues of downtown, while living in the suburbs
and using the press to endorse and plead for what they no longer
could vote for" (Ylvisaker, 1961: 107). Even electronic media, less
tied to the central city but still not dominant in the suburbs, are
challenged by the suburban support of "state's rights" and a strong
weekly suburban press, which opposes threats to suburban corporate
identity. The latter in turn is often attacked by the daily
metropolitan press, which paternalistically expects that central-city
goals will somehow unite all metropolitan interests. Thus the
headlines: Suburbs Are Told They Can't Exist Without Cities; Find

Suburbs Sap New York City's Growth Strength; Suburbs Need a Strong City; Suburbs Labeled Hasty In Opposing Metropolitan Rule, and How Suburbs Wound City.

Here, too, the media have sometimes tried to control the environment as well as report it. They often have viewed the federal urban renewal program as a means for the salvation of the central business district and its facilities. The downtown media have not only been rationally motivated, but have in urban renewal a crisis news story of dramatic self-interest, whose saleable sensationalized copy provides both a surveillance function and a signal that the seat of urban power may be endangered and the advertising and circulation base in peril.

Since the central city is the "heart" which is said to pump life blood into the metropolis, decay and decentralization are metaphorically seen as "sickness," an "illness" whose cure is renewal and rejuvenation of the old body, not the alteration of the total environment. The various "face-lifting," "save downtown" urban renewal programs have had the support of a small elite including "financial institutions, newspapers, department stores, owners of downtown real estate, academic intellectuals, city planners, city politicians, and others who have a strong stake in the maintenance and improvements of the city as they see it today" (Anderson, 1964: 218). Perhaps it is understandable that the federal urban renewal program has not been subjected to serious press scrutiny (Burd, 1968a), and that planning and urban renewal in the press tend to be considered good in themselves irrespective of the facts of the particular cases (Clark, 1959).

The media content is a combination of business-civic philanthropy and self-fulfilling prophecy, plus extensions of their rational economic self-interest offered as "social responsibility." It is presented through the mythology of editorial independence and the objectivity of reporters and crusading editors who guard the "civic conscience" in deciding on "all the news that's fit to print." The journalism textbooks describe the media's notion of the quality of urban life under such headings as "public service."

There has long been a "civic cold war" between city hall and the press, between those who control information and must maintain order, and those who sell information (which may bring disorder), between those in the media who often set the civic agenda, and those in government who are charged with carrying it out. Politicians must face the voters, whereas the media face only cancelled subscriptions, radio and television turnoffs, and rarely a boycott of goods, if citizens are unhappy.

Although the press sees itself as the people's "watchdog" against government, and much news does involve city hall, rarely does the personal war get exposure, except for surface contact when new television antennas are permitted, shabby downtown newsstands are charged with litter ordinance violantions, circulation trucks are ticketed for speeding, or extra police guards requested for newspaper plants. Not even the beatings of Chicago newsmen in 1968 at the Democratic convention was sufficient for news management to declare war on city hall. So there are areas of agreement between the two. As one editor put it:

> You and we [city hall and the press] are in the same general type of business—service to people. ... If we do not have much in common, something is wrong. ... All the money you have to spend is that provided by the taxpayers, and all the money we have to spend is that provided by the advertisers and the subscribers. And much in the quantity and quality of service each of us can give is determined by the dollar each receives [Futrell, 1960: 2].

MEDIA MANAGERS AND
COMMUNITY LEADERSHIP

News media personnel have both participated in and reported on city planning activity. With a combination of self-interest and civic devotion, they have not always been disinterested or "objective" bystanders. Three of four early Chicago mayors were publishers or part-owners of newspapers. Mayor-publisher roles are still not uncommon in small towns. One of Milwuakee's founders, Byron Kilbourn, used the city's first newspaper, the *Advertiser*, which he had established, to promote his improvement projects. A competing real estate man, Solomon Juneau, started his own weekly as a promotional vehicle for his own real estate projects. The paper became the city's first daily, the *Sentinel*.

Publishers have found the need to promote better transit to distribute goods and newspapers, and a beautiful and more attractive city for outsiders to see. In 1911, the president of the Dallas *Morning News*, George Dealey, selected the city's first city planner, and took part in negotiations for a new university campus and rail terminal unification as part of the "city beautiful" movement. In Houston, in the late 1950's, the editor of the *Chronicle*, M. E. Walter, served as chairman of the ad hoc city planning commission, and later brought in the first planning official.

Rucker and Stolpe (1960: 349) have observed:

> Certainly there is no better way for a publisher to put out a newspaper with perfect ease of conscience and with the respect of the community than through consistent initiation and sponsorship of needed community improvements. No one is in a better position than the publisher to sense important community needs or to lead in answering them; and the extent to which he gains good will through worthwhile services determines largely the value of his newspaper as a going concern.

It is often hard to separate city planning, the media personnel and the reporting function. Some journalists suggest that editors "may have a part to play in the social plan themselves. . . . [and] until the general public takes a greater social responsibility, press and radio may have to carry an abnormal share" (Campbell and Wolseley, 1949: 395). The editor of the Memphis, Tennessess, *Press-Scimitar,* Edward J. Meeman, a member of the board of directors of the National Conference on State Parks, once justified personal activity by the press on parks and other public facilities, saying, "A newspaper should sometimes make news as well as report it" (*Planning and Civic Comment,* December, 1958: 28).

In 1958, Amory H. Bradford, vice-president and business manager of the *New York Times,* was elected president of the Regional Planning Association in the New York area. The first executive director of the Kansas City Metropolitan Area Planning Council was also the director of WDAF-TV and radio in that city. In the late 1800's, the *Indianapolis News* hired a special reporter to "see that the organization was formed" for a Commercial Club, which later became the city's Chamber of Commerce (Grubb, 1940: 40). In Omaha, the publisher of the *World-Herald* in 1945 personally called together civic leaders to launch a city planning program, and appointed a committee to prepare a capital improvements program. (The *World-Herald* was later purchased by the owner of a large construction firm which erected many new downtown skyscrapers.)

In Chicago, the *Tribune* and one of its political columnists were personally influential in the selection of a site for the city's new convention hall, named for a former *Tribune* publisher, Robert

McCormick. In Los Angeles, the wife of the publisher of the *Times,* Dorothy Chandler, conceived of, and raised millions of dollars for, the city's new music hall, which was built near the Times building downtown and named for her. In Houston, the new performing arts center was named for the former owner of the *Chronicle,* Jesse Jones. In Kansas City, when the *Star's* founder, William Rockhill Nelson, came to the city in 1890, his first effort to improve the quality of urban life was to expose the "opera house" as a fire trap and work for a new one. (The opera house owner later guaranteed notes for new *Star* presses.)

Nelson's policy of "civic betterment" led him not only to crusade for charity, welfare, and cheaper gas, and against tenements, poverty, loan sharks, and fake land sales; he also took part in planning the city's street and park systems, and left his estate and art collection for the city's famous Nelson Art Gallery. His staff was told, "The *Star* has a greater purpose in life than merely to print the news. It believes in doing things. I can employ plenty of men to write for the paper. The successful reporter is the one who knows how to get results by working to bring about the thing he is attempting to do" (*Kansas City Star* staff, 1915). The *Star's* managing editor conceived the idea for a city convention hall in 1897. It was built, and although it burned, was rebuilt in time for a national Democratic convention.

Other examples of press involvement in city planning are numerous. The *Arkansas Gazette* helped finance the National Citizens Planning Conference in Little Rock in 1957. In 1959, the *Detroit News* paid twelve planners to make a study of downtown land use. The *Toledo Blade* publisher, Paul Block, Jr., after World War II, retained a man to draw up a city plan, financed a quarter of a million dollar city model, and helped form a port planning committee. Editors in Portland, Oregon and Fort Smith, Arkansas have served on committees to solve area water and flood problems. Tom Wallace, a one-time editor of the *Louisville Times,* was also vice-president of the American Planning and Civic Association.

A reporter for the *Pittsburgh Post-Gazette* and former editor of the local chamber of commerce publication became city planning director, and four city planning reporters became city planning administrators, after having been actively involved in civic betterment programs. Throughout the country, editors, publishers, and columnists write books of praise for their cities, serve on planning bodies and boards, set up programs, speak on local problems at civic luncheons, and accept awards for their campaigns to improve the city. Reporters do civic publicity on the side, help write model-cities proposals, and prepare articles for chamber of commerce magazines.

In many cities, press involvement in rebuilding is seen in the new or remodeled newspaper plants in redevelopment areas. Some plants have been demolished in urban renewal projects, and some have used urban redevelopment land for rebuilding. In Wheeling, West Virginia, the *Intelligencer & News-Register* in 1966 even announced plans to buy cleared downtown land and build shops, a plaza, and parking facilities on it.

Magazines, unlike all but a few newspapers, are generally directed to national rather than local audiences. Like newspapers, however, they not only report, but commonly participate in, city planning activity. Editors frequently speak to national urban conferences. Norman Cousins, editor of *Saturday Review,* was named by New York Mayor John Lindsay to direct the city's Committee on Air Pollution. Magazine writer Jane Jacobs, author of *The Death and Life of Great American Cities* (1961), has been active in the New York debate on urban renewal and open space. As far back 1942, *Fortune* magazine was experimenting with methods to implement city plans in Syracuse, and in 1957 *Life* sponsored a traveling wide-screen movie to educate civic leaders on city problems. *Look,* in 1952, started its "All American Cities" awards in cooperation with the National Municipal League, and in 1956 began its Community Home Achievement Awards. *Esquire* established a national "Business in Arts Award," and *McCalls* financed a campaign to improve Philadelphia backyards. The chairman of the board at *Life* was former president of ACTION (American Council to Improve Our Neighborhoods), which generated numerous books, magazine articles, and urban affairs journalists. Its successor, Urban America, Inc., began a magazine called *City* and spawned an Urban Writers Society.

Thus, those who have greatest authority within the mass media corporations not only behave as businessmen in the narrowest sense, but also as members of the local, and sometimes national, power elite. It would be ridiculous to claim that these roles have no influence on how managers and editors, and the staff closest to them, affect the content of what appears on the printed page, is heard from the loudspeaker, and is seen on the video tube. But it would be equally misleading to assert that this is the whole story.

THE NEWSMAN'S NORMS AND
URBAN REPORTING

Rules and guides for journalistic reporting and interpretation of urban problems appear to have grown out of the shared civic ideology of the press and other participants and promoters of the city. Owners and editors of newspapers appear to be "a species of the genus 'civic leader' " and it is difficult for them to be influenced unless appeals are made to the "public interest" (Clarke, 1959). But policy is rarely spelled out in the newsroom; it is transmitted by social osmosis. However, the business elite are protected by the media (Breed, 1958), and reporters are warned not to expose business except where government has become involved (Newton, 1961: 169). Probably the two main rules for urban reporters are to crusade for the "peepul" and the "little man," and to love the city. In the words of the motto of the *London Evening Mail,* the American tradition of urban journalism is like that of the oath of the young man of Athens:

> We will never bring disgrace to our city by any act of dishonesty or cowardice . . . We will fight for the ideals and sacred things of the city, both alone and with many. . . . We will strive unceasingly to quicken the public's sense of duty. Thus in all these ways, we will transmit this city, not only not less, but greater, better and more beautiful than it was transmitted to us.

As in the 1923 code of ethics of the American Society of Newspaper Editors, "The right of a newspaper to attract and hold readers is restricted by nothing but consideration of the public welfare." Similarly, the electronic media are restricted only by FCC requirements for some "public service" programming. Meanwhile, the media are expected to be (and often boast that they are) independent, responsible, honest, accurate, and able to entertain, sell goods, and educate the public to self-government. It is assumed that open exposure of ideas in the marketplace naturally creates a general public good and "social progress" by an informed public.

But what actually happens when the "city beautiful" remains the popular and press concept of the good city? Planning is still confused or mistaken for mere zoning, and the citizen vicariously participates in news of urban conflict and thereby avoids civic activity. Government becomes in the press a type of sideshow or sports news, where the citizen watches the players but avoids the issues. Drugged by news of conflict, he is not in the mood for making decisions; and excessive information (without explanation) brings communications overload, too many choices, and thus the rejection of any decision (Burd, 1968h).

Moreover, as gatekeeper of the civic symbols and custodian of the civic relics, the press is the city's civic salesman and press agent who points with pride or cries with shame and alarm. It reminds the public of the central business district as *the* city with civic superlatives: the centerpiece, the showcase, the crown jewels, the face and facade, the newest, the biggest, the tallest, the longest, the largest, the cleanest, the safest, and the greatest. Central area builders and buildings are idolized throughout the seasons and in all weather as news photographers capture the skyscrapers in the civic stereotype.

The news rooms are flooded with copy on announcements of new buildings, architectural drawings and sketches, planning decisions, groundbreakings, topping-out and bottoming-out ceremonies, dedications, the first sledgehammer for the old and the first tenants for the new. The politics and conflict behind the changes are often covered over by the superlatives, the civic promotion, and the routine news reports on budgets and appointments. For example, urban redevelopment and other building projects are popular in the news media stories because they are readily available, photogenic, and have the ingredients of news drama: the tearing down and rebuilding, the conflict and change of scenery, and the "before" and "after" scenes which create a civic advertisement for civic promotion.

Consumers of civic pride are informed via the mass media of the speed, volume, shape, length, mass, number, height, length, width, space, depth, and breadth of public projects. Glowingly, the media tell of the facilities which reinforce the central power of the core city: its medical centers and commuter colleges; libraries and museums; gleaming civic centers, ports, research parks; sports stadiums which generate news of civic pride and aid the economic base, attract urban leisure, and provide jobs for sports writers; fairs, conventions, and tourist attractions which boost "Main Street"; convention halls, auditoriums, and cultural centers, which carry the name of the central city and counteract the dirt and dreary grime of the nearby industrial areas; shoppers' malls near downtown luxury apartments which attract customers, camera clubs, and fashion shows; and acres of parking lots, where once were book shops, parks, restaurants, and small stores (Burd, 1968e).

Even prizes for mass media performance may merely reinforce the ritual of the urban facial. At the White House Conference on National Beauty in 1965, a London journalist, Barbara Ward, suggested that a highly competitive search for intercity beauty, sponsored by newspapers and taken up by television, "would arouse all those things that are best in free enterprise." Similarly, NBC correspondent, Nancy Dickerson, urged a national competition of local television stations in which an award would be given to the station producing the best documentary on a local beautification project. The winning producer, it was suggested, would receive "The President's Award" in the Rose Garden of the White House.

City planners often see the press as a vehicle for the planners' purposes, and planning texts usually refer briefly to "publicity" and "education" for putting a plan into effect and promoting it. Aiding the congeniality is the tendency for the press to treat civic improvements as "nonpolitical" (Burd, 1968a). Planners are reminded that their "most precious equipment is the admiration and respect for the values of the democratic process." The happy situation is one in which "both the planners and reporter talk the same language, [and] generally agree on what would be good for the community" (Wall, 1960: 5, 13). Despite whatever bad moments they have had with the press, planners as a group "rate fairly high with reporters, editors and publishers" (Wall, 1960: 12). What's more, the media tend to treat special planning authority groups favorably as the agencies lend themselves to advertising and civic promotion rather than regulation and enforcement. In such a case, it is hard for a city government to deny these agencies money when they are publicly portrayed in so good a light (Kaufman and Sayre, 1960). The New York press's protection of the ubiquitous Robert Moses is a choice example.

The mass media's definition of "the urban crisis" and its norms for newsgathering have, in many cities, frequently produced softer city hall and press relations at the local level and warmer Washington—to—city hall contacts suitable to a traditional national Republican press. Even the New Deal and Great Society programs were supported to the extent that the press saw its fate intertwined with that of government and business, all surrounded by the ghetto. In this situation, voters, shoppers, parishioners, subscribers, and taxpayers were in a similar boat, so that the politics of urban renewal created a sense of honeymoons between the media and political machines. Indeed, Washington aid has most often helped those with interests in the central business district and at the same time has made the "bad" political machine look better. The attractive photographs of new buildings in slum clearance projects not only impressed civic boosters and city editors, but for politicians they were cheaper than political campaign posters and more valuable because they appeared as "nonpolitical," yet appealed to the voters.

"The newspaper is both a leader in politics and a reporter of political phenomena," wrote Chilton Bush in his classic journalism text, *Reporting of Public Affairs.* Analysts of urban government have asserted that the mass media "are not merely contestants in the great game of the city's politics, they are also the principal channel through which other contestants reach the general public" (Kaufman and Sayre, 1960: 81). And one study of city hall reporters went so far as to say that reporters are "unwitting adjuncts to city hall," who "have been made, via their own role activities and the sources' advice-seeking behavior, a part of the governmental process" (Gieber, 1961: 296).

But to effect social change, as in urban planning, conditions have to be manipulated (a lesson learned by newspaper city planners). For a long time, the city planning beat was ignored by journalism textbook writers. The planned city or society was considered socialistic at least until the New Deal era. Even the urban crises of the 1960's have been viewed in the context of what might best be called the press's crisis of self-interest. Typical are these guidelines:

> ... the community publication that gives adequate attention to social plans in its territory builds belief in its own interest in the area it serves. The greater the civic loyalty of the readers and the greater the civic loyalty of the paper, the greater will be the loyalty of those same citizens and readers. . . . They are in the battle together. . . . Since the modern journalistic medium is first of all in need of sound finances, this pragmatic consideration is important. . . . Since social planning itself has grown markedly in the United States in the past fifteen to twenty years, it has become news, therefore of economic value to the press and radio. . . . Handling this will differ not a lot from covering speeches, business sessions of organizations, open meetings, banquets, or other standard types of assignments. Once again, the subject matter may differ, but the way to report and write the story does not differ in technique. . . . And it would do little or no harm, with the usual social planning story, for him [the reporter] to take sides. . . . Why should a newsman oppose an honest civic campaign to eradicate cancer? Why

should he be against systematic, planned steps to rid the community of some widely acknowledged evil? [Campbell and Wolseley, 1949: 395].

Thus, "selling information" becomes an extension of "free enterprise" in the marketing of ideas, while "objectivity" puts the news in a passive relationship to its market.

General assignments coverage of planning has been like covering a fire, with reports when it was announced (or broke out) and when it was carried out (or put out). Guides for newsmen have consisted of such criteria as the degree of interest the event arouses, attendance at meetings and the frequency of such gatherings, the number of persons active, and the importance of the project to the city's development. Newsmen are told there must be public support for the plans, "unless the publication, radio station, or magazine is financially independent and can risk creating antagonism by supporting or even giving publicity to some project which the more powerful forces of the community tend to disapprove" (Campbell and Wolseley, 1949: 395).

But unless the newspaper and its reporters are acutely sensitive to the need for planning, it is not likely that much planning news will develop as a result of the newspaper's initiative. Routine beat coverage will not bring to light the significance of the changes that are transforming our communities. "The planning-conscious newspaper, on the other hand, will seek out the significant. . . only when there is a fight does this kind of [meeting] news make big headlines. Rarely does news coverage of meetings contribute anything of lasting value to the public's comprehension of their urban problems" (Wall, 1960: 8-9).

Yet city planners also have expected the media to awaken public awareness of the fact of metropolitanism and its problems. The press, radio and television have a special role in this regard since they are metropolitan in scope and their economy, both in circulation and audience, are areawide. "They are needed in the vanguard of a modern crusade to put man in control of his urban destiny" (Sorenson, 1961: 170). It was also hoped, in the 1940's, that widespread use of opinion polls would permit "a democratic determination of the attitudes, desires and resistances of those who live in cities toward planning problems and proposed solutions" (Branch, 1942: 1). However, in the public mind, city planning is still seen as specific problems rather than an abstraction or image of a city.

How much—or how little—citizens actually learn from the mass media about the processes of changing the environment and institutions of the city is evident from the content and rhetoric of news reports on the face and the face-lift of the city: the breakthrough and the crash programs, the delays and the stalls; the go-ahead and the go-go, the up and the up-up, the onward and upward, the green light, the bogdown and the boom, the loom and the boon, the shot in the arm, the spark and the spur, the sag and the lag; the ugly eysore, the new shiny and shimmering facade. Such boosterism and its metaphorical self-fulfilling prophecy are wholly acceptable to the news sources and often supplied by them with an eye to the "city beautiful" and urban cosmetics, civic pride, and "other-directed" motives, without any change in the system which created problems in the first place.

There have been periods when the media's reportage about cities focused on matters other than zoning disputes, announcements of plans, and their specifications. Magazines, for example, in the 1885-1914 period had a positive notion about urban ills. *Scribner's* tackled the complexities of slums. *Arena* magazine opposed mob executions of Negroes. *Everybody's Magazine* and *American Mercury* discussed social ferment. *Cosmopolitan* crusaded on problems of the urban poor so enthusiastically that its articles provoked President Theodore Roosevelt to term such journalists "muckrackers." One of the more famous of these was *McClure's* Lincoln Steffens, whose "Shame of the Cities" articles became a casebook on corrupt urban government, often ignored by civic-boosting newspapers in the cities involved.

Newspapers as far back as the early thirties could count among their contributions "parks, better public school administrations, better police and fire protection, improved highways and streets, plans for development of public buildings according to architectural beauty and utility of layout, both in reference to the buildings themselves, their relations to each other, and their cost" (Thayer, 1934: 408). This concern for the quality of urban life is crucial in considering metropolitan solutions. "Quality as well as quantity is called for. Architecture and aesthetics count as much or more than water aqueducts; billboard control matters as much as boundary control; and the form and location of a freeway is as important as its traffic capacity" (Rehfuss, 1968: 408).

More recently, as already indicated, the mass media have become alarmed over decentralization in American urban cores and have formulated a national "save the cities" crisis. The metropolitan problem has been defined by the media as the task of finding ways and means to restore, rebuild, revive, and reconcentrate the power of the old central city and its central business district and maintain it as a single force dominating one unified metropolis. The aims of the city and the possible merits of decentralization allowed by modern communication and transportation are not thrown open to debate. Instead, the media picture the traditional city's problems as those of a bad business failure. Radio disk jockeys launch frantic "paint-up-fix-up" campaigns for old neighborhoods. Television specials appeal to a visually conscious audience with dismal slums needing repair. The downtown press constantly laments the exodus of white, middle-class taxpayers to the land of barbecue pits and crabgrass (which many journalists themselves view through cracks in the picture windows of their own suburban "split-level traps"). Almost daily, readers, listeners and viewers of the media are told of suburban "sprawl" and urban "decay," of the flight to the suburbs, the blight of the central business district, and the plight of the suburban commuters, whose numbers include newspaper publishers, reporters, and editors who must "return wearily to work each day in the central city as symbol manipulators and merchants of dreams" (Spectorsky, 1955: 10).

Whether the old central city *should* be preserved is not often debated in the press. The editorial "cure" for decentralization is similar to the civic ideology of those with a vested interest in the central business district: lure back higher-tax-paying groups to the downtown area and the approaching black ghetto, and halt what, the media often imply, is a devilish harrassment by some evil force which seeks to destroy the old, inner city, and replace it with a nonwhite, lower-income ethnic community. At a time when communications and transportation make the central city less necessary, the press urges people to go "downtown" for the opera or ballet, even though they can often fly to another city faster than they can fight local expressway traffic, or can stay home in suburbia and hear a symphony on their hi-fi.

But the news media's efforts to perpetuate old stereotypes of the city does not alter reality for the metropolitan citizen who finds the auto, the cloverleaf, and the commuter station more familiar than the trolley and the central business district skyline, and the shopping center mall more familiar than old Main Street or the courthouse square. Despite the increasing dominance of the highly advertised, decentralized suburban way of life, the media often see the *sub*-urbs as dependent and secondary to the old central city. Ironically, the civic effort to make the latter competitive in style with suburbia has led to the destruction of its unique aspects and made it merely one more shopping center in the metropolis.

The press has not been a harsh critic of the auto or expressway which may congest the central business district and force shoppers to suburbia, but which provides advertisement revenue, eliminates ugly downtown skid rows, cuts off the central business district from "bad" neighborhoods, provides new photogenic scenes of the skyline and news copy on accidents and traffic jams. Expressways facilitate the delivery of the latest editions to the suburbs, and some newspapers sell special expressway accident insurance policies to subscribers. Others campaign against the aesthetics of freeway billboards, which cut deeply into advertisers' budgets. Meanwhile, radio stations accept ads from commuter railroads which buy "public service" program time to relay helicopter and television information on expressway traffic jams to the radios of drivers while at the same time they urge commuters to go downtown, reading the newspapers as they travel by train.

POTENTIALITIES AND PROBABILITIES
FOR THE URBAN MASS MEDIA

The mass media, having built up a large audience, could take the initiative for improving the quality of urban life if they would try to demand an attitude toward this life as well as a demand for specific accomplishment, and if they would provide a critical aesthetic point of view rather than a simple solution to a given problem (Rehfuss, 1968: 103). But are there pressures for such change great enough to overcome the preoccupations and predelictions produced by the institutional commitments of the media as private business corporations, by their managers and executives as members of a community power elite, and by the perceptual and rhetorical traditions of newsmen generally?

Today the central-city-oriented mass media face the prospects of being surrounded by a black ghetto, and related white institutions face both a racial and social revolt. They still seek to hold on to the old "colonial" urban empire and to regenerate metropolitan leadership from the geographical center of the metropolis. Having helped build the "bigger" city and having learned that it was not necessarily "better," the monopolistic and rich mass media belatedly have turned some of their money-making machinery toward concern with the "urban crisis." Although often ex post facto, a sort of reportorial autopsy, news coverage in urban areas has slowly begun to move from a concern with quantity to quality; from how to "get things done" to "how good is it," whether public housing or the urban "neon wilderness"; from how to distribute news (and goods) to how to make the distribution equitable; from a worship of mere urban growth to an evaluation of the city's ability to inspire, elevate, and uplift.

But this comes at a time when the mass media establishment is still tied to the old city and its leadership and continues to seek control over urban communications, while masquerading self-interest as the public good. It comes in the midst of a crisis for media survival in the old central business district with the press tied to the old "objectivity" ideology. Meanwhile, new city residents seek new media forms in the midst of communications overload and a search for security and privacy. Ironically, understanding of urban issues is drowned in a flood of information, and the search for privacy and meaning is lost in the massive metropolis. We now have reason to believe that the mass media "have helped less to enlarge and animate the discussion of primary publics than to transform them into a set of media markets in a mass-like society" (Mills, 1956: 314).

The economic character of commercial media in a free enterprise society, as Klapper (1960: 42) points out, is such that they "appear destined forever to play to, and thus reinforce, socially prevalent attitudes far more often than they are likely to create or convert attitudes." Furthermore, by cultivating the common man, the press has put itself in a somewhat more vulnerable position than it occupied before. For "the news and editorial rooms stand in a passive relationship to the market-place" (Lindstrom, 1960: 77). Even saleable news on racial conflict is oriented toward the reader and listener market rather than toward a clarification of, or guidance toward, specific goals. J. Edward Gerald in his analysis of *The Social Responsibility of the Press* observed:

> When journalism abandoned politics for business as a way of making a living, it switched its personality from that of crusader trying to organize, teach and change the people and the community to that of

entrepreneur trying to make a good living with a minimum of trouble. . . . passive noninvolvement in critical problems is journalism's outstanding quality. Few of the editors of mass-circulation newspapers since 1830 have risked their careers to exert strong leadership. . . . The press is neutralist because its content grows out of preposession with the market rather than concern for causes or goals. . . . service to both the community and to selfish interest means compromises to minimize public interference with newspaper publishing [Gerald, 1963: 100, 117].

The mass media policy still reflects a middle-class mentality, which regards change and the problems of the "metropolitan 400" as inconvenient: the high cost of public aid for blacks and the need for new ghetto housing to "clean up" the slums; the poor taxicab service at the suburban-based airports and the pickpockets and shoplifters downtown; the embarrassing smoke and haze which cloud the views of new skyscrapers for visitors and suburban commuters seeking reassurance that a visible city *does* exist; the jet noise near suburban patios and the noise of central business district construction crews near downtown penthouses; the garbage "eyesores," dumps and pollution which create civic "BO," especially near the more affluent; the realization that the crime syndicate has moved to suburbia to taint its reputation as a "nice place to raise the kids." Debate is not opened up on the aim of the city, only on the means of maintaining the present one (Burd, 1968a). "It is no wonder that the New Jerusalem, as a rule, has less appeal to them [the media], than continued work on improvement of the present city. The scope of their concern for the welfare of the community is limited by interests of the firm . . . " (Gerald, 1963: 117).

Those minorities without economic power often lack access to communications, yet they are highly aware because of television, improved transportation, and the search for experience and action rather than passivity. The visual power of television has dramatized the contradictions of race and affected the expectations of minorities. Yet only gradually has the white media "integrated" blacks into the obituaries and society pages. Even this did not begin to occur until the media realized that they were surrounded by a new community and could not take their presses and radio booth equipment on the commuter train to escape to suburbia.

Cut out of urban assimilation, the black community has found a new identity in an increasingly thriving black press and radio, and in the entertainment aspects of television. Black homeowners and taxpayers, once crowded out by black criminals and prostitutes in the white sensational press, now have their own modern "immigrant" press (Burd, 1968i: 8). Similarly, the alienated youth, often whites, have their "underground" press, film, and radio; and for both minorities the search for a larger audience is by pickets, demonstrations, and other news-producing activity which attracts the Establishment media. This activity not only creates a temporary "community," but also provides a model for minority group press agentry and helps get items on the civic agenda which might otherwise be ignored because they were not highly visible (Burd, 1968f).

The "underground" press represents possibly a revival of alert journalism in America (Burd, 1967a). It is using new and imaginative makeup and display techniques; it is gutty, sarcastic, highly personalized, and subjective; it employs experienced writers who know their subjects from personal experience; and perhaps most significantly, it questions and challenges middle-class values, criticizes the Establishment press which upholds them and raises serious urban issues (Burd, 1967a).

It should be noted that criticism of cities and the press is not restricted to social minorities without power or prestige. Such criticism has emanated from Presidents Jefferson to Johnson, and from writers H. L. Mencken and Upton Sinclair to the 1947 Hutchins Commission on Freedom of the Press. An important difference, however, should be noted here: in most of these cases the attack on the media and society was devoted to their logistics and not their ends. The media were criticized for being sensational, controlled by "bad" men in the form of "press lords," and for failure to keep a close eye on the democratic machinery.

The press also has its own paid critics, but these are limited to covering the fine and popular arts—possible competition for the mass media. The coverage of literature, poetry, the dance, sculpture, and painting, and later movies, radio, television, photography, and recordings enhanced the advertisement potential of the press and gave it the opportunity for subjectivity in writing, and, perhaps more importantly, afforded it a means of guiding readers, influencing their taste, and bringing prestige to the press by an identification with quality.

But it was not until the late 1950's that the mass media began to realize the need for critics of the city and the quality of its environment. The impetus came from those concerned with the architectural design of cities, where the total impact of community rebuilding and growth was seen, and where the citizen sensed not only the function but the feeling of life in cities. The 1958 Conference on Urban Design Criticism at the University of Pennsylvania supported by *Fortune* magazine and the Rockefeller Foundation, and subsequent conferences in 1962 and 1965 on press coverage of cities by the American Institute of Architects were influential moves in this direction.

Out of this concern for the quality of reporting on urban design were to come the new magazine urban crusaders—some muckrakers—who by the early 1960's had helped formulate the urban problem agenda, and had inspired or precipitated literature critical of both cities and the mass media for the condition of the urban environment. Such books as *Fortune's The Exploding Metropolis* and Jane Jacobs' *The Death and Life of Great American Cities* had, by their critical outlooks, influenced newspapers to engage less in vulgar boosting and to utilize qualified urban critics.

One of the advocates of press criticism of city planning, Grady Clay, real estate editor of the *Louisville Courier-Journal,* called for an evaluation of the results and aims of redevelopment. He said that the press was so concerned with reporting *how* things happen that they neglected to appraise the results. "We concentrate on the legality and longevity ('Will the streets outlast the mortgage on the house?'), on durability and feasibility; on taxability and other such virtues. But we seldom have space and planners seldom have time to think about visibility, suitability, proportions, shape, colors, texture and scale" (Clay, 1960: 132). In the January, 1968, issue of *Landscape Architecture,* of which Clay also is editor, he argues persuasively for the establishment of more formal, effective, and organized systems to review and evaluate physical development and change. In his words there is "need to train an entire generation to evaluate good and bad environment . . . a built-in system of surveillance and review so that all major investments in public environment are followed by reliable evaluations." Clay has little hope that the professional press, caught by its dependence upon advertisers, can play this role.

The feeling that the American press was not taking a critical look at the environment was expressed in 1961 when FCC Chairman Newton Minow accused television of being a "vast wasteland" and in 1966, when CBS's Fred Friendly resigned over programming to suit mass taste. Later he wrote in his protest, *Due To Circumstances Beyond Our Control,* "Television's downfall is that it discovered a direct circuit between the box office and its production centers" and the "addicts of mediocrity have become the nation's taste-makers."

Further dissatisfaction with the performance of the mass media was revealed in early efforts to get Pay-TV, Educational Television, and the Public Broadcast Laboratory. The criticism of the media sky-rocketed after the three major political assassinations, the Kerner Report on violence, and the 1968 presidential election, as both the political Left and Right attacked the mass media, apparently with the agreement of wide sections of the general public (Burd, 1968b).

There is indication that the future will continue to display a growing discontent with the urban mass media. The metropolitan electronic media may be able to create a new urban image and a new outlook if they devise new terminology for new conditions, are not obsessed with the geography of cities, and abandon the philosophy of the central business district as the "mother" of urban centralization.

The decentralization of media is likely to continue as means of transportation and communication change. Although the "electronic city" may not arrive, television will cover spot news more and urban metropolitan dailies will become more regional and international. Suburban zone sections by metropolitan dailies will continue to be unable, except in a few cases, to cover the total metropolis outside the center city. In the latter, the community press will probably remain stable with the minority press thriving, while in suburbs, the rise of CATV and the suburban daily newspaper will aid further communications decentralization. Branch and mobile libraries (even museums), and the use of university extensions, office reproduction advances, satellite communication, and city magazines will also continue to decentralize communications patterns.

There is some indication that the suburban media will attract future journalists who seek escape from the central city and its media strikes, nepotism, and bureaucracy, and are attracted to new type and printing reproduction methods being used in suburbs. Additionally attractive is the fact that suburbia is the seat of future purchasing, voting, and decision-making power, and the locale of "big stories," whose quality and presentation may be dictated by the taste of the better educated (Burd, 1966: 3). New suburban dailies near larger central cities will seek regional monopoly with their own mechanical plants outside the old central business district, and some central-city dailies will try, if the federal government permits, to extend control over suburban area publications. However, any future zone editions will face the cost and circulation problems, production headaches, the artificiality of neighborhoods created by transportation arteries for distribution, and especially the stability of suburban weeklies and the function of shopper throwaways.

The specialized urban press will probably continue to thrive, as labor, religious, and political groups seek to reach their own audiences. Persons without group ties will continue to communicate with the metropolitan public through more "letters to the editor," more and more "action lines," talk shows, and polls.

Future urban journalism is likely to inherit present problems of professional conduct. Journalism schools will continue, at least for some time, to serve as trade and writing factories, tied vocationally to the Establishment press and isolated from other departments in universities, where journalism is held in low regard. Qualified graduates will continue to flock to public relations and promotion rather than reporting or investigating urban problems. Those few who enter the latter will continue to lack an enforceable code of ethics and will remain safe from public criticism. Journalism teachers will probably continue to evade critical analysis of press performance on city conditions, as research continues to be oriented toward a quantitative rather than evaluative stance. The former does not harm the scholar's standing with his "scientific" peers who neglect social issues; and the latter does not endear the journalism researcher to publishers or station management (Burd, 1966).

A disgruntled public in the future may be able to appeal to American press councils if current experiments succeed. The Mellett Fund for A Free and Responsible Press is carrying out projects in line with the recommendations of the Hutchins and Kerner Commission's criticisms of press performance on urban problems. Some magazines may continue sporadic criticism of the other mass media, and publications like *Nieman Reports* and *Columbia Journalism Review* are likely to continue to critique press performance.

One hope for the creation of an alert urban press is the rise of city magazines, which seek to maintain a metropolitan image of the city, but crusade for as well as boost, civic morale, and which appeal to a rather small, quality-minded elite who are influential in urban decision-making and move across political boundaries in the metropolis. Articles in these publications already indicate a depth and perspective often lacking in mass media accounts of urban crises marketed to reach the largest possible audience quickly and without offending mass attitudes and taste (Burd, 1968g).

Perhaps the greatest internal need in the metropolitan media, especially daily newspapers, is for a reorganization of the city room around specialized urban functions rather than around the traditional central business district ecology. This trend appears underway in some cities under the pressure of the "urban crisis." Such a reorganization might utilize corresponding knowledge from urban universities, bridge the gap between the masses and the specialists, and between journalism and related fields of study. It could also promote news departmentalization and reduce the fragmented jigsaw puzzle facing the urban citizen seeking information. The city room could become a prototype of the metropolitan area, with "civic detectives" rather than civic boosters, with writers and announcers interested in educating rather than promoting, with media creating an independent civic policy rather than merely reporting the poetic metaphor and propaganda of the civic ideology.

Such an arrangement might make the media more critical of the civic icons, and more anxious to engage in debate rather than to prevent it. It might also lead to a new civic image with the mental cameras aimed beyond mere geography. In such a setup, functional areas might be ecology and natural environment rather than the "outdoors editor"; transportation rather than the automotive and travel sections and the accident reporters; economic base rather than the business page; government and politics rather than city hall and the federal building; housing and community facilities rather than the real estate and home sections and the "school board" beat; health rather than the food section and the "death" beats at the hospital and morgue; crime rather than mass murders and the murder trial; human relations rather than the "church page," a black beauty queen, and news of pickets and demonstrations covered by general assignment; and city planning rather than zoning meetings, slum fires and the flash floods (Burd, 1967b). Only then will the news media be in a position to translate urban reality into a new metropolitan image and promote intelligent urban citizenship.

Finally, all of the media may recognize the possibility of direct joint participation with interests other than those of the present establishment and allies in the pursuit of meaningful innovations in urban institutions. This would transform the reading and listening orientation of many citizens, because they could then be perceiving *their media.*

In the words of Donald Webster in his book on *Urban Planning and Municipal Public Policy* (1958: 309): "Work of citizens' organizations should be carefully tied in with the use of such public information media as newspapers, radio and television. When citizen interest is fostered by direct participation, stories on planning have greater appeal from the standpoint of newspaper circulation, listening audience, and general civic interest." But only if that "use" goes well beyond "good reporting" and "public relations" in terms of contemporary conventions, will it be a symptom of a most important transformation of the media themselves.

REFERENCES

ANDERSON, MARTIN (1964) The Federal Bulldozer. Cambridge: MIT Press.

BRANCH, MELVILLE C., JR. (1942) Urban Planning and Public Opinion. Princeton: Princeton Univ. Press.

BURD, GENE (1968a) Voters, the Press and Urban Renewal. Milwaukee: Center for the Study of the American Press, Marquette Univ.

———(1968b) "Magazines and the Metropolis." Address delivered to the Annual Convention of the Association for Education in Journalism, Univ. of Kansas, August 27-28.

———(1968c) "Urban crises and criticisms of the mass media." Paper presented to Mass Media and Education Institute, Marquette Univ., July 24.

———(1968d) "Magazines and public images of cities." Address delivered to the Annual Convention of the American Association of Commerce Publications, July 15, Milwaukee, Wisconsin.

———(1968e) "The metropolitan press and urban problems." Paper presented to the American Political Science Association's Public Affairs Reporting Awards Seminar, June 26, Sun Valley, Idaho.

———(1968f) "Urban renewal in the city room." Quill 56 (May): 12-13.

———(1968g) "Urban alienation and the underground press." Lecture to Alpha Kappa Delta, April 16, Milwaukee, Wisconsin.

———(1968h) "Media in metropolis." National Civic Review 58 (January): 138-143.

———(1968i) "Media view the ghetto." New City 6 (January): 8-11.

———(1967a) "Pickets, political power and the urban press." Address delivered to Kiwanis Club, National Newspaper Week, October 10, Menominee Falls, Wisconsin.

———(1967b) "Cities and the press." Paper presented to the National Convention of the Association for Education in Journalism, Univ. of Colorado, August 30.

———(1966) "The suburban community press." Catholic School Editor 35 (June): 3-5.

CAMPBELL, LAWRENCE and ROLAND WOLSELEY (1949) Newsmen at Work. Boston: Houghton Mifflin.

CARR, STEPHEN (1967) "The city of the mind." Chap. 8 in W. R. Ewald, Jr. (ed.) Environment for Man: The Next Fifty Years. Bloomington: Indiana Univ. Press.

CLARK, PETER B. (1959) The Chicago Big Business Man as a Civic Leader. Ph.D. dissertation, Univ. of Chicago (unpub.).

CLAY, GRADY (1960) "Planning design and public opinion." Journal of the American Society of Planning Officials (Proceedings of the Conference at Bal Harbour, Florida, May 22-26: 129-139.

FUTRELL, ASHLEY B. (1960) "What the press expects of city hall." Southern City 12 (December): 1.

GIEBER, WALTER and WALTER JOHNSON (1961) "The city hall 'beat': a study of reporter and source roles." Journalism Quarterly 38 (Summer): 296.

GREER, SCOTT (1962) The Emerging City. Glencoe, Ill.: Free Press.

GRUBB, JEANETTE (1940) "A study of the editorial policies of the Indianapolis News and its relationship to the growth of the city of Indianapolis." (Master's thesis, Northwestern Univ., unpub.).

JOHNSON, EARL S. (1957) "The function of the central business district in the metropolitan community." Pp. 248-259 in Paul K. Hatt and Albert J. Reiss, Jr. (eds.) Cities and Society. Glencoe, Ill.: Free Press.

Kansas City Star Staff (1915) William Rockhill Nelson: The Story of a Man, a Newspaper and a City. Cambridge, Mass.: Riverside Press.

KAUFMAN, HERBERT and WALLACE S. SAYRE (1960) Governing New York City. New York: Russell Sage Foundation.

KLAPPER, JOSEPH T. (1960) The Effects of Mass Communications. Glencoe, Ill.: Free Press.

LAZARSFELD, PAUL F. and ROBERT MERTON (1948) "Mass communications, popular taste and organized social action." Pp. 147-153 in Lyman Bryson (ed.) The Communication of Ideas. New York: Harper & Row.

LINDSTROM, CARL E. (1960) The Fading American Newspaper. New York: Doubleday.

LYNCH, KEVIN (1960) The Image of the City. Boston: MIT Press.

MCLUHAN, MARSHALL (1964) Understanding Media: The Extensions of Man. New York: McGraw-Hill.

MUMFORD, LEWIS (1961) The City in History. New York: Harcourt, Brace & World.

MILLS, C. WRIGHT (1956) The Power Elite. New York: Oxford Univ. Press.

NEWTON, V. M. (1961) Crusade for Democracy. Ames, Iowa: Iowa State Univ. Press.

REHFUSS, JOHN A. (1968) "Metropolitan government: four views." Urban Affairs Quarterly 3 (June): 91-111.

RUCKER, FRANK and BERT STOLPE (1960) Tested Newspaper Promotion. Ames, Iowa: Iowa State Univ. Press.

SORENSON, ROY (1961) "A citizen looks at metropolitan problems." Pp. 167-174 in First Conference on Environmental Engineering and Metropolitan Planning. Evanston: Northwestern Univ. Press.

SPECTORSKY, A. C. (1955) The Exurbanites. New York: Berkeley Publishing.

STRAUSS, ANSELM (1961) Images of the American City. New York: Free Press.

——— and RICHARD WOHL (1958) "Symbolic representation and the urban milieu." American Journal of Sociology 63 (March): 523-532.

THAYER, FRANK (1934) Newspaper Management. New York: D. Appleton.

WALL, NED L. (1960) "The press, the public and planning." P. 5 of Information Report No. 134 for American Society of Planning Officials, May.

WEBSTER, DONALD (1958) Urban Planning and Municipal Public Policy. New York: Harper & Row.

WEISBERGER, BERNARD (1961) The American Newspaperman. Chicago: Univ. of Chicago Press.

WOLSELEY, ROLAND and GEORGE F. MOTT (1958) New Survey of Journalism. New York: Macmillan.

YLVISAKER, PAUL N. (1961)"Diversity and the public interst." Journal of the American Institute of Planners 27 (May): 107-117.

12

The Quality
of Urban Education

ROBERT J. HAVIGHURST
and DANIEL U. LEVINE

□ THE QUALITY OF urban education should be assessed in the
light of certain pervasive characteristics of urban society that help us
to see what the educational system should be expected to accomplish
in our metropolitan areas. The two most important of these
characteristics are that (a) technology is producing an abundance of
material goods and leisure time, and (b) knowledge-transacting
exchanges in modern society place a growing emphasis on
information-processing skills which most people acquire through
formal education.

The first characteristic recently has been taking on increasing
importance for educators, particularly those who work in the
suburbs and the outlying parts of the cities where the schools are
succeeding reasonably well in helping students acquire academic
skills but generally do very little to prepare students for living well in
the midst of an abundance of goods and leisure time. The second
characteristic becomes most prominent when one is concerned with
the inner cores of the central cities in which large numbers of pupils
from low-income homes are not acquiring the minimal level of
academic skills required for effective participation in our society.

The second problem is obviously a central part of the "urban" or "racial" crisis in the big cities and therefore is more immediately pressing then any other in education, but its solution has escaped us. For this reason most of this essay will be devoted to an analysis of the current situation with respect to the education of low-income students in schools in inner cores of the central cities. But before turning to this topic a brief comment should be made concerning the broader problem of education for an era of material affluence.

EDUCATING STUDENTS WHO SHARE IN THE ABUNDANCE OF URBAN SOCIETY

One inescapable test of the quality of education in modern urban society is its effectiveness with the majority of children of the metropolitan area who now grow up in middle-income families. What do the schools do for the children of middle-class families? These children do well on achievement tests against national norms, and they graduate from high school and enter college at a high rate. But is this enough?

With the average per capita income increasing in real purchasing power two or three percent annually, we are doubling our real income per capita every twenty-five or thirty years. At the same time, we do not have to work as many hours per week, per year, and per lifetime as we did in the past. This means new opportunity and responsibility for the educational system. New opportunity to serve adults with free time and to serve youth with a later school-leaving age. New responsibility to help students learn to use their growing leisure happily and wisely. Since the American economy is providing middle-class families with all the material goods they need and all that many of them desire, children who are growing up to become members of the middle class will have different problems to cope with than did earlier generations whose energies necessarily were expended in a continual struggle to attain a state of physical well-being and security. The functions of education in their lives should go far beyond teaching them the academic and social skills needed to make money. It should do at least the following for them:

(1) Teach them how to be wise citizens in an interdependent world.

(2) Help them to bring their minds and their hearts together—to make their reason serve their emotions—to develop themselves as expressive persons as well as instrumental persons.

(3) Help them *enjoy* themselves, and their fellows.

(4) Help them to learn on their own responsibility—to learn whatever they might like to learn, to read widely, to solve problems, to sing, to dance, to pray.

From this point of view it is not enough to compare urban schools today with urban schools of thirty years ago. We have a right to expect progress. The evidence of drug addiction, mental breakdown, meaningless drifting, and other signs of a general malaise and alienation among middle-income youngsters obviously is sufficient to indicate that the educational system must make much more progress in preparing students to function constructively in this new age of relative abundance and leisure. The public and private schools which middle-class children attend have the best equipment, the best-trained and best-paid teachers in the country. Yet, the problems—and the opportunities—posed by the challenge to help students live happily and wisely in an age of abundance are so enormous and exciting as to make the traditional goals of communicating the cultural heritage and developing mastery of academic skills seem almost pedestrian. It is probable that the schools have been progressing as rapidly as most other major social systems, but like these others the educational system must be greatly improved if urban institutions are to fulfill the promise inherent in a society of abundant technological and scientific resources.

EDUCATING STUDENTS FROM LOW-INCOME FAMILIES IN THE INNER CITY

One of the most useful ways to view the social system of modern urban society is as an almost infinite variety of human relationships in which knowledge is fast becoming or already has become the foremost commodity of exchange that determines the outcome of these interactions and their effects on the physical environment. As John Kenneth Galbraith, Kenneth Boulding, Buckminster Fuller, and others have pointed out in varied terminology, urban man lives in a knowledge-transacting society. More than any other resource, it is knowledge which gives individuals and groups access to their goals in private as well as public affairs. Knowledge and information have become keys to the satisfaction of desires for goods, status, and self-fulfillment.

Since the public school system is the principal instrument for ensuring that individuals acquire the skills needed to succeed in a society in which knowledge is increasing both in amount and importance, the demands made on the schools have been growing correspondingly greater in scope and importance. Given this increasing centrality of education in modern society, one is compelled to assess the schools in terms of their performance in helping the nation achieve fundamental social purposes which are attainable only if the educational system functions at a sufficiently effective level.

Today, one of the most urgent of these goals is to improve the status of disadvantaged citizens who no longer are willing to tolerate conditions that limit their access to the riches of a modern economy. Until quite recently, as J. Alan Thomas (1968: 1) has pointed out, ". . . knowledge has always been a scarce resource; hence, education has been rationed in such a way as to favor certain portions of the population." In the twentieth century, however, technological and scientific developments which have transformed society as a whole not only have made economic and social progress dependent on the continually improving utilization of intellectual resources throughout the population, but also have given the educational system the potential for providing many more students with more understandings and skills than was possible in the past. Though few could express it verbally, many disadvantaged citizens have realized that knowledge (i.e., educational attainment) need not be rationed in a manner that in effect condemns them to material and social deprivations which are anomalous in the light of our national ideals and unnecessary in view of the productivity of our economy. As a result, they are beginning to protest actively and vociferously against conditions which greatly limit their share in the urban economy. Since traditional information-processing skills are needed for productive participation in this economy, one obvious prerequisite to the solution of the serious problems which exist in the inner cores of the big cities is better education for economically and/or socially disadvantaged students who attend inner-city schools. In a knowledge-transacting society, in other words, one major criterion for judging the quality of urban education is the adequacy of the schools in educating socially or economically disadvantaged students who till now have had limited opportunities to become personally successful and socially useful. The fact that many of the children of low-income urban families do average or better work in school indicates that urban education is at least mildly satisfactory in their case, but a large proportion of these children fail to learn at an adequate level. This failure is the great unsolved problem in urban education today.

PROBLEMS IN ASSESSING THE QUALITY OF EDUCATION IN THE INNER CITY

There are two general ways of assessing the quality of education—through examination of the *process* and of the *product*. The tempting alternative is to examine the product. Why not simply test and measure the mental skills, the work habits, the civic attitudes, and the leisure-time activities of the young people in our schools, and thereby judge the quality of the educational system? This is similar to what we do in the case of the automobile. If the automobile is efficient and cheap we judge the automobile industry to be of high quality. That is, most of us do. A few, however, would also judge the quality of the automobile industry by the way it treats its employees, or by the amount of its profits—process factors.

Using the *product* method for measuring quality of education, some educational writers have recently advocated a definition of equality of educational opportunity as *equality of educational achievement*. They say that merely giving all children the same number of hours of schooling in the same size of class by teachers equally trained and in buildings equally equipped is not equality of opportunity. They point out that children have different degrees of opportunity in their homes and their local neighborhoods, and they argue that the function of the school system is to compensate for some of these differences by increasing the school's contribution to the education of those with less home and neighborhood opportunity.

In its basic thrust, this is an extension of the argument which has led to the granting by the state of extra money for the education of children with physical and mental disabilities that are either inherited or suffered through accidents. Proponents of this latter policy, however, generally assume that most children with such disabilities have handicaps so severe that they can only be somewhat reduced but not entirely overcome by extra education; hence it has not resulted in a demand for equality of educational attainment on the part of these handicapped children. Among many of the writers concerned with the education of low-income children, on the other hand, there tends to be an assumption that the handicaps of socially disadvantaged children can be overcome by enough of the right kind of education, and therefore that these children can equal the more advantaged group in educational achievement.

Provided the argument is not carried too far, we see much merit in using equality of achievement as a measure of quality in the product of education in the inner city. We would utilize this approach to the extent of setting a reasonable *minimum level of educational achievement* as a short-range operational definition of equality of opportunity. This minimum level would have to be established on the basis of what the modern society requires for vocational and civic competence as well as what is actually accomplished by the most effective educational methods we now possess. Possibly this minimum level would work out to or require seventh-, eighth-, or, in the near future, ninth-grade average reading and arithmetic achievement standards as they are presently defined. There are many inner-city pupils who go to school until they are sixteen, but still fall far short of these levels; from this point of view the educational system is not now turning out a product equal to the demands of urban society today. But if this initial goal were attained, the minimum standard of adequacy could then be raised on the basis of experience to a level closer to that achieved by students who are not economically and socially disadvantaged, thus continuing to assess urban education in terms of how well it succeeds in producing equal educational achievement among pupils of differing background.

However, in view of the known facts about hereditary differences in learning ability and the basic importance of family environment in the child's cognitive and emotional development, it is obvious that the argument for defining equal educational opportunity strictly in terms of equal educational achievement is absurd. The proponents of this argument are more or less aware that educational achievement is determined by a number of interconnected variables, but their analysis does not always give adequate attention to each of the major factors which determine the quality of the product of the educational system. Four of these factors are:

(1) The individual's family and local neighborhood.
(2) The inherited potentialities of the individual.
(3) The school system.
(4) The social context in which the schools function.

It is common for writers on this subject to ignore or play down one or another factor, thus leading to gross oversimplifications in discussions of the quality of education. In their most extreme form, some of these oversimplifications are:

(1) It all depends on the family and the local neighborhood. These influences are so powerful that the school system cannot overcome them if they are bad. Therefore the school system is doing as good a job as it can in the face of adverse circumstances.

(2) The population of the inner city is inferior, biologically, and becoming more so due to the out-migration of the white middle class and the in-migration of the nonwhite lower class. Though some exponents of "white supremacy" do use this argument to strike a responsive chord among other prejudiced whites, this position is not accepted by educated people who know that the great majority of experts on heredity believe there is no innate difference in potentiality for learning among the various racial and social groups in the United States.

(3a) It all depends on the teachers and the principals in the schools. They do not know how to teach children of minority groups. They believe, deep down, that such children cannot learn much, and consequently they do not try to teach effectively. Or, they are lazy, or malicious, or bad in some other way. We must get rid of these self-serving people and find a new kind of teacher with more faith in disadvantaged children, more of a service motive, and more knowledge of minority cultures.

(3b) It all depends on the size and complexity of the educational system. The big urban school systems are top-heavy with a rigid, unimaginative bureaucracy and a system of administration that prevents innovation and intelligent action by the classroom teacher and the school principal. This is a popular oversimplification just now, partly due to the publicity which the decentralization issue has received in the New York City schools.

For example, an article by a sociologist published in a reputable journal recently attacked the New York City school system as a "sick bureaucracy" (Rogers, 1968). Starting with the oft-repeated statement that the New York City school system is "caught in a spiral of decline," the article notes that:

> ... some of the most valid explanations are: demographic changes, increased poverty, and segregation ... interest-group fragmentation and the failure of a strong change-oriented coalition to emerge to force needed reforms; the fragmentation and administrative inefficiencies of New York City government and its failure to coordinate housing, welfare, antipoverty, mass transportation, urban renewal and other socioeconomic development projects with school planning; the dated curricula in the teacher training institutions; and the school system itself.

The article then focuses on the organization and administration of the school system as a major source of the presumed low quality of education in New York City:

> Notwithstanding the complexity of causes for the schools' failures, much of the responsibility should be placed on the school system itself. The New York City school system is typical of what social scientists call a "sick bureaucracy," a term for organizations whose operations subvert their stated missions and prevent any flexible accommodation to changing client demands. The system has the characteristics of all large bureaucratic organizations, but practices have been instituted and followed to such a degree that they no longer serve their original purpose. Overcentralization, fragmentation, the development of chauvinism and protectionist baronies within particular units, in-breeding, compulsive rule-following and role-enforcing, rebellion of field supervisors against headquarters directives, insulation from clients, and the tendency to make decisions in committees—making it difficult to pinpoint responsibility and delay in implementation of programs— are the institution's main pathologies. Add to this the poor training and negative attitudes of many educators toward pupils, parents, and the community; their limited expertise on what and how to teach; and the system's poor relations with outside institutions as well as citizen groups, and one has the makings of a markedly malfunctioning institution.

Perhaps this set of generalized opinions may be taken as an example of oversimplification. Admittedly there are relationships between teaching and learning in the classroom and the administrative organization of the big-city school system, but is there any evidence that a system one-twentieth the size of the New York City system would have performed better in the past decade? New York City will decentralize, but such decentralization is likely to create as many problems as it solves for the quality of education in that city. For example, it is difficult to see how a plan for decentralization in New York City can avoid creating some all-black school districts and some all-white school districts and thus make more difficult the task of racial integration in the public schools.

To those who believe that a "big institution is a bad institution," it may seem conclusive to merely assert that bigness per se makes it impossible to attain high quality education in a school system. But American business, cultural, religious, and educational institutions have all moved toward larger unit sizes in this century, and the result has not been altogether bad from the point of view of enhancing the productivity of these institutions.

(4) It all depends on the dominant power group. The American society is dominated by a middle class which believes that its own culture is best for all people; it therefore runs the schools so as to reward those who are apt to pick up middle-class behavior and to punish those who remain attached to their minority or lower class subcultures. From this point of view the discomfort with and dislike of school by many minority group youth is a sign that the schools should reinforce the minority or lower-class subcultures.

The wise man will not accept arguments which assess the quality of urban education in terms of a single, oversimplified ideological perspective. In our subsequent discussion of changes which tend to improve inner-city education, we shall use an eclectic definition of quality. We believe that the *product* of education in the inner city—the student—should have at least a minimum proficiency in the basic mental skills of reading and arithmetic. But just as educators are beginning to look beyond the obvious *product* of the predominantly middle-class school to assess the process of education there in terms of its potential for helping students *learn to like to learn* and to live happily and productively in an economy of abundance, a useful diagnosis of the quality of education in the inner city must look at what is known concerning the *process* of education offered in inner-city schools for cues to its improvement.

AN ASSESSMENT OF EDUCATIONAL OPPORTUNITIES AND PROGRAMS NOW
OFFERED FOR LOW-INCOME STUDENTS IN THE INNER CITY

Programs of "compensatory education," supported by substantial
federal government and private foundation funds as well as by local
school money, have been disappointing. For the most part,
compensatory education has attempted to expand services and
programs previously offered in the school, generally through such
approaches as reducing class size, increasing guidance services,
providing additional instruction after school, and conducting field
trips to furnish experiences thought to be necessary for
understanding instruction in the classroom.

Among research reports that are skeptical of the effects of
compensatory education, the one of greatest influence is the U.S.
Office of Education study on *Equality of Educational Opportunity*
(Coleman, *et al.*, 1966). Related to it is the Civil Rights Commission
report on *Racial Isolation in the Public Schools* (1967). The equal
opportunity report included a study of the variables associated with
the scholastic achievement of more than 645,000 students in 4,000
public schools and constitutes one of the largest and most significant
pieces of social science research ever conducted. In essence, James S.
Coleman, and his colleagues who carried out the study, concluded
that achievement is correlated more closely with family environment
and other nonschool variables than with school-related variables such
as expenditure and teacher experience. In confirming the general
conclusions of previous and concurrent research conducted in the
United States and Great Britain, the report came as no surprise to
many educational sociologists and research specialists, but it did
provide a substantial amount of needed documentation and greatly
enhanced our understanding of the factors which make for academic
success in the public schools. While a number of important criticisms
have been made concerning the study's methods and conclusions
(Bowles and Levin, 1968), in the judgment of the authors of this
essay none of them provides an adequate basis for questioning its
general emphasis on the importance of home and neighborhood
influences as determinants of school achievement.

The central findings of the report on *Equality of Educational
Opportunity* are summarized in the following:

> Taking all these [above] results together, one implication stands out
> above all: That schools bring little influence to bear on a child's achieve-

ment that is independent of his background and general social context; and that this very lack of an independent effect means that the inequalities imposed on children by their home, neighborhood, and peer environment are carried along to become the inequalities with which they confront adult life at the end of school [Coleman et al., 1966: 325].

Based on these conclusions, three sets of propositions can be identified as having particular relevance for an analysis of the quality of urban education:

(1) Regardless of where they are located or the type of student they enroll, public schools in the United States are not very potent institutions in the sense of motivating or otherwise helping students develop or achieve near the limits of their native capacity. Some students, particularly those from low-income families, come to school poorly prepared to achieve well in the instructional programs they are likely to encounter there and experience a progressive decline or cumulative deficit in achievement relative to national performance norms as they advance through the grades. Other students, particularly those from middle-class homes, come to school possessing language skills, work habits, and other "readiness" capabilities which enable them to perform adequately in the classroom. In general, as students pass through the school they continue to achieve about as well as one would expect, given their initial advantages and disadvantages.

(2) Economically disadvantaged students achieve somewhat better in schools in which they are in a minority than in schools in which most of their classmates also are from low-income families. Part of this differential probably is due to selective factors which suggest that low-income students for one reason or another enrolled in predominantly middle-income schools are not as educationally disadvantaged as students of similar economic and social status in the inner city slum school; but part of the difference also can be attributed to the influence of peer norms and standards which are least conducive to achievement when low-income students constitute a large majority in a school's student body as well as to a variety of difficulties teachers experience in attempting to provide instruction in inner-city schools.

(3) It follows that although ways must be found to provide more consistent and "powerful" learning environments for the disadvantaged pupil regardless of the type of school he attends, improvements in the achievement of low-income students can be brought about by

placing them in "destratified" classrooms in which conditions are not as detrimental to achievement as in the inner city. Since a relatively high proportion of black students in poverty areas in the big cities are from low-income families and since racial segregation also generates attitudes which impede the performance of minority group students, integrated education can be even more effective than destratified education per se in improving the academic achievement of low-income black students in the inner city.

In addition to reanalyzing the data of the equal opportunity study to obtain a fuller understanding of the relationships between race, social class, and school achievement, the Civil Rights Commission report on *Racial Isolation in the Schools* also assessed the impact of highly acclaimed special programs for disadvantaged students in New York, St. Louis, Philadelphia, and several other cities (U.S. Commission on Civil Rights, 1967: 113-114). Basically, it was found that compensatory education focused on inner-city schools had little discernible effect in raising the achievement of pupils enrolled in them, but that disadvantaged minority students who had been bussed to or otherwise enrolled in schools with a substantial proportion of middle-class students tend to make significant gains in achievement (U.S. Commission on Civil Rights, 1967: 115-140). Far from settling the issue, however, these results highlighted the need to find answers to three related questions before reaching definitive conclusions concerning alternative approaches for improving the achievement of disadvantaged students:

(1) Recognizing the pioneering nature of compensatory education programs conducted in the 1950's and 1960's, is it possible that programs funded more adequately and carried out more effectively than were those in the cities studied by the Commission might overcome educational and other disadvantages associated with low-income status?

(2) In view of the negative effects of stratification and segregation on low-income students, must even the best compensatory education programs inevitably fail as long as the students enrolled in them are confined to inner-city schools and neighborhoods?

(3) Recognizing that the low achievement of black students in the inner city can be attributed mostly to socioeconomic disadvantages but partly to segregation, what combination of school

improvement efforts and integration-desegregation efforts would ensure adequate achievement gains among these students through a maximally efficient expenditure of resources?

At present, then, it is possible to say that integrated and destratified education is a highly desirable and possibly even an indispensable element in programs to raise the achievement of disadvantaged pupils, but that racially and socially balanced education probably would overcome only a part of the "cumulative deficit" in the academic performance of low-income students in the inner city. In addition, while it is clear that most big-city school districts can and should do much more to eliminate segregation and stratification in the schools, it is probable that for the foreseeable future many and perhaps even a majority of low-income students in the inner city are going to attend schools whose enrollments reflect the segregated and stratified residential patterns which exist in the big cities. Thus educational programs for low-income students in the inner city must be appreciably improved if these students are to acquire the academic skills they need to participate productively in a knowledge-transacting society.

The evidence currently available on the status of efforts to improve educational programs for disadvantaged students indicates that educators have not yet found ways to bring about adequate achievement gains among low-income pupils in inner-city schools. The major national effort to improve these programs has been financed under Title I of the Elementary and Secondary Education Act of 1965, through which approximately one billion dollars a year in federal funds have been made available to public and parochial schools for the education of disadvantaged students. The most recent evaluation of the effects of Title I expenditures is for its second year of operation (1966-67) in which 9.2 million school children in 16,400 school districts were enrolled in programs supported with these funds (U.S. Office of Education, 1965). With respect to the achievement of disadvantaged pupils, the data on the performance of students in Title I programs indicate that though the situation varies a good deal from city to city and state to state, some Title I programs apparently are succeeding in bringing about statistically significant gains in achievement among the pupils enrolled in them. More precisely, compensatory education in some school districts has been made sufficiently effective to halt, though not reverse, the cumulative deficit which typically places the achievement levels of low-income students three or more years below grade level expectancy by the time they enter the ninth or tenth grade.

Title I programs have prevented many disadvantagedyoungsters from falling behind their more fortunate peers in scholastic progress. Where in the past they have lost ground each month, many Title I youngsters are now improving, sometimes gaining a full month of learning for every month spent in the classroom [U.S. Office of Education, 1965: 7].

As is noted elsewhere in the report, even these results must be treated with some caution since achievement gains, where they did exist, were registered in the spring of 1967 and might not hold up were students to be tested the following fall (as happened in New York City's controversial More Effective Schools Program. See Fox, 1967). In addition, many educators fear that the limited gains made to date may be due primarily to improvement in test-taking skills and attitudes; if so, these gains would have to be viewed as more easily attainable than subsequent improvements in the more abstract skills and understandings disadvantaged students must acquire to reach grade-level achievement standards. It is quite possible, in other words, that the uneven gains thus far reported under Title I programs may constitute the ascent to a plateau which is much easier to reach than surpass.

But we must recognize that Title I could hardly have great impact in making available only $1,011,761,000 for the estimated 8.3 million students who were enrolled under its provisions during the fiscal year of 1967, and that even then it was not possible to enroll more than a "small proportion" of the nation's disadvantaged students, particularly in the "inner cities . . . [where Title I] still reaches only a fraction of the poorly prepared, undereducated children" (U.S. Office of Education, 1967: 7). So one ought not to be surprised to find that the overall tone of the Office of Education's report on the second year of Title I is more bleak than encouraging. After taking note of "hopeful signs," the report concludes that:

. . . the Title I child is still far behind the average child. As many as 60 percent of the Title I youngsters in some districts fall in the lowest quarter on reading scores . . . . [even programs] that show signs of immediate success often are not, from a long-range point of view, really successful because the children begin so far behind more fortunate students and have so far to go to catch up (U.S. Office of Education, 1967: 7, 33).

To summarize the current situation, educational programs to help disadvantaged pupils acquire the academic skills they need to function effectively in a knowledge-transacting society are not now being carried out at a sufficiently effective level to overcome the educational disadvantages of pupils who suffer from the economic destitution and the social despair which are typical of the "crisis" social-class and racial ghettoes in the big cities. To improve the quality of big-city schools so as to achieve this objective will be an exceedingly difficult undertaking. It will cost a great deal of money, perhaps more than middle-class Americans are willing to pay to give low-income and minority group students the skills to compete on a more equal basis with their own children. But even more than money, better education in inner-city schools will require imagination, innovation, and careful evaluation, three characteristics which have been all too rare in American education. Nevertheless, despite the relative lack of progress to date and the slowness of the educational system in finding solutions to the problems of teaching the disadvantaged child, the outlines of a program which might improve the quality of education in the big cities are now becoming clear. Since no single approach or improvement can serve as a panacea for these complex and deep-seated problems, comprehensive action is needed in a number of varied directions. Some of the more important of the proposals and directions for bringing about substantial improvement in the quality of big-city schools are briefly described in the next section.

DIRECTIONS IN IMPROVING INNER-CITY SCHOOLS

There are two broad positions taken by people who are critical of urban school systems. One group can be called "anti-institutionalists" because they combine despair over the obvious shortcomings of present school systems with a general mistrust of social institutions. They perceive the present educational systems as being "self-serving, bureaucratic monsters that need replacement rather than reforms." (Schrag, 1968: 68). Their positive proposals for reform often have a certain freshness and air of originality. But these qualities seldom persist when these proposals are examined with a sympathetic eye which nevertheless looks for evidence that they are workable.

Some prominent anti-institutionalists, such as Paul Goodman, are as critical of middle-class schools as of low-income schools and argue that "minischools" or other educational settings which might better recognize the individuality and humanity of young people should be widely established in as many communities as possible. Other critics who argue for the establishment of schools alternate to, or competitive with, the public schools in the inner city appear to be primarily concerned with the present system's failure to raise the achievement of disadvantaged students (Clark, 1968).

The anti-institutionalists find themselves in a strange alliance with right-wing stalwarts, such as the editors of the weekly business newspaper *Barron's* who favor a "whiff of competition" to achieve "denationalization of the education industry," partly out of nostalgia for a "happier day and age" when minority groups had no opportunity to demand "costly . . . boondoggles" (e.g., Headstart) of the public treasury (Barron's, 1968). Regardless of the differences in their specific orientations, however, these critics share a deep dissatisfaction with a system which they believe has allowed the schools to achieve a "near monopoly position as gatekeepers to social and economic advancement" (Schrag, 1968).

One of the most detailed of the anti-institutionalist proposals for alternative school opportunities is described in a recent article by Theodore Sizer, Dean of the Harvard School of Education, and Philip Whitten (1968). Starting with the widely accepted and hardly arguable premise that ". . . *we must discriminate in education in favor of the poor* [original italics]. We must weight the education scales in favor of the poor for the next generation and commit a major share of our resources to providing superior educational programs for them" Sizer and Whitten propose ". . . quite simply, [to] give money in the form of a coupon to a poor child who would carry the coupon to the school of his choice, where he would be enrolled" (Sizer and Whitten, 1968: 60-61). Using a sliding scale which, after taking family size, regional differences, and other variables into account, would provide an average annual subsidy of $1,500 to each poor child, the plan would cost an estimated $15 billion per year. By thus giving low-income parents the power to choose the schools their children would attend, existing schools would be forced to compete with each other to provide more effective instruction for disadvantaged students, and new schools would be founded to improve the quality of education available to the poor.

Sizer and Whitten are aware that the success of their proposal would depend on the accuracy with which parents assessed the educational opportunities available to their children and that many parents would make unwise choices and decisions, but they have even less confidence in "the present monopoly of lay boards and professional schoolmen" and they maintain that "accreditation by regional professional groups and the states" as well as "new mechanisms" for systematically assessing "the quality of the 'output' of the schools" could provide the information most parents need to make adequate choices. Also cognizant of other problems such as the difficulty of persuading middle-income parents to send their children to school with low-income students, the authors acknowledge that their proposal would require not only a "federal school-building program established to house imaginative and integrated educational programs," but also a vastly improved level of public understanding of education. Though the task of implementing such a proposal would be extremely difficult, Sizer and Whitten conclude that "It can be done. It must be done." (Sizer and Whitten, 1968: 63).

Unfortunately, however, there is no convincing evidence to support the conclusion that competing schools, such as have been proposed in the Sizer-Whitten "Poor Children's Bill of Rights," would function any more effectively than do inner-city schools today. To our knowledge, for example, none of these proposals have given even the slightest hint as to what might be done within "competing school systems" to raise teacher salaries or give teachers reasonable protection with respect to pensions or tenure of position; yet questions such as these cannot be disregarded if one hopes to ensure high level performance on the part of professional employees. More important, the "free market" model school could be expected to advertise its program, thus compounding the tendencies which already exist in the public school system to give more attention to appearances than reality and to use data selectively and sometimes irresponsibly to make favorable claims. The difficulties involved in evaluating the results of any educational enterprise, and the fact that accrediting groups seldom if ever have succeeded in probing below the surface of conditions in the schools, suggest that one cannot lightly dismiss the problem of assessing the quality of education offered by many competing institutions which would be tempted to falsify product descriptions, give kickbacks to parents on the government coupons, exclude relatively "nonmalleable" input (i.e., hard-to-teach students), and engage in other profit-oriented practices detrimental to the purposes of their well-intentioned advocates. Similarly, it is not difficult to find instances in which high quality instructional programs have convinced some middle-income white parents to keep their children enrolled in schools with disadvantaged minority pupils as long as these schools have remained predominantly middle class in character; but it is another matter to assume that improved facilities and programs supported through federal or other sources can be widely successful in counteracting the segregating and stratifying tendencies inherent in a plan based on the exercise of free choice by parents. In short, it is possible that proposals for improving the education of the disadvantaged by breaking the "monopoly" of the public schools not only would fail to achieve their goals, but would accentuate some of the most dysfunctional and undesirable features of the present system.

In the final analysis the priority assigned to the anti-institutionalists' proposals for competing school systems depends more than anything else on an explicit or implicit reading of the possibilities which exist for improving the present system. School districts—like every other major organization in urban society—are highly bureaucratic. The bureaucratic characteristics of standardization, centralization, and impersonal operation often serve to inhibit the public school district from adapting to solve the most pressing demands in its social environment. However, following the lead of Weber and other social scientists who have analyzed the growth and problems of organizations in modern society, it also should be recognized that large bureaucracies offer many advantages for the democratic achievement of important social and individual goals in an age of technological and societal complexity. Unless one believes, therefore, that the public schools or other large bureaucracies are so irredeemable as to be totally incapable of reform, it is possible to agree with the anti-institutionalist critics that the present public school "system which blames its society while it quietly acquiesces in, and inadvertently perpetuates, the very injustices it blames for its inefficiency is not in the public interest" (Sizer and Whitten, 1968: 62), and still give highest priority to vigorous and comprehensive attempts at improving the quality of public education offered in the inner city and throughout urban society. In terms of general organizational functioning, much might be done to increase the effectiveness of the public schools by utilizing the ideas and suggestions of Warren Bennis (1966) and other leading students of modern organization who argue that it should be possible to move "beyond bureaucracy" by administering organizations in ways which retain the advantages but minimize the negative tendencies of bureaucratic practice.

Since it is questionable whether even the most ingenious proposals for competing school systems would lead to improvements rather than further deterioration in the educational opportunities available to disadvantaged students in the inner city, active efforts looking toward reform within the present system are more justified than would be the diversion of time, energy, and other resources toward the implementation of the radical changes proposed by the anti-institutionalist critics of the present system.

In contrast to the anti-institutionalist critics there is a group of "activist reformers" who would work within the present institutional structure but seek thoroughgoing reforms. Some of these reforms are concerned with change in the organization and governance of education and some with improvement in instruction in the school. Since space limitations allow for only a few brief illustrations in each category, it should be clear that the examples cited on the following pages by no means constitute an exhaustive analysis of the many steps that should be taken or are worth trying as part of a comprehensive effort to improve the quality of urban education in general and inner-city schools in particular.

Decentralization

Although educators and laymen in big cities throughout the country are seeking to achieve some form or degree of decentralization in the public schools, demands and proposals for decentralization may refer to at least three different goals or changes which are conceptually distinct though not mutually exclusive with one another. The three most fundamental are as follows:

Transference of decision-making authority within the administrative hierarchy to officials closer to the individual building and classroom level than are the superintendent and the members of his staff in the central office. The most common approach to achieving this kind of decentralization is to divide a large school district into administrative subdistricts under the immediate authority of administrators who report directly to the general superintendent of schools. The Chicago public schools, for example, now are divided into twenty-seven administrative units under an equal number of district superintendents and three deputy superintendents with appreciable authority to make decisions formerly reached in the central office. Another potentially valuable form of hierarchical decentralization has been partially achieved in the Kansas City, Missouri, Public School District where inner-city schools receiving federal assistance for educating disadvantaged students have been organized into a separate division under an assistant superintendent of schools whose staff supervises and initiates instructional programs in these schools and speaks for their special interests within the larger district bureaucracy.

The desirability of these kinds of decentralization in larger school districts is almost self-evident. Just as is true of other kinds of organizations which are faced with the task of providing nonstandardized services to a variety of clients in a rapidly changing environment, urban school districts must distribute administrative authority so as to ensure that as many decisions as possible are made by officials who are close to the operating level of the organization and therefore have a chance to understand the nature of the problems which exist there.

Transfer of powers from the district-wide board of education and its administrative staff to boards of education partly responsible for the operation of one or more public schools within the larger district. In addition to moving decisions closer to the level at which they are implemented in the schools, the principle arguments in favor of establishing and transferring decision-making powers to local boards of education are that this type of decentralization may serve to (a) help residents of the city, particularly low-income and minority citizens, acquire a greater sense of control over their lives and destiny; (b) encourage citizens to become more supportive of the efforts and more understanding of the problems of professionals who staff the schools; and (c) ensure that the schools will be more accountable to the needs and desires of the people they serve.

There is not as yet very much agreement on how powers should be divided between central and local boards of education or between local professional staff and local board members. Through trial and error as well as conflict and confrontation, decisions now are being made to distribute administrative and policy-making functions more widely within the big-city school district, but it will be a number of years before anyone can speak confidently concerning the optimal distribution of power between and within local and district-wide centers of authority. At the present time (September 1968), for example, the New York City Board of Education is proposing to give thirty-three local boards of education greater powers with respect to recruiting and assessing teachers, preparing budgets and allocating funds, and determining and modifying curriculum in the schools; but disagreements between community organizations, teacher organizations, and other groups may force substantial changes in the proposal. In fact, the actions of one local board have brought on a city-wide teacher work stoppage.

Most readers will be aware that major experiments in this type of decentralization are underway in the Ocean Hill-Brownsville, I.S. 201, and Two Bridges sections of New York City, the Woodlawn neighborhood in Chicago, and elsewhere in the country. Several of these experiments involve universities or other specialized institutions in a position to contribute a great deal to the solution of the urban crisis, and many of them have considerable financial and political support from foundations as well as from state and federal government agencies. The potential importance of such experiments is obvious. Not only do they provide an opportunity to draw together many community and specialized resources which might help improve the education of disadvantaged students in the inner city, but the purposes to which they are addressed, clearly focus on central problems which will have to be faced and dealt with in any attempt to improve the quality of life in the central cities of nearly all our metropolitan areas. As city-planner Edward Logue has pointed out, for example, the problem of making the performance of government functions accountable to the people of the city is evident with respect to housing, urban renewal, and other crucial urban services (*City*, 1968: 30). Because they signify a deep determination to ensure a better education for students in the inner city, the demands of low-income and minority citizens for a voice in controlling school affairs generally should be welcomed as an encouraging awakening of local interest in the problem of improving the quality of inner-city schools.

Mechanisms to involve larger numbers of citizens more directly and intensively in local school affairs. If decentralization involves only a handful of citizens representing the interests of inarticulate masses on a local school board, its impact may be mostly limited to the small number of persons who actually sit on the board. Just as is done in many middle-income suburbs, school officials should work in a variety of ways to bring about widespread citizen participation in the affairs of the inner-city school, but it is doubtful whether such participation can be achieved or maintained unless citizens are given a meaningful role to play in educational decision-making, i.e., unless some greater measure of decentralization is introduced in the educational system. President John H. Fischer (1968) of Teachers College at Columbia University recently described one possible approach to encourage citizen involvement in urban education when he argued that

> We shall need a truly radical conception of decentralization, for what is involved is creating means by which principals and faculties can obtain from their communities far more regularly than they now do both their signals and their rewards.

> One way to bring this about would be to establish in every school a group of parents and other citizens to work with the principal and teachers ... [to] advise the school staff on educational priorities and objectives, on curriculum development and on the types of services most likely to aid the students. It could submit to the local board at least annually its appraisal of the school's success in meeting the problems the community considers important....

All three kinds of decentralization described above offer hope for overcoming the negative aspects of bureaucracy in the modern school district. To the extent that influence is transferred closer to units in the field and to the clients served by an organization through a combination of these approaches to decentralization, opportunities are gained for counteracting bureaucratic tendencies toward overstandardization and rigid, universalistic determination and application of policy.

Improvement in Administration

Reforms related to curriculum and instruction in inner-city schools will have little chance to succeed unless prior efforts are made to secure administrators who are more competent, independent, and creative than the larger number of administrators now employed in big-city school districts. As regards the individual building level, for example, informed observers working on the team which surveyed the Chicago public schools reported that the leadership of the school principal seemed to be "the vital factor" differentiating between successful and unsuccessful inner-city schools (Havighurst, 1964: 67). Similarly, the author of a recently completed study of Chicago elementary schools reached the same conclusion in attempting to account for the difference between "well functioning" and "poorly functioning" inner-city schools (Doll, 1968), while the author of an unusually comprehensive survey of developments in educating the disadvantaged recently made the following observation:

> In some school systems, an isolated school will be identified as doing an exceptional job of educating disadvantaged children as evidenced by community support, academic achievement, or other criteria. These objectives are achieved far beyond those attained by comparable schools without direct reference to special funds, although special funding is at times in evidence. In each of these schools, one finds a dynamic, determined, and competent principal who has inspired children, parents, and teachers to join in the successful venture [Jablonsky, 1968: 3].

The long-range prospects for improving inner-city schools also depend on the quality of the administrators and supervisors who hold positions either in the central office or intermediate between the individual school and the central office. The actions of administrative and other specialized staff members in these positions will do much to determine whether school district resources will be focused productively or wastefully on programmatic improvement in the inner-city classroom. Unfortunately, however, these positions often are filled by educators with little or no knowledge of, or experience with, disadvantaged children. Many of them earned their promotions by going along with the most dysfunctional features of the big-city school bureaucracy and are not much inclined to initiate the deep-seated reforms needed to bring about constructive change in big-city schools.

Given the central importance of administrative leadership at all levels of the urban school district, it is disheartening that very little has been done to train, select, and promote more competent administrators for city school systems. In the past year, however, several developments have occurred which indicate that this critical problem may begin to receive the attention it deserves. The new Education Professions Development Act, for example, now makes substantial federal funds available for the training of city school administrators. Fordham University and the University of Wisconsin are conducting intensive training programs to prepare principals and personnel administrators, respectively, for positions in big-city school districts. With respect to the actual administration of such districts, budgetary problems have prompted the superintendent and school board to release 381 "old-line" central office and middle-management personnel in the Philadelphia public schools (Carr, 1968: 1, 3), thus opening the way for their eventual replacement by personnel specifically chosen for their talents in meeting the emerging challenges of urban education. Such developments now represent little more than "stirrings in the wind," but it also is possible that they may point the way toward significant reform in public school systems in the big cities.

School Board Policy

Through a variety of additional policies involving the organization and operation of big-city school districts, boards of education can do much more than most are now doing to improve the educational opportunities available to disadvantaged students.

Superintendents and school board members who understand that segregation and stratification serve to depress the academic performance of low-income minority students in the inner city should work out a comprehensive plan to desegregate and destratify the urban school district. Most of the possible components in a desegregation plan, such as bussing to underutilized schools, pairing of elementary schools, and closing old school buildings in the ghetto, are well known and require no additional exposition in this essay. The manner in which school administrators and school board members act to implement a desegregation plan and to obtain public support for it is equally or more important, however, than the specific elements in the plan. For example, among the steps which might be taken in working toward the goals of desegregation and destratification are the following:

(1) Written policies should explicitly recognize recent research indicating that the surest way to raise the achievement of disadvantaged minority students is to place them in desegregated and destratified schools which also provide compensatory services for youngsters who need special help in overcoming learning problems.

(2) Similarly, school officials should make sure taxpayers understand that a combination of destratification and compensatory

education is the least expensive way to raise the achievement of disadvantaged students, and hence that desegregation is highly desirable on purely "academic" as well as on moral and social grounds.

(3) The school board should sponsor and publicize studies showing the effects of local integration plans on the achievement of middle-income white students. As long as the latter are not placed in predominantly low-income schools, such studies invariably show that middle-income students do not suffer and sometimes gain in achievement after desegregation.

(4) Through its written policies, the school board should publicly commit itself to do everything possible to help stabilize interracial or transitional neighborhoods in the big city. Among the actions which should be taken to help implement this policy are: (a) provide additional resources to schools in interracial neighborhoods and direct that these resources are expended on highly visible and well-publicized school improvement programs; and (b) as is being done on a limited experimental basis in Chicago, counteract resegregation by guaranteeing that bussing, pupil reassignment, or other methods will be used as necessary to maintain racial balance in schools in desegregated communities.

(5) Since parents are deeply concerned with the conditions in local public schools, board policies and actions can influence the development of housing patterns in the city. Thus, rather than sitting back and bemoaning the obvious fact that segregated housing makes it difficult to desegregate the schools, school board members can take affirmative action of the kind recently recommended to the Board of Education in Columbus, Ohio:

> The Board of Education should take immediate steps to place all plans for new school construction or additions to existing facilities under pre-construction open housing agreements hammered out in advance. The Board of Education can work with state legislatures in the passage of state wide legislation calling for such agreements to precede all public service developments in Ohio. . . .

> Imagine, for example, that an undeveloped tract or tracts become available for housing. The area or section is zoned for residential purposes. Before public services—water, sewers, schools, police protection—can be extended to this area, plans for area development must be presented to local governments by developers. Plans must provide evidence of the following: that the development will (1) include single and multi-family dwellings; (2) include rental as well as sale property; (3) will provide

services of realtors and financial institutions for people regardless of race, color, religion, and national origin [O.S.U. Advisory Commission, 1968: 86-87].

In Berkeley, California, Evanston, Illinois, and a few other cities where plans for district-wide desegregation have been introduced and apparently are working relatively well (Sullivan, 1968), school officials are demonstrating that a great deal can be accomplished to reduce segregation and stratification in urban schools. Admittedly, desegregation is much more difficult to achieve in larger cities and in cities where community influences are less favorable, but, by the same token, the success attained by school administrators and school board members in the school districts cited above indicates that the obstacles to desegregation are not as insurmountable as school officials with less resolution and ingenuity would have them appear.

Early Childhood Education

Headstart and other preschool programs for disadvantaged youngsters have won widespread and deserved support for their goal of reaching the student in his formative years before he falls far behind in the basic cognitive and social skills needed for later learning. Although the gains made to date by students in Headstart and similar projects generally have been disappointingly small and impermanent, these results can be attributed largely to the relatively short duration of these programs, to lack of intensive training for their professional staff, and to other deficiencies in implementation rather than to fallacies in the theory responsible for their initiation. In carefully planned and controlled longitudinal experimental projects being conducted in several parts of the country, educators and researchers are beginning to identify the instructional approaches that are likely to be most effective in developing the intellectual skills of educationally disadvantaged children, and in some cases very encouraging results already are being reported in programs focusing particularly on methods designed to raise the measured intelligence of low-income students (see, for example, the report in *The New York Times,* September 1, 1968).

In general, most concerned and knowledgeable observers now agree that (a) preschool programs should enroll disadvantaged children by the time they reach two or three years of age; (b) gains made in these programs often will be lost unless instructional conditions and approaches in the primary grades are appreciably better than presently can be found in most inner-city schools; and (c) preschool curricula must be organized so as to constitute a systematic program for developing the perceptual, cognitive, linguistic, and other skills which may be inadequately developed among a given group of children. In addition, many school officials responsible for the conduct of preschool classes are becoming aware of the contributions which can be made to their programs by social and racial balance, parent involvement, and other goals that need to be sought throughout the educational system.

One important technical question which has not yet been resolved involves the relative efficacy of preschool programs as against projects attempting later remediation, but within the next few years it should be possible to estimate the degree to which resources expended on the former might reduce the costs of compensatory education for older students whose intellectual and social behaviors are much more difficult to modify. All in all, the concept of preschool training for the disadvantaged remains the single most promising component in any comprehensive program for improving the achievement of students in the inner city.

Teacher-Training and Retraining

Both common sense and recent research indicate that teachers need to be unusually skilled and competent if they are to work successfully with disadvantaged students in the inner-city school and that the capabilities and attitudes of teachers are an even more important determinant of the achievement of disadvantaged students than is true in schools in which most of the pupils have been well prepared to succeed in the classroom (Coleman *et al.*, 1966; Burkhead *et al.*, 1967; Rosenthal and Jacobson, 1968). Yet, whereas teachers in middle-class schools have had two, three, or more years of relevant professional training for working with students in relatively favorable instructional situations, few teachers in the inner city have had more than one or two courses which have given them any practical and concrete guidance for solving the much more difficult problems posed by the educationally disadvantaged student. Until such training is widely available, a large proportion of the billions of dollars being spent to improve educational opportunities in the inner-city school inevitably will be wastefully expended to finance inappropriate instruction; as a result it will be impossible to assess the limits of the gains which eventually might be made in raising achievement levels among disadvantaged students.

Since adequate progress in teacher-training and retraining is not likely to be made until laymen and professionals explicitly recognize the extent of the need, it is encouraging to note that a recent survey of the public schools in Washington, D.C., suggested that fifteen to twenty percent of the time of teachers in big-city school districts should be set aside for in-service training. Other encouraging signs of an awakening to the need for providing in-service training on a much more systematic and intensive basis than has been possible through the traditional provision for monthly faculty meetings and occasional workshops can be seen in the Atlanta schools where specially trained teams of teachers have been formed to free regular staff members for participation in continuing in-service training programs (Martin and Henson, 1968), and in Kansas City where four teachers will spend the 1968-69 academic year at the University of Missouri at Kansas City preparing to become in-service training specialists upon returning to the public schools.

Although the scale of the in-service efforts now being made to improve the competence of inner-city teachers is still much too small to have a significant national impact on the achievement of disadvantaged students, many educators have recognized the enormity of the demand for this training and some have made a good beginning in finding ways to organize and conduct successful training programs.

Improvement of Instruction: The Emerging Emphasis on Motivation

The primary point which must be made in discussing the improvement of instruction in the inner-city classroom is that no one has found a simple formula which might magically raise the achievement of disadvantaged students in big-city schools. The conventional way to work on the problem of teaching disadvantaged children of school age is to work directly on the mental skills of the child—his vocabulary, reading, writing, arithmetic: *teach, teach, teach* with all the energy, time, patience and techniques necessary to implant the desired academic skills. This has not worked very well. It has led some critics of the schools to contend that the school principals and teachers are stupid in their insistence that the key to effective teaching of disadvantaged children is to give them more of the same materials and experiences that have not proved useful—an extra hour of drill after school, a Saturday morning class, a reduced class size so that the teacher can give more attention to the individual pupils.

Some of the activists appear to have discovered methods which do work with disadvantaged children. Their methods appeal to motives. If the pupil wants to learn, he will try to learn. Boys who do not learn in school nevertheless learn to play basketball, and girls who do not learn in school nevertheless learn to dance very well. Both accomplishments require practice, as well as bodily coordination. These boys and girls spend hours practicing what they want to learn.

Small, informal schools and classes are springing up in the inner city that appear to be accomplishing more than the conventional schools with disadvantaged youth. For example, the "street academies" of New York City appear to be working successfully with some dropouts and failing students from the high schools. These are described in an article by Chris Tree in *The Urban Review* for February, 1968, and are now a part of the Urban League's Education and Youth Incentives Program. Herbert Kohl taught a sixth-grade class in Harlem with a kind of freedom and spontaneity that seems to have motivated many of his pupils to care about their school work. Perhaps it is significant that he did relatively little drilling, and did not bother to correct spelling and grammar. In fact, he drew criticism from his supervisors because he did not emphasize the mental skills in the usual way. And Jonathan Kozol, in Boston, made friends with his pupils, took them on trips with him, visited their homes, but did not seem to stress the conventional training.

A recent experiment in tutoring seems to have succeeded through its motivational value, in spite of the fact that the wave of tutoring projects of a few years back has been a disappointment. The conventional tutoring project puts college students or middle-class adults in the role of tutor to inner-city pupils. But the experiment undertaken by Robert Cloward of Rhode Island University used tutors only a little bit older and more skilled than the pupils being tutored. He used eleventh-graders of below average reading ability in slum areas as tutors to middle-grade pupils in slum schools. Both tutees and tutors gained in reading achievement more than their controls did in a carefully designed experiment. These results can best be understood in terms of a "will to learn" that was increased in this situation. Certainly it was not a matter of superior methods of teaching used by these untutored tutors.

Educational games are being used as part of the school curriculum, especially in the middle-class schools. Games are motivating to most players, and they try to learn in order to win the games. James S. Coleman and his colleagues of Johns Hopkins have been working out games for high-school students. The Mecklenburg Academy in Charlotte, North Carolina, has a number of games available for junior-high schools. This approach may prove to have great value for improving achievement in the low-income school.

Social psychologists have underway some important researches on academic motivation which are beginning to suggest that most children of the lower working class can be taught more effectively by somewhat different methods of organizing lessons, giving approval, and correcting children's work than the methods that work best with middle-class children. It should be possible soon to show teachers of disadvantaged children how they can best teach, with methods no more difficult than the methods that are best used with middle-class children (Katz, 1967).

Assuming that we can and will learn more effective methods of teaching disadvantaged children through research on motivation, what chance is there that these methods will be quickly and widely adopted? Here we meet the obstacle of bureaucratic resistance to change which we have already discussed.

Accepting the fact that innovations meet resistance in city school systems, some of the activists are now working on ways to overcome this kind of resistance. An example is the Consortium School in Utica, New York. After a successful experience with Upward Bound summer programs, the principal people working in this program resolved to try to put their methods to work in the regular school program. Five colleges in the neighborhood of Utica joined the Utica School Board in a Consortium School Board Agreement, by which the colleges help to staff an experimental secondary school within the school system, supported by ESEA Title III funds. The plan provides for continuity between high school and college through a combined high-school-college staff (Doremus, 1968).

CONCLUSION

When we find the good and effective ways to teach disadvantaged children and youth, we will still have to solve the problem of social integration of ethnic and poverty-plagued minority groups into the economic and political life of our large metropolitan areas. The solutions will go hand in hand. Big-city and suburban governments as well as big-city and suburban school systems will be remade in this process of social urban renewal.

The quality of urban education depends on what is done for disadvantaged children and youth, and the amount of space we have devoted to this problem denotes its immediate importance, in our eyes. Yet this is essentially a short-range problem. As we solve this problem we shall become more and more aware of the long-range problem of improving education for the growing proportion of children, youth, and adults who are free of the curse of poverty. In an economy of abundance can we teach ourselves to live an abundant life?

REFERENCES

Barron's National Business and Financial Weekly (1968) "Little red school house? Community control of education invites mob rule." (Vol. 48, September 2): 1.

BENNIS, WARREN (1966) Changing Organizations—Essays on the Development and Evolution of Human Organizations. New York: McGraw-Hill.

BOWLES, SAMUEL and HENRY M. LEVIN (1968) "The determinants of scholastic achievement—an appraisal of some recent evidence." The Journal of Human Resources 3 (Winter): 3-24.

BURKHEAD, JESSE et al. (1967) Input and Output in Large-City High Schools. Syracuse: Syracuse Univ. Press.

CARR, JOHN (1968) "NYC shakeup ordered: Philly's Shedd fires 381." Education News 2 (May 27): 1.

City (1968) "Analysis: Should the schools decentralize?" (Vol. 2, March): 30-32.

CLARK, KENNETH (1968) "Alternative public school systems." Harvard Educational Review 38 (Winter): 100-113.

COLEMAN, JAMES S. et al. (1966) Equality of Educational Opportunity. Washington, D.C.: Government Printing Office.

DOLL, RUSSELL C. (1968) Types of Elementary Schools in a Big City. Chicago: Ph.D. diss., Univ. of Chicago (unpub.).

DOREMUS, J. C. (1968) "Upward bound in transition." Educational Opportunity Forum 1 (No. 2): 51-61.

FISCHER, JOHN H. (1968) "Fischer on decentralization." Education News 3 (August 5): 16.

FOX, DAVID J. (1967) Expansion of the More Effective School Program. New York: Center for Urban Education.

HAVIGHURST, ROBERT J. (1964) The Public Schools of Chicago. Chicago: Board of Education of the City of Chicago.

JABLONSKY, ADELAIDE (1968) "Some trends in education for the disadvantaged." IRCD Bulletin 4 (March): 1-7.

KATZ, IRWIN (1967) "The socialization of academic motivation in minority group children." Pp. 133-190 in Nebraska Symposium on Motivation. Lincoln: Univ. of Nebraska Press.

MARTIN, JOHN S. and E. CURTIS HENSON (1968) "Atlanta takes the initiative." NEA Journal 57 (May): 28-29.

Ohio State University Advisory Commission on Problems Facing the Columbus Public Schools (1968) Recommendations to the Columbus Board of Education.

PASSOW, A. HARRY (n.d.) Toward Creating a Model School System: A Study of the Washington, D.C. Public Schools. New York: Teachers College, Columbia Univ.

ROGERS, DAVID (1968) "New York City schools: a sick bureaucracy." Saturday Review 51 (July 20): 47-49, 59-61.

ROSENTHAL, ROBERT and LENORE JACOBSON (1968) Pygmalion in the Classroom. New York: Holt, Rinehart & Winston.

SCHRAG, PETER (1968) "The end of the common school." Saturday Review 51 (April 20): 68.

SIZER, THEODORE R. and PHILIP WHITTEN (1968) "A proposal for a poor children's Bill of Rights." Psychology Today 2 (August): 59-63.

The New York Times (1968) "Special training program at City University's center in Harlem sharpens children's perceptions." (September 1): 46, cols. 4-5.

SULLIVAN, NEIL V. (1968) "Discussion." Harvard Educational Review 38 (Winter): 148-155.

THOMAS, J. ALAN (1968) "Modernizing state finance programs." Paper presented at the NEA School Finance Conference held in Dallas, March 31-April 2.

United States Commission on Civil Rights (1967) Racial Isolation in the Public Schools. Washington, D.C.: Government Printing Office.

United States Office of Education (1967) Title I/Year II: The Second Annual Report of Title I of the Elementary and Secondary Education Act of 1965. Washington, D.C.: Government Printing Office.

13

Determining the
Quality of Education

A Political Process

MICHAEL DECKER
and LOUIS H. MASOTTI

☐ IN THE FOUR years since Leonard Britton surveyed the literature on educational quality in America, the number of serious studies
of this subject has probably doubled (Britton, 1964). We do not
propose to add yet another piece devoted primarily to defining
quality or trying to measure it. Neither is it our purpose to map the
distribution of educational quality in urban America or to identify
its determinants. We view the provision of educational quality by
formal organizations—public and private, elementary through university—as a social problem. We propose, therefore, to examine the
demand for, and supply of, this quality in an effort to forecast the
most probable solutions upon which urban America will settle.

The demand for educational quality is not unique to contemporary America. It has existed in all societies which have developed a
literate class, for as Ralph Pounds notes, formal educational organizations are designed by such classes as a means for transmitting the
ability to write (Pounds, 1968). What distinguishes the contemporary
concern with and demand for educational quality is: (1) the increasing number of individuals and groups who demonstrate concern for
educational quality, (2) the increasing intensity of that concern, and
(3) the demand for a greater amount of quality. What follows is a
description of the shape and sources of that demand. Since much of
the conflict over educational quality stems from differences in definition, we first describe the location of various meanings of quality.

SOURCES OF QUALITY DEMANDS

School personnel can be divided into three principal interest groups: teachers and other operations personnel, line administrators such as building principals, central office or staff administrators, and policy-makers (boards of education). That such a classification can reasonably be made, is a function of increasing pressure on the schools to produce quality, as these groups variously define it. The creation and increased militancy of separate formal associations which work for changes, often contradictory, in educational resources and processes is ample evidence that they see themselves as separate interests.

Teachers, of course, contend that they are the primary determinants of educational quality. They demand adequate teaching conditions, low pupil-teacher ratios, modern plants, reasonable wages, and the like. These demands are defended on the grounds that better educated students can be produced, in the final analysis, only through improvements in the quality of instruction. Such improvement can be achieved only if schools provide adequate terms and conditions of employment for their teachers. The latter, in particular, assume a high correlation between instructional quality and the acquisition by students of a good education. In effect, they are pressing schools to improve the educational service to students, indirectly, through them.

Line administrators often fit the stereotypes of the man in the middle; they are in somewhat the same position as plant managers in large manufacturing concerns. The important difference is that the schoolmen usually have less control over their operations than plant managers. Their primary problem is to maintain some measure of balance between the demands of staff administrators, on the one hand, and of their employees, on the other. Thus, while school principals aspire to an experienced, well-educated and competent staff, excellent physical plants, and powerful curricula, their aspirations are often blunted by their organizational positions.

Staff administrators are—both by law and their own perceptions— the guardians of public and private investment in education. It is this perception which causes them to be characterized by many as conservatives, a description which is usually appropriate. Change requires additional resources, and administrators normally require compelling evidence of the need for improvement before supporting modifications of any magnitude. Because of their position at the boundary between the school and its environment, their primary concern with the maintenance of adequate resource inputs, administrators tend to conceive of educational quality in terms of minimizing conflict. It is, indeed, very difficult for communities experiencing substantial conflict over their schools to attract administrators who meet the desired standards.

Yearly reports provide evidence concerning specific administrative versions of quality. These reports typically describe additions to the school's physical plant, increases in the educational level of teachers, efforts to maintain a stable tax rate and, in some cases, attributes of the student body (including the percentage of students going to college, the number awarded scholarships, and such performance information as comparisons between local reading levels and national averages) (Vincent, 1965). All these illustrations are designed to maintain the flow of money, staff, and students at acceptable levels.

Boards of education occupy the boundary between citizens and the central administration, and to them are directed most heavily the demands of interest groups of all sorts. Given this position, boards are in the delicate circumstance of having to satisfy seemingly insatiable appetites for quality in the face of more and more limited resources. Their concerns are thus mirrored by those of their closest associates—central office administrators.

All educational organizations are concerned with the quality of students coming to them. We find this particularly true of post-secondary schools, whose admissions policies generally reflect the indications that students' performance at this level is best predicted by their performance at the end of the secondary experience (Astin, 1962). The output performance of universities and colleges is usually defined in terms of test scores, higher degrees attained, and positions secured; while their input is defined in terms of the general intelligence, academic aptitude, and past school performance of the students. The care and resources devoted to the task of admissions in colleges and universities provide clear evidence of the stress placed on the abilities (quality) of incoming students.

Business firms are also concerned with educational quality. As we might expect, firms emphasize student characteristics in their conceptions of quality (i.e., product quality) since they use former students in achieving their objectives. Not only are particular vocational skills required, but students must be schooled in the work culture. A second source of business concern for the human products of the educational process is the requirement of political stability in the business environment. This concern is a major factor in current business involvement in central city education. The relationship between Michigan Bell Telephone Company and the Detroit public schools is a good example. Finally, firms demand process quality of the local school systems (new physical plants, experienced and well-educated teachers, and the like) in order to attract employees to their geographic areas.

The distribution of firms by type is not even throughout urban areas, however. Business establishments which are required to remain in central cities separate into two classes: heavy manufacturing industries which are immobile because of their large investment in capital plant and, therefore, are primarily concerned with labor pool (product) quality; and small businesses which are unable to move for lack of capital and are primarily concerned with the maintenance of civil order, hence product quality. Those firms that are more mobile, on the other hand, tend to follow their labor pool not, of course, without cost (Commission on Civil Rights, 1967). They are, therefore, principally concerned with process quality as a means of attracting desired human resources to their areas.

Nonschool governmental organizations also exhibit concern for educational quality. This is particularly true of federal agencies which invest large sums of money in support of education at all levels—both public and private. In general, federal conceptions of quality reduce to two. One involves the labor pool, or attention to vocational skills and work-culture socialization. The second is input quality, focusing attention on preschool children and the skills they need in the school environment (Rowland and Wing, 1967).

At the local level, municipal government demands civic order and a degree of political stability, a demand it transfers to the local educational system in several ways. Following the assassination of Martin Luther King, for example, the superintendent of one major-city school system declined to close the schools, in order not to contribute to the confusion. And in many urban areas the use of school facilities by the community has been increasing, in part as a result of summer programs designed to lessen potential instability.

The second clearly significant demand of municipal government focuses on what we have called process quality. Municipal officials know that a school system which enjoys a reputation for having a high-quality plant, faculty, and curriculum, helps attract a population requiring relatively low service levels and producing relatively greater revenues for municipalities.

Parents also regard educational quality in a certain light. In part, their definition relates to their aspiratons for their children, for parents expect the educational system to prepare their offspring for the future. Those of upper-class professional status are interested in having schools serve a preparatory role for the university and later professional life, while those in the nonprofessional categories expect preparation for the world of work. These parental expectations are reflected in school programs (Project Talent, 1962).

Two other parental conceptions of quality are worthy of note. One is the aspect of predictability. Since many parents arrange their daily activities around school schedules, events like unannounced closings are met with severe criticism. So, also, have attempts to introduce university-type class scheduling met with resistance from parents as well as merchants. The second conception relates to the fact that parents pay, in one form or another, the cost of education —whether public or private. Thus, cost is one of the criteria of school quality often used by them (Norton, 1966).

Students provide an additional source of definitions of school quality. Only very recently has behavioral evidence become available from which to infer such definitions from this source, and then only at the secondary and university levels. The child-centered culture apparently does not extend to listening to the young about their educational aspirations. Recent student activity indicates, however, that those in secondary schools define educational quality in terms of the educational process—the conditions under which they attend school. Most of the student violence has occurred where conditions for the education of secondary pupils are worst (central cities). This is understandable, given the conditions as described by Kozol (Kozol, 1967) and others; it is evident in these situations that students are exercised about the conditions of their education, not about its content.

The university student presents a more complex picture in his definition of educational quality. As the Berkeley and Columbia incidents have made quite clear, large numbers of college students are strongly disenchanted with the process of higher learning, the relevance of its content, and in some cases the community roles played (or not played) by their institutions (Cox Commission, 1968). A more traditional concern of the college student has been the relative absence of skill learning congruent with the demands of contemporary life. This latter point, of course, raises questions about the role of a university as dispenser of wisdom or post-high school vocational center, questions which are at this time receiving serious reconsideration.

Our final source of quality definitions is the body of researchers on whom educational institutions depend for new knowledge. Almost without exception, educational quality is defined, by members of this group, whether they be school psychologists, consultants, or educationists, as measured student performance (Burkhead, 1967). Probably the best illustration of this conception of quality is to be found in the now well-known "Coleman Report" (Coleman et al., 1966). Although the educational research group has had only a marginal impact on school operations in the past, we expect its role to enlarge as the demand for quality increases. And, as its impact on urban education becomes greater, so its definition of quality will take on greater significance. One clear illustration of this increasing impact is the connection between the research establishment and the New York City schools relative to decentralization. Whether or not the knowledge base provided the New York system by behavioral research can be institutionalized is still not clear.

Our description of various meanings of quality has proceeded from what we assume to be the most powerful groups with respect to the operation of urban educational organizations (teachers, administrators), through users of educational quality (firms, government, and those most closely connected with the educated (parents, students), to the education research establishment. When the quality landscape, thus delineated, is replete with various definitions, several general characteristics emerge.

First, the specific definition of educational quality held by any group is closely related to the conditions required by it to achieve its own objectives. Second, the closer a group is to the educational process, the more likely is quality to be defined in terms of attributes of that process; while the farther away it is, the more likely it is to be concerned with educational outcomes (product) as evidence of quality. Third, the skills and abilities of students are universally used as indicators of quality (though sometimes indirectly, to be sure). Finally, educational quality is assumed to behave as an economic entity. It is real, if abstract. It is used by administrators to attract teachers, money, and good will; by firms in location decisions in addition to the considerations of production and marketing of goods and services; and by parents in making residential location decisions. Quality is produced and consumed; hence it exists in greater or lesser quantity. It is the economic character of educational quality which permits us to examine the growing demand for it as well as the programs designed to satisfy this demand, a task to which we now turn our attention.

PROGRAMS FOR QUALITY EDUCATION

A number of social processes have stimulated the demand for educational quality by all the aforementioned groups. One such process is the development of an increasingly complex communications network. Making status comparisons is a very much easier task now than it was in earlier generations. It is now, more than ever, possible for residents of inner cities and rural areas to compare, for example, their educational services with those of more affluent people. This comparison has resulted in an increase in relative educational deprivation felt by disadvantaged groups and an attendant increase in the quality of education demanded by residents of these areas and their agents.

The process of urbanization, dating at least from the closing of the frontier, has increased local population densities in metropolitan areas and has had the same sort of consequences that have flowed from the availability of modern communications (Dentler, 1967). The most widely recognized aspect of urbanization, however, is the striking and growing isolation of the poor, the black, and the disadvantaged in core cities and the segregation of their white and more affluent complement in suburban areas. Isolated as well in central cities, are school districts which are underfinanced, understaffed, overpopulated, and under heavy pressure to accomplish the impossible dream.

This separatism has led to added visibility for the urban black and urban poor. Their aspirations to become part of the national life, coupled with their high social and economic visibility, has produced demands for substantial quality increases—immediately, not at some future date. Even if black Americans generally move to adopt their own separatism, it is unlikely that their demands for educational quality will abate.

A third social force producing demands for more quality is the drastic upward trend in skill requirements—in both urban and rural areas. Technological factors account for this upgrading. It is all very well to claim that cybernation will increase the number of jobs available in America, as indeed it may, but those jobs will demand, as they have, higher and higher skill levels. Because of this trend, firms increase their personnel standards, and colleges, universities, and vocational schools stiffen their entrance requirements. The users of educational quality, in other words, are themselves under pressure to demand more of it as a condition of their continued "success."

Fourth, the costs of education are spiraling, due in part to inflation but principally to increasing costs of personal services. Education is so highly labor intensive at this time that small increases in the unit cost of labor produce substantial resource requirements. Estimates by School Management indicate that, even discounting inflation costs, the nation's average current expenditures per student increased approximately 25 percent between 1957 and 1965 (School Management, 1966).

Fifth, services provided by educational institutions have greatly expanded since World War II (Rowland and Wing, 1967). Kindergartens have come into widespread use; school lunch programs are the rule today; and even the idea of the school as a community center has been resurrected.

Finally, international competition has intensified national concern for educational quality. One has only to date the first Conant Report (Conant, 1959) to recall that Sputnik I went up in 1957. And what middle-aged parent can forget the development of the "new math," the "new science," and the "new social studies" which began in the late 1950's?

The effects of these last three social processes have been twofold. Each of them has served to call the matter of the quality of American education to the attention of large segments of our society. At the same time the cost-service-competition squeeze has stimulated the search for ways to increase educational quality.

We observe the specific consequences of these social processes, as they affect the demand for educational quality, in (1) additions to the population of the concerned, (2) heightened intensity of concern, and (3) the development of programs to satisfy the growing demand. Probably the single most significant change in the demand for educational quality has been the translation, by all segments of our population, of their demands into programs for action. The number of programs designed to increase educational quality boggles the mind. As Rowland and Wing point out, there are at least 237 different opportunities for local schools alone to take advantage of federal funds for improving their operations. And this does not begin to exhaust the array of available sources of assistance.

Since we intend to provide some insight into the possible out-comes of the search for ways to increase quality, our classification of current programs having that goal is based on our conception of the variables of quality production. We assume that these variables include money, knowledge, technology, curriculum, time, teachers, students, and organization; and that the application of resources to the education of students adds value to them. This allows us to describe programs as they are directed at changing specific resource applications.

Money is first on our list of the factors of quality production because it buys all the other factors and is, therefore, one of the most critical conditions necessary to increase quality. At the local level, money flows, in the main, to the districts which need it least. This circumstance reflects state aid distribution formulas, local property tax procedures, and the encapsulation of the poor in central cities. Money produced at the local level tends to be hard won—through the electoral process in most instances—and used almost exclusively to meet increased enrollments and/or demands for salary increases. State aid to local schools has leveled off in many areas of the country, and higher education seems to capture a substantial share of state monies allocated to education. According to the United States Office of Education, revenue receipts for local schools in the United States as a whole show, since 1920, substantial de-creases in receipts from local sources, increases leveling off about 1950 from state sources, and increases also leveling off from federal sources (U.S. Office of Education, 1966). Federal aid, had been steady at about 4.5 percent of total local revenues from 1953 to 1963—a contribution fifteen times that of 1919. The federal share has again increased with the passage of bills such as the Elementary and Secondary Education Act of 1965. Clearly, new money is coming and will probably continue to come from this level.

Two sorts of knowledge have received the major share of research attention in recent years. On the one hand, there is research of the type reported by Bruner in *A Study of Thinking* (Bruner, 1956) and Bloom in *Stability and Change in Human Characteristics* (Bloom, 1964). The objective of this kind of research is to provide accurate information concerning the learning and development processes in human beings so that effective educational programs can be devel-oped for them. Research along these lines has served as the knowl-edge base for a wide variety of programs as diverse as Head Start and curriculum changes in the physical and social sciences.

The other kind of knowledge produced has been directed at improving the management of local schools. Most active in this work has been J. Alan Thomas (Thomas, 1964), whose research on the determinants of educational outcome follows in the tradition begun in the 1950's by Paul Mort and his associates (Mort, 1952; Ross, 1958; Mort and Furno, 1960). It is more difficult to trace program design to this sort of research than to development research. One reason for this is that administrative research has failed to identify factors controllable by school managers which, if changed, would affect educational quality. Certainly the findings of development research allow us, for example, to direct our attention to the early years as a good place to improve schools and provide students with opportunities for discovery. No such comparable epigrammatic knowledge derives from administrative research.

Related closely to the development of knowledge about the learning process, and depending heavily on the work of B. F. Skinner, have been efforts to introduce new technology into the formal educational process (Skinner, 1958). This technology has supported the curricular innovation we call programmed instruction (Bushnell and Allen, 1966). In the early 1950's, programmed instruction technology was the book reorganized. But with the development of large computing installations, we have seen the development of computer assisted instruction (CAI), a development greatly aided by the work of Suppes and Atkinson at Stanford University (Suppes, 1966). While CAI is not yet generally available, undoubtedly it will have important consequences for the future growth of educational quality. Nor may we neglect developments in the construction industry which permit the adaptation of buildings to curriculum instead of the other way around.

The mechanization of tasks which previously required heavy investment of labor—student drill, class scheduling, payroll, assessing the effects of alternative means for improving education—is now proceeding rapidly due to electronic computers. We expect, with John Pfeiffer, that advances in production technology will continue with obvious implications for the development of quality (Pfeiffer, 1968).

As indicated above, the curricular factor has received much attention by Skinner and other proponents of programmed instruction for well over two decades. There is no doubt that programmed instruction can increase the quality of education in America. The doubt arises when one considers its use in the schools. Of greater operational moment have been the spate of Sputnik-induced curriculum content changes for which we have the usual abbreviations—PSSC, GCMP, ITA (Commission on Civil Rights, 1967). These changes, however, have generally altered the course content rather than the teaching process.

The factor of time is singled out here partly because of the so-called dropout problem which has implications for the supply of educational quality. Students who are not in school cannot be treated through formal educational organizations, and will produce a negative influence on product quality indicators. And it is precisely among those students who enter schools from educationally deprived environments that most dropouts are produced.

The other development which is most closely related to the time factor is the development of federal manpower and education programs. Head Start extends the school life of various disadvantaged groups. The addition at the beginning years can be as much as two years, during which children are prepared for public schools. At the other end of the age scale are programs to provide a place for students who have taken themselves out of formal education and require vocational training in order to obtain jobs. While these and other federal programs often include new curricula and technology, it is apparent that the primary effort is to increase the amount of time during which society, through formal education, has control over the development of disadvantaged students and citizens intending to increase their skills and capabilities.

Teachers, the "delivery system" of the educational process, have long been considered central to the quality production process. Programs dealing with them have been of two types. The first, included in the recently passed Education Professions Development Act, has been directed at increasing the supply of certified teachers. While some parts of the country find themselves oversupplied in some teaching specialities, there is an undersupply of teachers nationwide, especially in central cities and particularly in the subject area of the physical sciences. The unionization and increased militancy of teachers will probably increase the supply as wage rates rise relative to other occupations.

The second teacher-related program focuses on teacher education. With the development of new curricular materials, National Science Foundation summer institutes have been funded to help teachers utilize these materials. The policy has spread to the social sciences and humanities. Increasing emphasis on the role of guidance counsellors has led to summer institutes for them as well. Moreover, not all teacher education programs have been public or adaptive oriented. For example, General Electric has sponsored economics institutes for several years in an attempt to increase the kit of skills available to social science teachers. States and local districts have also encouraged professional growth for many years, through stepped salary schedules and certification procedures. Thus, there is manipulation of the teacher factor on the supply and quality fronts alike.

Students are a factor of quality production in a rather special sense because we associate increases in quality with changes in them. But they can be moved about in an effort to effect those changes. While the drive for school integration is rooted in the civil rights movement, integration and de-stratification can, according to the Coleman Report, increase quality independently. If this is true, then integration plans can be said to affect the supply of educational quality.

Racial Isolation in the Public Schools reports three basic variants of desegregation programs in common use (Commission on Civil Rights, 1967). Pairing merges schools such that two K-6 schools might become one K-3 and one K-6. Central schools are sometimes established to serve all students of one level in a single district. This program is exemplified by the current interest in campus secondary schools and educational parks. Finally, attendance area changes are in common use in large cities where population mobility is of high magnitude. Thus far, metropolitan integration has not been tested on any sizable scale.

At the other end of the integration-segregation continuum is the matter of school decentralization. Advocates of decentralization argue that large centralized districts are not responsive enough to the needs of a heterogeneous student population (Mayor's Advisory Panel on Decentralization, 1967). The way to increase quality, they claim, is to turn the factors of production over to neighborhood control. If decentralization occurs in large central cities, we will be able to assess the validity of this claim. It is important to note here, however, that the conflict between integration and decentralization as a way to increase quality is in its early stages and will probably become more general and more intense with time.

Compensatory programs like Head Start and Project Follow-Up are attempts to increase the level of student input to the central schools by, as we have noted, increasing the levels of quality of all the other factors of quality production.

The final factor of educational quality postulated—organization change—has a long history. Efforts to incorporate all students in sufficiently large K-12 districts has resulted in a decrease in the total number of school districts in the United States to approximately 28,000, a decrease of almost 80 percent in the last thirty-five years. The number of one-teacher schools has experienced an even sharper decline. Spurred by rising costs, many proposals for dividing teacher labor have been advanced, including team teaching and the use of para-professionals.

Two recent organizational proposals concerned primarily with central cities are currently being advocated as a means for increasing both process and product quality. The first pertains to school decentralization, which we discussed above in terms of its implications for student separation or institutionalized segregation. But decentralization also has obvious organizational implications for education in the unwieldy school systems of large central cities, (Gittell and Hollander, 1967; Rogers, 1968a and 1968b). Decentralization plans range from the highly controversial neighborhood control experiment in New York City (which resulted in open conflict between neighborhood parents adamant about control over curriculum and personnel, and the powerful teachers' union concerned about job protection), to the decentralization of the existing "downtown" school administration (e.g., the Cleveland plan). All have the intent of shifting educational decision-making closer to the parent-student level, but they differ significantly in degree, and thus in effect.

There is a second movement, hardly noticeable at this time, in which inner-city parents, mostly black, seek to afford their children a private education within the community as an alternative to the public schools. Private schools in the ghettos may provide low-income parents with the kind of alternative until now only available to the affluent, an alternative that has the advantage of increasing control without the inevitable struggles with organized public school teachers and administrators (Jencks, 1968).

CONSTRAINTS ON THE DEVELOPMENT
OF QUALITY

Historically, the supply of educational quality in the nation has increased along many dimensions. A larger proportion of our population is in school, school years completed have risen, illiteracy has declined and skill levels have increased. But the increases have not been distributed equally in all geographic areas nor among all segments of the population. (For example, the median years of school completed increased 3.4 years from 1940 to 1965 among whites over twenty-five, and 3.2 years for non-whites, this from respective 1940 figures of 8.4 and 5.8). Educational quality is lowest in rural and central-city areas, and among the poor and the black. It is no surprise that the most vocal demands for increases in quality come from these areas and groups. The quality problem confronting urban America has come to be recognized as a distributional problem. And most of the current quality related programs, whatever their source or target, are designed to decrease the differences in levels of educational quality along geographic, class, and ethnic (racial) categories of our population.

The outlook for success in this effort is not a happy one—not, at least, given our present educational system. We cite Project TALENT and the Coleman Report in this connection, which comport with many similar studies:

> The remaining unaccounted-for variations (in student mean test scores) must be split among hundreds of school characteristics. . . . the chances are that the real unique relation of the (school) practice being studied accounts for only a fraction of one per cent of the output measure being studied (Project Talent, 1962).

> The school appears unable to exert independent influences to make achievement levels less dependent on the child's background—and this is true within each ethnic group, just as it is between groups (Coleman, *et al*, 1966).

In short, it is not at all clear that the public schools, as we now know them, can solve the pressing social problem of satisfying the growing demand for educational quality. The demand for such quality in urban areas is increasing faster than its supply and the system is becoming unstable, a circumstance pervasive in the quality of urban life. Changes can only come through modifying or removing constraints on the production of educational quality in these areas, which is to say that the factors of production must be applied in increasing amounts.

Knowledge and technology, in our opinion, have developed to the point where the shape of quality distribution is not significantly affected by constraints on them. We know from Bloom and others, for example, that environment most powerfully affects human development when growth is at its highest rate, and that people grow fastest during childhood (Bloom, 1964). This knowledge is part of conventional wisdom. We know also from many studies that teacher quality has some independent effect on the production of educational quality and that peer relations affect educational quality. Over 5,000 years of experience in formal education have, in short, generated a substantial body of knowledge. Flowing from it is a growing stockpile of sophisticated educational technology, which, with intelligent application, can increase the quality of urban education. However, much of this knowledge and a great deal of the technology is not utilized extensively. We must look for other constraints, and we need not look far, because they abound.

It is becoming ever more difficult for local schools to generate local tax monies by vote of district residents (Miner, 1963; James, 1963; James *et al*, 1966). The distribution of resources in metropolitan areas favors suburbs, in general, and the same may be said of state aid formulas. Nor is federal assistance limitless. The current allocation of resources at the federal level is heavy on defense, light on domestic programs for urban education, and not easily changed. These income constraints create, as we have indicated, constraints on all other factors of production. CAI programs cost on the order of $300,000 per course to develop. Teacher wages are increasing and their supply is limited, particularly in the largest school districts and in the field of higher education. Institutional reorganization is costly and time consuming beyond the work life of any change agent. The Mort studies indicate, indeed, that about one hundred years are required to complete the diffusion of a single educational innovation in a single state. But the large question remains: Whence come the constraints? Their sources reside in the distribution and use of values and power. The process of producing educational quality is fundamentally constrained by the values held by people and by their effectiveness (real power) in transforming these values into public policy. In pursuit of clues as to the future of the supply of educational quality in urban America, we now consider the politics of education.

THE POLITICS OF EDUCATION

Unless it can be assumed, and we suggest it is an unrealistic assumption, that there is widespread consensus not only on the need for more quality in urban education, but also on the *specific* goals and means for achieving it, education, like all other areas of public policy, becomes "political," i.e., it is involved in a process for resolving conflicts among contending values and participants. The basic ingredients for a politics of education were outlined in the opening paragraphs of this chapter: an increasing number of educational interest groups who pursue their often conflicting definitions of quality with growing intensity (Eliot, 1959; Minar, 1967; Masotti, 1967). These conflicts among values and groups will only be resolved through the techniques of the political process: influence, bargaining, persuasion, accommodation, compromise. They should be recognized, even if not adopted, by participants who hope to compete effectively in the contests for advantageous policy outcomes. The purist and the idealist may opt to reject these techniques in favor of extolling the merits of his proposal, but he must be willing to pay the price for principle. As long as he is unaware of the rules of the game he is playing and the techniques likely to be used by other participants, he has a lower probability of achieving policy success and is in a relatively weak position to seek changes in the decision rules which will benefit his program. Raymond Moley made the point well:

> Politics is not something to avoid or abolish or destroy. It is a condition like the atmosphere we breathe. It is something to live with, to influence as we wish and to control if we can. We must master its ways or we shall be mastered by those who do (Moley, 1964: 142).

Politics, defined as a process of conflict resolution, is clearly applicable to proposals for educational change. The central focus of the political process is conflict over goals and/or means and it abounds in public education, including the deliberations and debates over both process and product quality.

Education has always been political in the sense that we have used the term. However, until relatively recently the politics of education has been an "intra-institutional" politics dominated by professional educators—the "schoolmen." The current "quality-equality" educational revolution, set in motion by the 1954 school desegregation decision of the U.S. Supreme Court and the 1957 launching of Sputnik, has had the effect of diminishing the relative influence of the schoolmen and made education politics more "public" as it shifted into the mainstream of the legislative, administrative and judicial processes at the local, state and national levels. This politicization of public education has increased significantly as a result of the mobilization of both sincere and crackpot critics of progressive education, a focus on the schools by the civil rights (integration) movement, and more recently by the black power (decentralization) issue. Education politics is no longer an intra-institutional phenomenon; there is now an increasing concern for and awareness of the extra-educational determinants of educational systems and policy.

There has been a general recognition that education is a powerful force in determining the social, economic, and military welfare of the nation, a recognition which has made it an obvious battleground for contending forces who disagree on goals or means. What distinguishes the contemporary conflict in public education from those which preceded it is its scope and intensity: the current battle over educational change involves an enlarged set of participants on a wider front. The problems of providing the quantity and quality of education demanded by a rapidly growing and technologically oriented society have produced concomitant increases in the range and intensity of the conflicts over attempts to cope with these problems. The task of physically accommodating the spiraling student enrollment has been complicated by the needs for qualitative improvements in education articulated by its designers and consumers— teachers, administrators, policy-makers, students, parents, and the commercial-industrial recruiters.

Quality education for all is clearly in the public interest as well as being privately functional. However, until some method of obviating differences among definitions of quality is found, conflicts will be resolved through a political process. But politics is a slow and cumbersome process as it seeks to accommodate differences. Qualitative changes in urban education will come but they are likely to continue to be incremental, and neither massive nor rapid, unless those who possess the knowledge to effect such changes also achieve sufficient power to implement them. This is the price of a pluralistic society in search of quality.

REFERENCES

ASTIN, ALEXANDER W. (1962) "Productivity of undergraduate institutions," Science 136 (April): 129-135.

BLOOM, BENJAMIN S. (1964) Stability and Change in Human Characteristics. New York: John Wiley.

BRITTON, LEONARD M. (1964) Report of Visitation and Sources Investigated. Cleveland: Martha Holden Jennings Foundation.

BRUNER, JEROME S. et al. (1956) A Study of Thinking. New York: John Wiley.

BURKHEAD, JESSE et al. (1967) Input and Output in Large City High Schools. Syracuse: Syracuse Univ. Press.

BUSHNELL, DON and DWIGHT W. ALLEN, [eds.] (1966) The Computer in American Education. New York: John Wiley.

COLEMAN, JAMES S. et al. (1966) Equality of Educational Opportunity. Washington, D.C.: National Center for Educational Statistics.

Commission on Civil Rights (1967) Racial Isolation in the Public Schools. Washington, D.C.: Government Printing Office.

CONANT, JAMES B. (1959) The American High School Today. New York: McGraw-Hill.

The Cox Commission (1968) Crisis at Columbia. New York: Random House.

DENTLER, ROBERT et al. [eds.] (1967) The Urban R's. New York: Frederick A. Praeger.

ELIOT, THOMAS H. (1959) "Toward an understanding of public school politics." American Political Science Review 53 (December): 1032-1051.

GITTELL, MARILYN and T. EDWARD HOLLANDER (1967) Six Urban School Districts. New York: Frederick A. Praeger.

JAMES, H. THOMAS (1963) Wealth, Expenditures and Decision-Making for Education. Stanford: Stanford Univ. School of Education.

——— et al. (1966) The Determinants of Educational Expenditures in Large Cities of the U.S. Stanford: Stanford Univ. School of Education.

JENCKS, CHRISTOPHER (1968) "Private schools for black children." New York Times Magazine (November).

KOZOL, JONATHAN (1967) Death at an Early Age. Boston: Houghton Mifflin.

KRUYTBOSCH, CARLOS E. and SHELDON L. MESSINGER [eds.] (1968) The state of the university: authority and change. The American Behavioral Scientist 11 (May-June).

MASOTTI, LOUIS H. (1967) Education and Politics in Suburbia. Cleveland: The Press of Case Western Reserve Univ.

Mayor's Advisory Panel of Decentralization of the New York City Schools (1967) Reconnection for Learning: A Community School System for New York City. New York: Ford Foundation.

MERVIN, JACK C. and RALPH W. TYLER (1966) "What the assessment of education will ask." Nation's Schools 78 (November): 77-79.

MINAR, DAVID W. "The politics of education in large cities." (1967) Pp. 308-320 in Marilyn Gittell (ed.) Educating an Urban Population. Beverly Hills: Sage Publications.

MINER, JERRY (1963) Social and Economic Factors in Spending for Public Education. Syracuse: Syracuse Univ. Press.

MOLEY, RAYMOND (1964) The Republican Opportunity. New York: Duell, Sloan and Pearce.

MORT, PAUL R. (1952) Educational Adaptability. New York: Metropolitan School Study Council.

——— and ORLANDO F. FURNO (1960) Theory and Synthesis of a Sequential Simplex. New York: Columbia Teachers College.

National School Public Relations Association (1968) Computers: New Era for Education. Washington, D.C.

NORTON, JOHN D. (1966) Dimensions in School Finance. Washington, D.C.: National Educational Association.

PFEIFFER, JOHN (1968) New Look at Education. New York: Odyssey Press.

POUNDS, RALPH L. (1968) The Development of Education in Western Culture. New York: Appleton-Century-Crofts.

Project TALENT (1962) Studies of the American High School. Pittsburgh: Univ. of Pittsburgh.

ROGERS, DAVID (1968a) 110 Livingston Street. New York: Random House.

——— (1968b) "New York City schools: a sick bureaucracy." Saturday Review (July): 47-49.

ROSS, DONALD H. [ed.] (1958) Administration for Adaptability. New York: Metropolitan School Study Council.

ROWLAND, HOWARD S. and RICHARD L. WING (1967) Federal Aid for Schools. New York: Macmillan.

School Management (1966) "The national cost of education index – 1965-66." (January): 115-120.

SCHRAG, PETER (1967) The Village School Downtown. Boston: Beacon Press.

SKINNER, B. F. (1968) The Technology of Teaching. New York: Appleton-Century-Crofts.

SUPPES, PATRICK (1966) "The uses of computers in education." Information edited by Scientific American San Francisco: W. H. Freeman.

THOMAS, J. ALAN (1964) Administrative Rationality and the Productivity of School Systems. Chicago: Midwest Administration Center, Univ. of Chicago.

United States Office of Education (1966) Digest of Educational Statistics: 1966. Washington, D.C.: National Center for Educational Statistics.

VINCENT, HAROLD S. (1965) Quality Education. Milwaukee: Milwaukee Public Schools.

14

The Quality
of Urban Leadership

CHARLES R. ADRIAN

☐ THESE ARE TIMES of rapid change in American urban communities. The demands for effective leadership at the local level have probably never been greater. Yet, the danger implicit in despair confronts us, for when we hear intelligent, knowledgeable people say that so indispensable a world center as New York City "can't be governed," we begin to realize the extraordinary nature of the times. Changing economic patterns, life-styles, ecological patterns, and social expectations mean changing patterns of political power and hence changing leadership patterns.

It is difficult to assess the quality of urban leadership at any time. Indeed, it cannot be done other than intuitively. Community leadership is drawing today from among the best educated, most experienced, and most aware people in our history, people who have lived for a longer period of time in an urban environment than did their parents or grandparents. It is a leadership that is, with some significant exceptions, willing to tackle the formidable tasks of race relations, housing shortages, traffic congestion, smog, police-community relations, poverty, changing education needs, beautification, jobs, and whatever. Most in this cadre also recognize that they cannot lead alone, but that all the important issues require complex patterns of intergovernmental leadership.

But there is another aspect to the picture: today's urban leadership, is also, in part, inexperienced, naive, angry, uncooperative, uncertain, and frustrated. Like the other components of the "power structure" in all but the smallest cities, it is internally competitive, it lacks consensus on policy goals and procedures, and it is sometimes too strife-ridden to make decisions. Yet it is an American leadership, for it comes from nearly every segment of society, though not necessarily in optimum or equal amounts.

THE MOTIVATION TO LEAD

The quality and the source of urban leadership vary according to the method by which offices are filled. The type of person who seeks elective office differs from the type who accepts an appointed position and both of these, in turn, differ from the individual who plans to devote a career to community work through the merit system of selecting public servants.

Those who become political leaders through either election or appointment, I have described elsewhere as including professionals, hobbyists, advertisers, status-seekers, and ideologues (Adrian, 1968).

PROFESSIONALS

The professionals at the community level represent a group that is dwindling in supply and perhaps declining in quality. (The few professional politicians we have today tend to concentrate on state and national offices.) As community leaders, their importance has declined each year since the end of World War II. Prior to that time, the urban political machine and its highly visible "boss" represented a colorful part of American politics. In an age when the bureaucracy was based upon a patronage personnel system many young people entered the political arena, seeing it as a possible avenue for social and economic advancement. The traditional routes out of the ghetto by persons with leadership ability have been through the clergy, organized crime, professional sports, and politics. In the last generation, another route, that of education, has been added. This new pathway has gained greater use, as a result of the increasing number of public institutions for higher education with low tuition charges, and the expanding opportunities being made available to educated members of racial and ethnic groups.

Until a generation ago, the opportunities for movement into leadership positions and out of the ghetto were generally not connected with educational factors. Even in the case of the clergy, where some training was necessary, learning was narrowly restricted to a technical and, usually, not a very difficult sequence of topics. Whether Catholic or Protestant, the intellectual demands were relatively few, although they were sufficient to exclude those who lacked motivation. The avenues of crime, sports, and politics, however, required only certain personality types or natural skills, not formal training. Opportunities for higher education during the last generation, therefore, have tended to reduce the number of persons motivated to enter politics as a route toward social and economic betterment. Throughout American history, members of the middle class have found professional politics of little interest as a means for advancement. This path has been the preserve of the lower classes and the aristocracy. The traditional pattern will probably persist and, with the continued expansion of the middle class—an expansion certain to occur despite critics who insist nothing will be done to help the poor—the supply of persons committed to political activism will decline.

ADVERTISERS

As the percentage of Americans educated beyond the high-school level has increased, so has the size of the group of leaders who have emerged from the category of "advertisers." Most of the members of this group have been attorneys, insurance men, and realtors. They often can make important contributions toward community leadership. They are usually intelligent, highly motivated, and possessed of valuable skills. Characteristically, however, their participation in politics is short-lived. Their principal interest is to establish informal contacts with political leaders of the community which can strengthen their professional effectiveness. They are also desirous of making their names known to the general public so as to attract clients or customers. Once this has been done, they have little motivation to continue in political leadership positions. Indeed, if their advertising efforts are successful, the business attracted will be such that they can ill afford to spend much time away from their professional or commercial endeavors. Only a small proportion of them—generally those who hope to move on to higher office—continue in politics after a few years. Even so, their overall performance and importance are not insignificant.

HOBBYISTS

The third category is that of the "hobbyists." This group replaces, in part, the professionals and is a reflection of the fact that ours is increasingly a middle-class, educated society. Perhaps the most significant difference between members of this category and the old-time professionals is one of sex. The latter were invariably men, while the predominant portion of the hobbyists are women, generally well educated themselves, and married to persons holding professional and technical jobs in private and, to a lesser extent, government employment. These are the "do-gooders," a class of activists that has existed as long as there have been persons who have had leisure time as a result of their high-income levels. Such individuals have generally not been interested in gainful employment through public positions, but have been willing and even eager to accept appointments to boards and commissions or to other nonsalaried positions for which they can gain recognition and community acceptance. Another group of persons who must be included as hobbyists are members of the aristocracy. They are few in number, but they often become involved in community politics as leaders out of a sense of noblesse oblige. They have no interest in private gain through such activity. John Lindsay is an example. Many others could be pointed to, such as Theodore Roosevelt, who served as police commissioner in New York, Charles Taft as mayor of Cincinnati, and members of the Cabot and Lodge family, who once were active in the politics of Boston. Their counterparts in the middle-sized city were described by the Lynds in *Middletown* (1929), and by Williams and Adrian in *Four Cities* (1963). They continue to be found in the politics of small and middle-sized cities where members of the local aristocracy assume that community leadership is part of their responsibility to society.

The efficiency-and-economy movement of the second and third decades of the twentieth century sought to bring into community politics leaders who previously had avoided participation in such activities. By eliminating the partisan ballot in local contests and by changing the role of the councilman from that of errand runner to political and community leader (by electing him at large), additional numbers of upper-middle-class citizens from business, industry, and the professions were brought into local politics. These individuals, in other words, were willing to become involved in a political process with a relatively small amount of "politics" as that term has conventionally been interpreted by members of the middle class. A half century or so later, they are still contributing to community leadership. They now, after the conflict level has risen sharply from the days of the earlier decades of the century, find such participation perhaps less comfortable, but an established part of their community roles.

The hobbyists, separated from working-class people and ethnic-group members by income, education, and life-style, once tended to work largely in isolation. Today, they are seeking to improve their awareness of issues and their effectiveness, by joint action with various groups and classes. Hence, in 1967, high-status businessmen, labor and Negro leaders, educators, and others joined in forming the Urban Coalition, to find and pursue common goals for improving urban life.

STATUS-SEEKERS

"Status-seekers" are another important source of local political leadership. Generally, these are persons who have little education and modest, anonymous jobs, but who have a strong psychological need for recognition. Politics offers them an opportunity that does not exist through any other channel. Thus a truant officer or a fry-cook who has some political leadership talent may be elected councilman or mayor, though he is not likely to become a department head.

IDEOLOGUES

"Ideologues" constitute a fifth category on the local political scene. These persons have either strong commitments to specific public policies, or general perspectives on policy. Because the American political system makes the brokerage function a particularly important one for the role of political leader, ideologues do not usually rise to the higher positions of councilmen or mayor. Even in the lower echelon, they are likely to be something of a nuisance and embarrassment to a political organization, given their closed-minded approach to issues and contests. They usually feel greatly frustrated by the system, but are unable to ignore it because of their high awareness of political events. Their greatest leadership asset is their willingness to commit personal resources of time, effort, money, and skill on behalf of a given cause. The principal task of the senior leaders is often to channel these energies toward organizational goals (McClosky, 1960).

MINORITY-GROUP MEMBERS

A long-time source of political leadership has been the minority groups. The top city-wide leaders of the old-time machines were characteristically, though by no means invariably, Irish-Americans. This was partly because the Irish culture seemed to lend itself particularly well to the American political role, and the Irish-American politicians were skillful and plentiful. But the pattern also developed in part because the Irish were often able to serve as the lowest common denominator. While Polish-Americans, for example, would often refuse to support a German-American for the office of mayor, they would be willing to support a candidate with an Irish name. But the Irish were not the only group to advance through political channels. The political machine characteristically was operated by a coalition of different ethnic leaders. The ward or precinct leader would be a member of the nationality group dominant in that particular area of the city. Ward and precinct boundaries commonly were drawn along ethnic-group lines. (In this sense, the traditional practice in the old machines was akin to what some of the contemporary Black Power advocates seek.)

During the last quarter of a century, the old style of urban politics has been disappearing. Most of the big-city machines fell apart in the decade following World War II. By the close of the 1960's, the classical model of the machine could be found only in Albany and Chicago. With its decline came also a decline of the ethnic leader as an important figure in the community. There were many reasons for this development, among them the often successful efforts of efficiency-and-economy reformers to secure the election of the city council at large rather than by wards; the lessening of ethnic awareness and of ethnic subcultures, particularly after World War II; the professionalization of urban services and the consequent loss of available patronage positions; and the diminishing importance of the activities of city governments to the life-styles of the grandchildren and great-grandchildren of immigrants.

In the contemporary political system, the equivalent of the old-time ethnic leader is increasingly the black leader, especially the one leaning in the direction of black nationalism, or at least emphasizing pride of race, the importance of black leadership for black areas in the city, and of black spokesmen for the Negro community.

Although ethnic leadership has not completely disappeared, it is frequently blended into racial leadership. Puerto Rican and Mexican-American spokesmen, for example, often have difficulty in deciding whether they are primarily ethnic or racial leaders, with the contemporary vogue leaning toward the latter.

BLACK LEADERSHIP

The most talked-about aspect of urban leadership today is that of the black community. The long period of despair, apathy, and alienation relative to municipal politics came to an abrupt end in the 1960's with the rapid rise of not only a black leadership but also a new type of militant radicalism that some have criticized as racist.

When I began to make some preliminary inquiries in the early 1950's into Negro leadership in Detroit, a large city with a sizable Negro population, what I found was so routine and uninteresting that I abandoned the thought of making a systematic study of this nature. The leaders at that time were those traditionally involved in the white-black nexus for communications purposes. They were principally attorneys, other professionals, morticians, and insurance men. All of these played something approximating an "Uncle Tom" role, although it is likely that some of them did so only because they felt they had no other choice if they wanted to provide some meaningful communications links with the white community. Neither Detroit nor Los Angeles elected a Negro to the city council prior to 1953, and this was true in the latter despite the fact that all fifteen of its councilmanic seats are filled by election from single-member wards.

The transition in Detroit from traditional Negro political leadership to the contemporary pattern was a relatively smooth one. The

first black councilman was Charles Diggs, whose father had been a
mortician and political leader. Diggs later moved on to become the
first black congressman from the predominantly Negro thirteenth
district of Michigan in Detroit. He still serves in this capacity, interested in the welfare of his people and in civil rights; but he is not in
the forefront of Negro militancy. He previously played a similar role
in the Detroit common council when he was a member of that body.
Although Detroit was later to have other Negro councilmen, it had
none in 1968, even though Negroes made up nearly thirty-five
percent of the city's population. Los Angeles, on the other hand, had
three black councilmen at that time.

The development of Negro leadership participation within the
regular governmental structure in the United States has been
inhibited to some extent by the formal rules concerning councilmanic elections. These rules were not devised originally to minimize
Negro leadership and representation on councils, but rather were a
part of the efficiency-and-economy movement's strategy for weaken-
ing the representation of ethnic groups and, in general, the working
classes. The use of at-large elections, or the selection of councilmen
by districts much larger than traditional wards, had this effect. In the
second and third decades of the current century, when "reforms" of
this character were particularly popular, Negroes were not generally
viewed as much of a political threat or force. But a structural provision that had weakened the power of Polish-Americans or Italian-
Americans also worked to exclude or underrepresent Negroes on city
councils. The few opportunities for political leadership in public
office had the effect, of course, of discouraging the development of
strong Negro political leadership.

A study of Negro representation on the councils of nine of the
nation's largest cities shows the effect of formal structure upon black
leadership opportunities (Wilson, 1966). In March, 1965, Negro
representation on the Cleveland city council was proportionate to
that of Negroes in the total city population, while in Los Angeles,
Negroes were actually somewhat overrepresented, having about 13.5
percent of the population and 20 percent of the seats. (Both cities
employ election by wards or districts.) In the seven other cities
(Boston, Chicago, Cincinnati, Detroit, New York, Philadelphia, and
St. Louis), however, Negroes held less than a proportionate number
of seats. The greatest deprivation of blacks appeared to take place in
large cities with small councils elected at large. At the time of the
study, Boston, Cincinnati, and Detroit all had nine councilmen

elected at large and of these cities, only Cincinnati had Negro representation on the council and this consisted of one person, or one-half of what the black community would be entitled to on a proportionate basis.

This pattern of systematic underrepresentation on councils and on various other governing or advisory bodies, such as school boards and recreation commissions, caused growing dissatisfaction in black communities, particularly after 1960 when the quiet Eisenhower years ended. A third generation of black leaders began to emerge. The first had been the "Uncle Toms," often morticians and insurance men. They profited from their ties with the dominant white community and its financial institutions as well as from members of the black community, who were their customers, clients, and patients. The second consisted of those who were interested in civil rights, passive resistance, and peaceful coexistence. Congressman Diggs and, more prominently, the Reverend Martin Luther King, Jr., represented this generation.

The third or present generation of Black Power militants are reacting against nonrepresentation or underrepresentation just as have various ethnic groups in the past. They are, moreover, unwilling to accept virtual or stand-in representation by whites who are sympathetic to their problems and goals. In this point, also, their demands are similar to those made earlier by ethnic groups. While the latter were often willing, if necessary, to go along with a WASP or an Irish-American as a mayor, they would accept one of themselves, only, to represent their part of the city.

Much of the rhetoric of the Black Power movement has been expressed in provocative, threatening, and even insulting terms, but only a few of the leaders appear to be genuine advocates of insurrection or rebellion. In most cases, whatever language they may use, their goal is simply one that has long been familiar in urban American politics. (Or at least white persons committed to the traditions of liberal democracy desperately want to believe so.) They want black persons to be represented by blacks on councils, boards, and commissions, and they want that representation to be powerful enough to have some actual effect upon governmental decisions. And they are demanding that members of the black community be given a fair share (something that probably means different things to different people) of the benefits of governmental services and activities.

We are, thus, now confronted with a leadership that is seeking the same general payoffs from the system as have, in the past, been sought by leaders of the most recent and lowest-status immigrant group. The principal difference lies in the fact that Negroes are not recent immigrants (although they could be so viewed if we accept the idea that they have only recently been accorded the right to join the mainstream of American society). In addition, this effort by Black Power advocates comes at a time when the formal structure of city government and the general outlook of its leaders is such as to make the task of the new leadership more difficult than was that of earlier ethnic group spokesmen. The passing from favor and importance of ethnic-oriented politics has caused Black Power demands to seem more threatening and deviant than they probably, in fact, are. Furthermore, the absence of an old-time political boss with his sympathetic understanding or resigned acceptance of ethnic-group demands upon the system and his replacement by a middle-class mayor with "organization-man" values has made the task of communication and conviction even more difficult.

In the last few years, Negro leadership has begun to emerge at the very highest community levels, with the long-expected election of Negro mayors in middle-sized and large cities. In 1966, Negro mayors were chosen in both Cleveland, Ohio and Gary, Indiana. It seems highly probable that these are only the first of a large number of cities that will eventually elect Negro mayors. Cincinnati, Chicago, Detroit, Philadelphia, and St. Louis, for example, are likely to choose Negro mayors in the 1970's or 1980's.

Some Negroes (along with members of other previously apolitical or excluded groups) have been naive upon first entering into leadership positions involving cooperation with others. John W. Gardner (1968: 267) has noted:

> In the course of the collaboration the *less* powerful elements in the community learn some interesting lessons. Typically, they begin the collaboration with some astonishingly oversimplified notions about power. They are apt to think, for example, that there is somewhere in the community someone so powerful that he could press a button and solve most current problems if he were sufficiently generous of spirit. So the game is to find the man near the button and persuade him (or force him) to press it. They soon discover in the course of open collaboration with the so-called power structure that all power operates under constraints, and that there isn't any button that can be pressed to solve our most serious problems.

The coalition principle requires that minority groups be represented in the effort to solve community problems. And such representation is itself a step toward solving the toughest problem of all—effective dialogue between the black and white communities. When a crisis strikes, it is too late to begin the long process of building effective channels of communication. If there is to be fruitful collaboration between black and white leaders, it must begin and be tested in a noncrisis atmosphere. Then when trouble strikes, if it does, men who have worked together and trust one another can go into action together.

Evaluating the quality of community leadership is a game at which anyone can play, since there are no universally acceptable criteria or any quantitative methods for measuring it. However when a racial or ethnic group which has previously been apolitical and alienated comes to believe that political activity is efficacious in terms of its values and goals, the persons who make themselves available for leadership in this arena are likely to be of high quality, and probably person-for-person of higher quality, than are those who enter into similar roles in stable communities in which there is relatively little challenge in the political arena. In the latter case, the intelligent, upwardly mobile individual is usually attracted to other types of activities, such as leadership in educational or corporate institutions. If this is so, the probability is that politics is going to attract some of the ablest and most pugnacious members of the Negro community throughout the remainder of the twentieth century and perhaps beyond. Aspirations, a sense of efficacy, occasionally disappointing but not discouraging setbacks, and a route for status advancement, all offer reasons for politics to become an important and worthwhile activity for the most able black leaders.

YOUTH AS A SOURCE OF LEADERSHIP

Young people have received a great deal of attention in recent years as a result of their concern for the problems of the contemporary urban community. While both youthful and older writers have commented upon the feeling of young activists that the problems of the contemporary urban community are not being adequately met by existing or even proposed public policies, the leaders of various youth groups do not appear to have anything approximating a specific set of proposals to overcome these problems. It is easy to state what one believes to be unsatisfactory conditions; it is

vastly more difficult to provide leadership for those who would seek
to develop effective attacks upon such conditions.

Although many of those who are likely to be among the leaders in
seeking to deal with the problems of urban society in future years are
already active and "aware," they have not yet progressed beyond the
point where they can provide more than criticism. Some of them
will, no doubt, eventually be able to make useful contributions
toward societal accommodation to human problems. At the present
time, this is not the case. Those who list themselves as being among
today's "concerned youth" have reached the stage of recognizing the
difference between existing conditions and traditional American
ideals; they have not yet progressed toward identifying meaningful
methods by which the two may be brought together. In this sense,
today's youth, like that of past generations, represents a potential
pool of leadership talent, but not one that can yet replace existing
leaders by right of superior ideas, skills, or even motivation.

COMMUNITY MODEL AND
COMMUNITY LEADERSHIP

The source and quality of leadership in a community depends, in
part, upon the image that community leaders have of that community and of the purposes of local government. Oliver Williams and I
have developed a typology of local government based upon images of
its ideal function (Williams and Adrian, 1968: ch. 1). We have suggested that the principal images of government in the United States
today are those emphasizing the booster, amenities, caretaker, and
brokerage functions.

BOOSTER FUNCTION

"Boosterism" is the characteristic approach of chamber of
commerce managers and local businessmen's service clubs. The
emphasis is upon the growth of the size of the community population through an expansion in the number of jobs available in the
community. Governmental policies favor business and industry, and
the measure of success is in the population and tax base growth rate.
Leadership in this type of a community tends to be drawn from

among businessmen and industrialists, or from among lower-status persons who are willing to support the causes and interests of business and industry in return for a large tax base that will keep public charges low for working-class and lower-middle-class home-owners. The more affluent industrialists and businessmen often do not live within such communities, but only work there. A small or middle-sized city will often accept the booster approach to local government policy as an alternative to the higher taxes implied in a policy of keeping a community primarily residential. City managers, newspaper publishers and others may support the booster image because it is financially advantageous to them—a growing population means higher pay for the manager and increased circulation and income for the publisher.

AMENITIES FUNCTION

A commitment to the provision of individual and group amenities as the primary function of local government is found primarily in middle- to high-income suburbs. The emphasis is upon consumer rather than producer services. The goal is to secure amateur elective policy makers from among middle- and higher-income, well-educated persons. These individuals are expected to contribute their time and efforts to community advancement and, hence, to spend a term or two on a city council or other governing or advisory board. Leadership, in other words, is provided by persons whom Whyte has described as "organization men" (Whyte, 1956). Such men are themselves members of complex bureaucracies, either public or private, and they expect administrative leadership to come from a cadre of professionals. Typically, communities of this sort are operated under a council-manager plan. Professionalism is the hallmark, with the chief administrative leader being the city manager, specially trained for his job. He chooses professionals for department heads and other high-ranking positions. Other jobs are filled by professional and technical personnel who are selected through the merit system of civil service. Under such a system of leadership, communities tend to be neither reactionary nor very progressive in their approaches to public policy. Middle-class values predominate.

CARETAKER FUNCTION

Those committed to the desirability of a caretaker government in the local community believe that governments should perform only traditional functions at a minimal level of service. This approach is most commonly found in towns and small cities. It is the image held by many working-class home-owners who can barely afford to continue the payments on their mortgages, and by elderly retired persons living on fixed incomes. It is, however, also the approach to local government of many small retail businessmen (as distinguished from organization men in larger commercial organizations or in industry).

Political leadership in such communities tends to come from among the operators of small retail business, retired persons, or from career politicians who are willing to eke out a modest living through serving the community faithfully, honestly, and to the best of their rather modest skills and talents. The lifetime city clerk in a small- or middle-sized community is the prototype of such a leader. He has had a modest education, has few skills but a likable personality, is conscientious though noninnovative, and is dedicated to the community, its traditions, and its political style. His power as a leader stems from the fact that (a) he works full time on the local political scene, one of the very few persons who enjoys that advantage; (b) he knows a great deal about the traditions and past history of the community; and (c) he has a greater knowledge about laws, policies, and procedures than almost anyone else who is involved in the local decision process.

A community dominated by the caretaker ideology is likely to have little administrative leadership. The personnel system for non-elective individuals is usually based upon patronage personnel practices—the recruitment of professional and technical personnel through the merit system is likely to be viewed as too expensive to be afforded. On the other hand, payroll padding and schemes by which public funds are diverted to private use, both characteristic of the old-fashioned political machines, are not tolerated by the tax-conscious individuals who support this ideology. As a result, no more administrative and clerical personnel than are minimally necessary are hired, and those who do work for the municipal government generally serve with low pay and, consequently, are persons of modest training and ability. In general, they do not qualify as community leaders, but given the ideological commitment of those supporting this approach to community government, few leaders of any kind exist, for in the model the ideal is to have virtually no change and hence little need for innovation or leadership. The decisions which occur make only the most marginal increments of change, involving minimal risks and hence little talent in order to find the most desired choice.

BROKERAGE FUNCTION

The image of city government as a broker of services, negotiating among the various, often conflicting, interests and values of the community is the one most commonly found in large cities or in a city of any size in which there is considerable class, ethnic-group, or racial conflict. In this type of community, elective and appointive leadership must often be of a professional quality, with the career politician being highly aware of his responsibilities as a broker or diplomat and having relatively low levels of commitment to a particular ideology. In the past the brokerage image usually included the acceptance of a patronage approach to personnel administration, with the distribution of administrative leadership positions and professional and technical jobs based upon a proportionate allocation among the various classes, groups, or races active within the political system. Today, the brokerage concept is increasingly being amalgamated with a merit system of personnel administration and of recruitment for administrative leadership positions. This has occurred because of the increasing need for skilled personnel with specialized training to administer the ever more complex functions of local government.

The type of leadership a city or town enjoys also varies according to its age. An older municipality is likely to be dominated by either a broker or caretaker ideology, while the newer communities are more inclined to be dedicated to booster or amenities images. The most recent communities are also likely to benefit from the "Hawthorne effect," a euphoric, all-in-the-same-boat syndrome, which tends to produce a larger leadership supply than is to be found in older cities. This syndrome also encourages a greater effort to seek consensus in approaching the multifold problems of a new city (Roethlisberger and Dickson, 1939). The "Hawthorne effect," however, is temporary and, hence, has only a marginally permanent effect upon the political style of the community.

The rate of change of a community is an additional factor affecting patterns of leadership recruitment. Slowly changing communities tend to be dominated by the caretaker or brokerage images, while those undergoing more rapid change are likely to endorse either the booster or amenities images, both of which involve a commitment to a substantive set of policies implying growth, change, and "progress." In a slowly changing community, moreover, the most intelligent, best educated, and generally most able members are unlikely to devote their energies and talents to community leadership. After all, there is little reason to call upon the best minds and most forceful personalities if there are few questions to be decided. On the other hand, a community with a rapid rate of change is invariably confronted with a large number of controversial issues, and the decision process is likely to be highly complex. The issues are also likely to be highly visible and to be regarded by many residents as of great importance. The number of persons of leadership ability and the amount of time they are willing to devote to public policy issues is, in general, proportionate to the rate of social and economic change in the community.

PROFESSIONAL ADMINISTRATIVE LEADERSHIP

The most important source of professional administrative leadership at the local level continues to be from among those persons who have a middle-class orientation and who, sometimes unconsciously, desire power without being attached to the heavily publicized roles associated with the holding of public elective office. It is possible that individuals with this kind of interest are declining in numbers today. A serious shortage of qualified professional, administrative, and technical personnel at the local level has long been forecast (Municipal Manpower Commission, 1962). Certainly such persons are more commonly interested in serving in the federal bureaucracy or in those of the states that are most professionalized in their personnel systems. For the individual oriented toward professionalism in public service, the civil services of the federal government and a few states (particularly California, Connecticut, Michigan, Minnesota, New York, and Wisconsin), are far more attractive than at any municipal level.

Members of the municipal professional bureaucracy are accepted as leaders today by only a relatively small number of community citizens. Educated members of the middle and upper-middle classes tend to defer to the judgment of such persons when it comes to the development of community public policy. These individuals are themselves accustomed to working within a bureaucracy, whether public or private, and they understand the way in which it works. They recognize that professional specialists ordinarily have the most meaningful understanding of how to approach a particular problem and they accept the specialized and impersonal character of a bureaucracy because of their own experience within one.

Although professionalism in local administration was once generally accepted by educated middle-class persons of all ages, such acceptance has been on the decline in recent years. The professional bureaucrat has been criticized particularly by individuals who identify with the extreme left or right. Those on the right, largely committed to a politics of nostalgia, who see the ideal America as existing in the small town of fifty or one hundred years ago, are not able to accept a professional bureaucracy; instead, they continue to believe that the community can best be led by amateurs who devote odd moments to the administration of local affairs. In the small town of yesteryear America, the community leader was a jack-of-all-trades with relatively little formal training and no permanent commitment to public service. Right-wing extremists have never been able to accept the idea that such a role is now outmoded and cannot be made appropriate to the needs of contemporary urban society.

Members of the New Left, similarly, do not accept the professional bureaucrat as a suitable leader of the community. Their interpretation of this functionary in a mature social system is, in general, similar to that described by the late nineteenth-century German social scientist Max Weber: one who is technically competent but unable to innovate and who tends to see established procedures as ends in themselves. The assumption is that emerging problems will have to be met by persons outside of the professional bureaucracy or by those who are within it but refuse to accept its basic value pattern.

Young people continue to be interested in community public service and leadership. A few years ago they tended to avoid entrance into business bureaucracies, although that had been a popular avenue for advancement after World War II and through the Eisenhower years. Today, they are less interested in, or willing to enter, the bureaucracy at the federal level, a type of activity that was extremely popular a generation ago. They do, however, seek to move into positions of leadership in government service in relation to the anti-poverty programs at all levels. The young idealist, in other words, is attracted to federal service when his contribution can be made in the urban ghetto, and particularly in the black ghetto.

The municipal personnel officer seeking to recruit young persons into potential leadership positions in municipal government is finding applicants of a very different type from that of a decade or more ago. Current applicants are, with increasing frequency, members of the New Left, or the black militants, or those who are essentially committed to the basic attitudes of such groups. They are fundamentally unsympathetic to the traditional values of a bureaucracy and, indeed, are often hostile toward the Weberian bureaucratic values. They are entering into the system dedicated to changing it. The fact that Weber did not believe the ideology of a mature bureaucracy could be changed does not deter them from their determination. Those who are today being recruited into junior leadership positions are asking increasingly for a new type of professional administrator, one who is committed to the development of new ideas and procedures. Whether such a community leader is desirable is perhaps secondary to the question of whether or not he is a possibility. The overarching question in this regard is whether members of the most recent generation of persons entering into community public service will be able to repeal the well-established rules of bureaurcracy. If they cannot do so—as seems likely—will they continue to attempt to make meaningful contributions to the problems now confronting urban society? Irrespective of what disillusionment some may suffer, it seems likely that the concern and high motivation of many members of the current college generation in respect to the problems of urban society will provide a reservoir of leadership talent for the administrative machinery needed in the next generation.

FUTURE PROSPECTS

The pool from which we are today drawing our urban leadership talent is broadening and deepening—but so are the problems confronting that leadership. This chapter has presented a discussion of the potential sources of talent for this role of service at the community level. To be complete, a discussion of the quality of urban leadership in America should also include an examination of the manpower sources for strategic offices to be filled in state governorships and legislatures, in the White House and on Capitol Hill. It is perhaps in these places that the really significant policy decisions of the future will be made and the most able leaders will be needed. But it seems reasonable to conclude that leadership at the local level will continue to provide the final specifics in the shaping of urban public policies and to dominate the all-important responsibility of deciding upon the local agenda of items for action or inaction. The quality of this leadership in the future seems likely to be high. Whether the character of the problems to be faced is such as to make accommodation to them beyond the grasp of the leaders is a question of the greatest importance to contemporary society. Unfortunately, the question is one that the social scientist is not equipped to answer.

REFERENCES

ADRIAN, CHARLES R. (1968) "The autonomy of local political leadership." Pp. 1-9 in Richard W. Taylor (ed.) Political Leaders in Action. Iowa City: Sernoll.

GARDNER, JOHN W. (1968) "City Hall can't go it alone." Public Management 50 (November): 266-267.

LYND, ROBERT S. and HELEN M. LYND (1929) Middletown. New York: Harcourt, Brace & World.

McCLOSKY, HERBERT et al. (1960) "Issue conflict and consensus among party leaders and followers." American Political Science Review 54 (June): 406-427.

Municipal Manpower Commission (1962) Governmental Manpower for Tomorrow's Cities. New York: McGraw-Hill.

ROETHLISBERGER, F. S. and W. J. DICKSON (1939) Management and the Worker. Cambridge: Harvard Univ. Press.

WHYTE, WILLIAM H., JR. (1956) The Organization Man. Garden City: Doubleday.

WILLIAMS, OLIVER P. and CHARLES R. ADRIAN (1963) Four Cities. Philadelphia: Univ. of Pennsylvania Press.

WILSON, JAMES Q. (1966) "The Negro in American Politics." Pp. 431-457 in John P. Davis (ed.) The American Negro Reference Book. Englewood Cliffs: Prentice-Hall.

15

The Quality
of Urban Management

NATHAN D. GRUNDSTEIN

☐ WHAT FOLLOWS IS a conceptual essay. It is an endeavor to link the urban environment to urban management. The linkage is developed in a manner that demonstrates how the former generates certain design problems for the latter. These problems are indicators of the quality of the urban environment as it is encountered by urban management. The approach visualizes urban management as searching for certain attributes in relation to its environment. In effect, there is a quality of urban management that can be delineated in relation both to the quality of its environment and to its design. The environment of relevance for the quality of this management is the social. The resort to an equilibrium model of the latter provides the reference base for an explanation of the quality of the urban environment that is of concern to urban management. The quality of the latter can then be expressed as a distinctive type of management which can be given design components and assigned social functions with respect to the environment in which it operates.

THE CRITERION PROBLEM

No pretense of any analytic treatment of the quality of management can be sustained without first disposing of the criterion problem. For this purpose, I shall begin by specifying the criterion and then probe both the logic in support of it and its relevance for urban management.

The quality of management is a phrase that refers to something other than technological efficiencies in the performance of functional activities, or to measures of performance in program management, or to the rationality of problem-focused organization decisions. The social function of management is to achieve desired environmental states. Its quality lies in its capability to select and attain these states.

The parallel to the logic supporting this criterion is that found in the distinction drawn by Barnard between *effective* and *efficient* action by individuals and cooperative systems (Barnard, 1947: 19-20, 32, 56-57). Human action is accompanied by consequences that are either intended (desired or sought) or unintended (unsought or undesired). An individual action is "effective" when it attains "a specific desired end." It is "efficient" when its unintended consequences are "unimportant or trivial" in relation to the desired end. Human action can be efficient but not effective, a situation which occurs when the unintended consequences of action that do not achieve a desired end nevertheless satisfy the desires of the actor. What we have here is the idea of a net effect of human action that can be described in terms of dual parameters of efficiency and effectiveness.

Barnard, at the same time, interconnects environmental states and the purposes of human action (both individual and social). In actuality, he is able to transform teleology (purpose) into nonteleological environmental states of existence-action. These states are time-related constraints on the achievement of human purpose. The physical environment is always a perceived environment; that is, it is contingently human. The individual actor decomposes it into mentally connected constraints on the achievement of his purposes. Action has as its object (but not as its purpose—i.e. action without teleology) the achievement of those "changes in the physical environment" that the individual actor recognizes "as limiting the accomplishment of purposes." (Teleology is not part of the physical environment of the actor.) A change in an environmental state is a change in a perceived set of time-related constraints on the achievement of individual purpose. The dimension of time is not in the physical environment but in the actor. The environment is present or future only in relation to the remoteness from the present of the purposes of the individual actor. As against the physical environment, the individual actor does not have an infinity of purposes. His purposes are contained within the range of actions that are required

THE QUALITY OF URBAN MANAGEMENT [397]

for types of environmental changes. Purposes remote from the present are achieved only through "roundabout action."

Referring back to the criterion of the quality of management, but now assuming the existence of organization, the capability attributes of management are not the equivalent of its organizational forms and operational procedures. While management can proceed through a variety of organizational forms—function, program, project, decision —these are but choices of social forms for managerial manipulation and control of a present environmental state in relation to a desired, time-related state of the urban environment. Organizationally based capability attributes have to do with managerial effectiveness in the selection and achievement of desired environmental states. One might ask whether assertions such as the following can be made about the quality of management:

A capability to move from lag-time to lead-time in dealing with undesirable environmental states;

A capability to cut down on response time to environmental changes;

A capability to bound ill-structured problems;

A capability to maintain a field of choice over time;

A capability to adapt the structure of organization to the strategy of management;

A capability for cumulative learning and for auto-intelligence.

Organizations are a type of cooperative system and as such they may require certain capability attributes as a necessary condition for being effective. Again the logic of Barnard is relevant. For him, a cooperative system to exist as such must incorporate distinctive types of action that represent three different capability attributes: (a) distributive action, (b) cooperative facilitating action, and (c) cooperative maintenance action (Barnard, 1947: 32-33). Here the quality of management is focused on the selection and achievement of an internal environmental state of the organization itself. The relationship between individual action and cooperative effort constitutes this environment. It is, however, an unstable one.

Instability, in fact, turns out to be the heart of the matter, and coping with it the test of the quality of management. There are multiple sources of instability: that generated from within the internal environment of organization; that generated by managerial efforts to introduce adjustments within this environment; that generated by changes of the physical (external) environment of organization; and that generated by the very success of cooperative effort itself—namely, a reconstitution of purpose and a reperception of action possibilities.

CRITERION RELEVANCE TO
URBAN MANAGEMENT

The relevance of the stated quality of management criterion to urban management will be dealt with here by examining the Sayre-Kaufman in-depth study of New York City in which they present an equilibrium model explanation of the governance of that community (Sayre and Kaufman, 1965). With this model as the base for an explanation of the administration of the city, three possibilities for the quality of urban management emerge: a zero capability, a negative capability, and a positive capability. In the first instance, environmental states emerge independently of urban management. In the second, an environmental state emerges which is either not sought or counter to that sought by urban management. In the third, an environmental state emerges which is in some measure the causal product of the actions of urban management and one sought by it.

Given a tendency toward an equilibrium state (a counterpoise of social forces), the persistent and enduring tendency in New York City is toward a zero level in the quality of urban management, one that is without any capability to select and achieve a desired environmental state. The urban managers are not free agents; and the manager who is not a free agent is in danger of extinction. To avoid such a fate, he expends his energies on getting control of his organization (the internal environment) and manipulating the environment external to his organization. His universe of action is ambiguous and full of uncertainties. And as we know, risks are inherent in uncertainty. The root uncertainty is whether to adopt a personal strategy of leadership or one of accommodation and subordination in the management of the urban agency. Each course has its dangers, its inconveniences, its costs and rewards. Everything, moreover, works toward a counterpoise of forces. There is much calculation; there may be high levels of activity; there may be a great searching for options; there is a diversity of social strategies for a variety of situations—but cumulatively the tendency is toward an equilibrium state, a zero-level of capability as the quality of urban management. The foregoing exists whether line or overhead agencies of the city are involved. There is a ceaseless oscillation from one social equilibrium to another.

The phenomena of administration in New York City disclose that there is in actuality a double game taking place in urban management. And here we get to the distributive processes of cooperative systems discussed by Barnard (1947: 22-23, 253-256). The individual contributions which sustain cooperative effort rest on the satisfaction of personal purposes by organizations. These purposes are satisfied through a variety of utilities which organizations create and distribute among their members. Such utilities have value in relation to the motivations of the individuals who are organization members. But these are not the only recipients of utilities or the only utilities generated and distributed by organizations; and herein lies the double game.

The agencies of management are a source of utilities or benefits (contracts, appointments, promotions, status, salaries, cash benefits, policies, etc.) that serve as motivational inducements for the maintenance of political parties, electorate constituencies, and the organized bureaucracies of the city. The motivational satisfiers for the personal purposes which sustain individual contributions to cooperative effort are generated by the agencies of urban management for the benefit of members of those organizations that are part of the external environment of urban management. There is an inversion of the capabilities for selection and achievement of an environmental state, because there is an inversion of the incentive base of organization. As it works out, the agencies of urban management in New York City generate more negative incentives than positive inducements for their own membership.

Does the ceaseless oscillation from one social equilibrium to another that marks urban management in New York City mean there is no managerial capability for movement as between environmental states? It is clear from Sayre and Kaufman that there is only a contingent and residual capacity for innovation and change on the part of urban management. What takes place is more like perturbation around a base point. Innovation increases uncertainty, while accommodation within the established framework of things decreases uncertainty. And the avoidance of uncertainty dictates the choice of management strategies. "The path of innovation is rocky, twisting and full of pitfalls." Although innovations do occur, they are limited in effect, narrow in scope, selective in application, and isolated in occurrence. Can we even say that change takes place where nothing cumulative, simultaneous, extensive, nontraditional, profoundly different, or of great magnitude in the way of innovation is to be expected from urban management?

If change comes, it does not occur because urban management is the dominant influence in the system of social choice and action. In fact, as Sayre and Kaufman put it, "nobody 'runs' New York." The city sort of emerges, accretes, and secretes. There is a randomness to it; but where is the intelligence? The former is easier to explain than the latter. The explanation is of an aggregate kind. The community (that is, the most comprehensive and inclusive external environment of urban management) generates "proposals for new activities and programs" out of its infinite diversity. These proposals go through a lot of social screening and revisory and obstacle courses, but there are too many of them to be completely filtered out. A residue remains of which change is the product. Urban management thus functions as an adaptive response structure to the external environment. There is a direction of causality in the relationship: it is the external environment that compels adaptive responses by urban management.

In the end, therefore, urban management is left in an ambiguous and unstable position so far as its capability to select and achieve an environmental state is concerned. The shifting diversity of the external environment (an open system) prevents the maturing of the tendency towards an equilibrium state within the structure of urban governance (a closed system). Urban management will on occasions exhibit a negative capability in some degree, and at other times a positive one. These are not equal possibilities; the degree of the former is likely to be larger than that of the latter. Whether urban management is in a position to exhibit the one or the other is more the outcome of contingency than of choice.

DESIGN PROBLEMS OF URBAN MANAGEMENT

At this point of analysis it is possible to relate the quality of management to fundamental design considerations. A large number of contingency elements are present within the structure of urban management, within its external environment, and among the interfaces between the two. Contingency elements introduce a probabilistic quality (Monte Carlo effect) into urban management, with very complex interdependencies among a large number of individual instances of management decision. An equilibrium model explains these interdependencies in terms of the interrelationship of the actions of social aggregates (party, constituency, agency, organized bureaucracies, etc.) to the decisions of urban management. Thus the interdependencies have a particular social form, one which tends to constrain the field of choice of urban management and to reduce the net effect of its decisions. Not only is there an inertial effect toward a zero capability state, but any positive capability state on the part of urban management exists in an unstable condition. Individual human motivation provides a base that underlies and supports all of this.

What we have is a demonstrable tie between the design of urban management and its quality. To change the latter it will be necessary to change the former. The reference criteria for the design changes are known. They have been provided by the analysis of the quality criterion of management and criterion relevance to urban management. As they have emerged in the analysis, these reference criteria are: (a) reduce the contingency (probabilistic) aspects of urban management; (b) enlarge its decisional field of choice; (c) introduce stability into its positive capability states; (d) lessen the magnitude of the inertial effect toward a zero capability state; and (e) remove the dysfunctionality between its cooperative effort and the motivation of its manpower.

To deal at the level of design rather than at the level of administrative procedures, organization revisions, and discrete instances of functional adaptations is to fix on a strategy of change in the quality of urban management. As we have previously noted, the criterion of quality is a capability to select and achieve desired environmental states. At the level of design we are dealing with effectiveness, which is the nub of the quality criterion. At the nondesign level, where changes in administrative procedures and revisions of organization dominate, an environmental state is beyond the reach of urban management. The technological efficiencies that yield incremental improvements within the paths marked out by existing management activity, while desirable, relate to the efficiency of urban management. It is possible, however, for the technological efficiency of urban management to become better while the environmental state becomes worse. The latter exists in actuality as a matter of common knowledge, for it is a matter of widespread urban experience that whole sociophysical substates of an urban complex enter into decline and decay independently of the technological efficiencies of urban management. On the other hand, the regeneration of these undesired sociophysical substates is not possible without an effectiveness capability on the part of either urban management or a substitute for it. The quality of management is not without its technological content, but this content is not the efficiency technologies of management.

A resolution of the design problem involves the working out of another type (or style) of urban management. It also involves a reperception of it within another structural form. The latter imposes a differentiation in the manpower capabilities for urban management. It also leads to different occupational mobility paths, and it is these that can be related to manpower motivation in the design of urban management.

TIGHTLY COUPLED URBAN MANAGEMENT

At one time urban management had the option of existing without urban planning. It is not so much that the option no longer exists, but that urban planning in its contemporary form requires a type of urban management, a type that is to be found in the coupling of the two. Essentially what has taken place, from an historical standpoint, is the emergence of the stage of managerial urban planning. It is a state in the historical development of urban planning that requires a certain capability level (quality) on the part of urban management. What has come about is a reversal in the positions of urban planning and urban management in which the capability levels of the latter have been outstripped by those of the former. There is now a very difficult linkage function to be worked out between the two that is critical for the quality of urban management.

Some historical aspects of urban planning are relevant to the foregoing. Since 1900 there has been no radical change in problem awareness with respect to the city. Put another way, by the beginning of the present century nearly all of the urban problems of which we are aware today were recognized by those who comprised the social movement favoring city planning. The actual social recognition of urban planning as a formally needed function of city government was generated over a twenty-year period extending from about 1890 to 1910. In the early literature on the city written during this time span, all, or nearly all, of the problems that we currently talk about as associated with the city were recognized and listed as problems for planning. Looking back, we can say that urban planning was the first of the professions confronted with what we know as contemporary urban dilemmas. Its difficulty as a profession was that it lacked the analytic base and science-related technology for dealing with these problems at that time. What planning has done is not so much change the problems of which it has spoken, as change both their analytic base, the science-related technology to be applied to them, and the perceptions of them as technical problems of urban planning. In short, it has been engaged in a period of developing its science.

After city planning became a recognized urban function (about 1910), the question of who would be its patron became critical. In the first stages of its development, it was decided that urban planning would not be allied with urban management. This was not a consciously and centrally determined choice; it just developed this way in terms of the values of the period. Since then we have witnessed one of the long-term trends of structural change in urban planning: a shift of its patron. Urban planning, in other words, started with a general independence from urban management, that is, with a focus on the general urban community and a civic base, and then moved toward a gradual identification with urban management.

Although aware of a series of urban problems, urban planning did not have a science base adequate to deal with them. Its science base is one of urban change and development, but in the early years of urban planning there was no science that could be applied to these factors or forces in the city. Calls from within the profession for the building of such a science date back as far as 1902. Since then, and particularly since the 1950's, efforts have been made to construct a science base for urban planning capable of dealing with urban change and development.

Initially, urban planning was viewed as consisting of a set of a priori, design-based components that in fact represented very static ideas about the city. These included the civic center notion, the neighborhood notion, and the parks and recreation notion. These concepts were not tied to urban dynamics but to the problems of social reform of the period. The critical influences on planning in its early days were the sad state of municipal government, the underdeveloped character of the social sciences, and the very low state of local politics, namely, its corruption. Urban planning, in its early days, sought for civic stability and the advancement of those social reforms based upon what we would characterize today as elevating concepts of human character. That is to say, it sought to change the motivations of persons in relation to urban government and to their city. However, it possessed no arsenal of concepts and technological tools for accomplishing this.

It did advance the idea of urban rehabilitation, which was then meant to be a change in the environment according to a preconceived plan. This emerged later as the comprehensive plan for the city. But it had not thought out a way of devising such a plan in its early days, nor has it ever conceived a way of making such a plan operational. The notion of regulation of development is also to be found among the ideas of early urban planning. This might mean either the regulation of private development according to an ideal plan, or the regulatory control of functional land use through zoning and eminent domain. The question of control is one on which urban planning has foundered. The tools of zoning and eminent domain have never proved adequate to this purpose, and the tasks of regulatory planning still remain one of the large undeveloped science areas of urban planning itself. The history of zoning is really the history of the ineffectiveness of urban planning. The end result was the development of legal tools for the control of city development which were ineffective, and the development of the independent city plan commission as the structural form for relating planning to the municipality. Both of these proved inadequate. The task of building a science of urban planning had to wait upon the development of analytic technologies which, in turn, had to await the development of social science theory and the bringing of this theory to bear upon the phenomena of the city.

Meanwhile, throughout this period, a parallel movement relating to the governance of cities had also been in the making. Efforts to achieve a science basis for urban management had been taking place simultaneously with the attempts to develop such a basis for urban planning. By 1887, the notion of a science of management, or of administration, as it was then called, was fairly well established as a feasible social and professional aspiration. What happened to the science of city government? It was first conceived of as operational routines set in the framework of the politics of administration. The two things that had to be done in the early years of city administration in the twentieth century were inadequate for the tasks of urban planning as we know it today. One was the routinization of urban operations. Unless one reads the early literature, he can have absolutely no idea of the disordered, and what I would call the pathological, state of urban government during those years. A great deal of time has gone into achieving the routinization of certain operations. The introduction of this functional routinization into municipal administration represents the dominant management technology of a period of city government that is now past. Cities have learned the technology for routinizing a whole list of operations and of constructing functionally constrained bureaucracies around them. This, however, has not been sufficient for the tasks of urban planning. Second, while it was historically necessary to find room for urban management in the corrupted political wilderness of municipal government—and for this task a great deal of structural reform had to be infused into the government of the city—it was an error to set urban management within the intellectual framework of the politics of administration. This is a framework without a science orientation and an analytic base that can contribute to the advancement of any science-related content for urban management. The historical revenge of management science is that it has developed independently of public administration, and has turned out to be the dominant science upon which both urban planning and urban management have jointly drawn.

The social phenomena, the structures, and the behavior identified with the domain of the political in the governance of the city have always been real for both urban planning and urban management. They cannot, however, be placed outside the science-technology base of either. Every science has certain social conditions that are necessary for it to be used effectively, and the problem of working these out is part of the problem of the science itself. It is possible to build the analytic content of a science independently of the human actors within the user institutions, but the implementation of that science brings to bear certain social conditions upon its utilization. The tighter coupling of urban planning with urban management compels a reconsideration of the social conditions generated by the politics of city governance in relation to the implementation of those respective sciences.

The start of a research and development period of urban planning began in the late 1950's, when it initiated a stage of conscious building of its science base. A result of this R & D activity has been an effort to fit urban planning into the structure of science and to relate it to patterns of human organization for the use of science. Fundamental ideas of science were brought to bear upon the content and the activity of urban planning, both to determine to what extent they applied, and to discover to what extent urban planning required its own special set of science-related concepts. This new stage in the development of a science of urban planning coincided in time with a stage in the development of management science itself. What took place, as both began concurrently to focus upon the city and to attempt to deal with the dynamics of urban change and development, was the establishment of analytic relevancies between the concepts used by each and the discovery of joint problems and foci of mutual interest. Urban planning was linked with management science in the creation of an intellectual infrastructure for its own science in order to deal with urban change and development, but this linkage was not the equivalent of a coupling with urban management.

THE STRUCTURE FOR MANAGERIAL URBAN PLANNING

The coupling of urban planning with urban management for purposes of urban choice followed upon legislation by the national government's subsidizing urban planning and fitting it into a federally coordinated program for urban change and renewal. Within the structure of the local economy, the two were coupled to assume a social function (assist in a managerial strategy of urban development) assigned by an external economy. The cash flows to the city generated by the federal financing of urban planning, and the interdependence of planning and urban management within the framework of national policy, forced the former into a closer relationship with the latter than had previously existed.

Essentially what has taken place is that the professor has entered into the stage of managerial urban planning. In the historical development of urban planning, we can speak of a sequence of stages or changes in the distinctive focus of its professional content—physical planning, land-use planning, comprehensive planning, socioeconomic planning, and the like. The stage of managerial urban planning is the latest in the growth of the profession. It involves changes both in its professional content through the spillover effects of management science on urban planning, and in its relationships with urban management. These changes provide a point of entry for altering the capability levels of urban management, (or they represent an alteration in the capability levels of urban planning). The difficult linkage function that now has to be worked out with urban management involves the following.

CHANGE IN THE SOCIAL LOCATION OF URBAN PLANNING
IN RELATION TO THE STRUCTURE OF COMMUNITY CHOICE

Managerial urban planning should be able to get in between the social components of an urban complex and the political and economic components that act as determinants of community change and development. It can be assumed that each of these is characterized by social structure. Managerial urban planning should have the competence to cross the boundaries of this structure and enter into processes of deliberation as a dispassionate resource for each of these sets of components. It is not without a teleology, but its real value as a resource is that it has an analytically based view of the city and a large inventory of diagnostically oriented information about it. It thus can be utilized as a fulcrum for the social exploration of action possibilities. Within the structure of community choice a clarification is needed between multiple social perceptions of the urban environment and the achievement of future oriented purposes. The utility of managerial urban planning with respect to the structure of community choice rests in its capacity to change the order of intelligence brought to bear on the decisions of urban management.

A CAPACITY TO ASSEMBLE AND APPLY A TECHNOLOGY OF SEARCH AND
INFORMATION ANALYSIS TO FACILITATE URBAN CHOICES

Managerial urban planning supports a change in the allocation of the resources of urban management as between its thinking (intelligence) and its operational activities, and an increase in the level of investment in its analytic decision capabilities. It is a matter of building an adequate diagnostic and prognostic base for the decisions of urban management. The criterion of adequacy imposes the constraints arising from a rule of economy of managerial effort. At any point in time there will be limits on the magnitude, simultaneousness, and purposes of what can be done by urban management. It is not, therefore, the ideal of a search and information analysis system of infinite capabilities that is relevant. Managerial urban planning can neither avoid looking at the city as a whole system nor escape doing some across-the-board studies of it. Predictions will force it to clarify the time-distributed interactions of a variety of factors and dimensions associated with urban phenomena. Specific action recommendations will require a depth of information detail.

CAPACITY TO CHOOSE WITH REFERENCE TO THE WHOLE CITY AND WITH REFERENCE TO A SET OF ITS TERRITORIAL SUBUNITS

A dual capability is involved here. Managerial urban planning cannot deal with the city only at its most aggregate level; it must also be capable of dealing with it at levels of subaggregation. Social system differentiations which are territorially distributed within the city compel a strategy for dealing with these differences. The distributive effects of decisional consequences generate problems of equity as between territorial subunits. Particular physical and economic characteristics of an area vary in relation to development possibilities (investment and land use) and this fact also supports territorial differentiation at a subaggregate level. Three real life illustrations are given below:

(1) Area 2 of City P:

The area was in obviously poor condition, blighted, and in need of clearance. The master plan indicated that it should be developed industrially. A steel mill was located in the area and the owning corporation wished to expand the plant. Thus a land developer was ready before a planning decision for the area was made. There were possibilities for improvement of the economic base of the city and for considerable slum clearance through industrial plant expansion.

(2) Area 10 of City P:

This was a tremendous and complex physical area characterized by a downgrading of its commercial portions. The original master plan had set aside an excess of territory for commercial development; however, residential use had been creeping in. Subsequent market studies showed that a large commercial area could not be supported. It was therefore necessary to revise the master plan classification of land use.

Local business had been pushing for area redevelopment and there was no problem of private moneys being available for this purpose. With a smaller business area, however, all of the merchants who had been in the area could not now be accommodated. Moreover, the land values of the developed area rose to a level where they were no longer affordable by the smaller merchant.

To deal with the physical magnitude of the area, the development agency tried to break it up into smaller sectors and build in phases, but found it was not practicable to do so. It consequently had to revert to doing the whole area.

(3) Area 16 of City P:

The district was created especially for a projected sports stadium, although the master plan designated this area as industrial and cultural. However, there were now revised economic base considerations involved. The change in the land use and the stadium project were part of longer-term plans for the development of the waterfront of the city. Accommodation to an expressway location had been worked out so that there were no conflicts with transportation plans. The changed land use meant problems of business relocation from the area.

In relation to the city as a whole—at its most aggregate level—managerial urban planning is confronted with the task of generating and adhering to an overall plan, rather than merely conforming to a master plan. Comprehensive planning plus a strategy over time in relation to a normative conception of the city is part of the capability of managerial urban planning.

CAPACITY TO BUILD A PLAN-RELEVANT, PROGRAM-RELATED ARRAY OF ALTERNATIVES AS OPTIONS FOR MANAGERIAL CHOICE

Urban management is faulted not so much by a lack of options as by the kind of options which are lacking. It tends to be without means of escaping an opportunistic approach to urban development. There are always some immediate needs in the public sector pressing for disposition, and for these urban management is open to opportunistically based alternatives that are politically persuasive. Moreover, developer entrepreneurs who are also out searching for options pressure urban management for projects that will realize their perceived gains.

Managerial urban planning can generate a shelf of project alternatives with information regarding the consequences of each, as well as the relation of each to known problems and goals of urban change and development. While not programmed in time, these projects can reflect a plan for the city. For urban management, the heart of the matter is to be able (a) to identify the particular areas which require public action to meet community objectives, (b) to design the appropriate action for these areas, and (c) to evaluate such action as an alternative.

THE STRUCTURE FOR URBAN MANAGEMENT

Managerial urban planning supplies the reference base for a type of urban management that is characterized by its coupling to urban planning. What is of concern is the level of capability at which urban management must function in order to be effective when so coupled. A coupling that enables it to initiate, maintain, or reverse a tendency in relation to a desired environmental state will represent an operational expression of its quality. In relation to an equilibrium model of urban choice, such a capability involves an alteration of the direction of causality as between urban management and the external environment. The fundamental question involves assessing the possibilities of the coupling between managerial urban planning and urban management for minimizing the contingency aspects of the latter and for enlarging its decisional field of choice.

Managerial urban planning does not reduce the structural complexity or simplify the external environment of urban management. Whatever effect it might have on the relation of urban management to its external environment must be explained in terms of its contribution to the internal environment of urban management. Its function is to help tie together the analytic and social resolutions of questions of urban choice for urban management. An interrelation of structure, information, and manpower is involved, together with the changes that will have to be made in the internal allocation of resources to achieve whatever level of investment in the foregoing may be required for tightly coupled urban management.

The structure of urban management at which the analytic and the social aspects of urban choice are worked through can appropriately be designated as the politico-managerial. The relationship of this structure to planning is explicable in terms of information and decision processes. Planning data are part of an information flow. Managerial urban planning breaks through the isolation of earlier planning from the social components of the external environment. Through expanded search techniques, it enlarges its touch points with this environment, and can engage in an information exchange with components within it. Planning data, however, have to be built into politico-managerial decision processes. In this connection several questions arise: Are these data in a form suitable for this purpose? How far is the politico-managerial structure of urban management prepared to use them? What are the implications of their utilization as a reference base for choice?

The politico-managerial decision-makers are free to decide to what extent their choice should be influenced or changed by the data and alternatives of planning. Do these decision-makers have any clear and explicit notions or criteria of what is good or bad, or useful or reliable, information for decisional purposes? Put differently, can the decision processes at the politico-managerial level also be said to be essentially analytic in their orientation, so that in this respect they are not different in kind from the processes associated with managerial urban planning? The object of the coupling structure is to facilitate a conjunction of two varieties of analytic thought about the management decisions of the city. These will exhibit differences in terms of their respective information content, information sources, modes of analysis, and, above all, valuational bases and perceptions of social reality. At the level of the politico-managerial, we are beyond that of problem analysis. We are at that of reality formation, where all of the judgments that precede a constrained, problem-focused mode of analysis take place. (Vickers 1965: ch. 4). At this level, all of the information that managerial urban planning can provide—its normative reference base, its diagnostic and prognostic data, its plan-related options, its inventory of alternatives—does not cover the total range needed for managerial decision-making. The point is not that managerial urban planning should provide more information, but that it is not privy to the other kinds of data essential for decision at the politico-managerial level.

At this level the political encompasses knowledge of the dispersion of power, of the conflicts of personal and social interests, and of the social systems in existence within the external environment. Knowledge of this kind represents information about a set of variables in the environment of urban management which, while they are nonquantitative (leadership, influence, values, goals, etc.) in the form in which they are known, are often the most significant for decision-makers at the politico-managerial level. These variables supply criteria by which the distributive effects of urban choice are assessed; it is in relation to them that the weighing and evaluation of planning information takes place. For any aggregate of decisions, there is some mosaic of reference which is a source of generalized guidance, constraint, and value. In short, there is an inescapable analytic logic of choice that must be engaged in at the politico-managerial level of decision.

Urban management needs something better than on-line trials with urban choices. The advantage of tightly coupled management is that it enables planning to operate as part of an internal feedback system required by the politico-managerial level of urban management. What the latter needs from managerial urban planning is (a) some relevant numbers and a decision base for assessing the non-quantified variables that are deemed significant for politico-managerial choice; (b) the transformation of planning data into information significant for decision-makers; (c) assistance in the formulation of some decision rules about intervention in the urban economy; (d) help in predicting policy-relevant consequences of decision options; and (e) the generation of program/project proposals as feasible alternatives. Urban management is without quality if its decisions simply mirror the contingent opportunities for choice opened by the external environment.

There are two other structural components of tightly coupled urban management: one is the general management; the other is the program operations. The latter is another level of relationship between urban management and its external environment while the former relates to its internal environment. Cities are today engaged in the management of a deck of multifunctional and cross-functional programs. For purposes of program management, cities are in need of a redesign of their managerial structure. The managerial level at which functions are aggregated for program objectives is the level of general management. It is the level at which the internal environment of urban management is simplified for purposes of control and at which scale and diversity of activity are brought within the boundaries of manageability. It is the level for managerial control in relation to the whole city and the strategies of urban management. It is the level for program goals, program balance, and program design. It is the level for resource allocation among program claims, for fixing the magnitude of program activity, and for choosing program instruments. It is the level for avoiding the introduction of fiscal and budgetary instabilities into urban finances. And finally, it is the level for the scheduling of program activity, for its horizontal coordination, and for its appraisal in terms of norms of performance. The contribution of the general management component is to the rationalization and maintenance—and thus to the stabilization—of the positive capability states of urban management.

The program operations component reintroduces a relationship between urban management and its external environment. The character of the relationship at this structural level is dominated by the wholly new meaning that has now permeated the spatially associated aspects of urban management. Spatial decentralization has become a vehicle for the territorial politicizing of urban management. In the past, the decentralization was organizational rather than spatial, and its significance for management was in terms of organizational efficiency. Spatial decentralization, however, is not concerned with efficiency considerations in the rationalization of organization. It is concerned more with radical alterations in the capability state of urban management, and it is associated with the collapse of all of the urban services parameters that once were taken for granted.

The implications of spatial decentralization for tightly coupled management open possibilities for new strategies with new social decision structures as instruments for these strategies. This type of decentralization provides urban management with the territorial base for a strategy of disaggregating the city for purposes of development choices and resource allocation. It thus opens the way to a whole new set of users and conditions of use of managerial urban planning.

The latter is the vehicle for introducing science-oriented information and technology into the bases of urban choice and development. Neither local party structures, nor the politically dominated city councils, nor the political level of urban management have built or tied into social mechanisms for the utilization of the information and methods of science in their own decision processes. It is through the coupling to managerial urban planning that this structural defect, which is critical for the capability state of urban management, is to be remedied. The objective is to establish an urban use structure and an urban clientele or patron for science. The use structure can include both the management and the decision structures of the city's social components. Tightly coupled management can not only establish the comprehensive plan relevancy of any option for urban choice, but it can also permit and supply the information for selection from among a set of plan-relevant alternatives by territorially based social decision structures. These kinds of choices can be linked into the processes of resource allocation. In effect, a powerful organizational and resource base is introduced to support motivational foundations for changes in an environmental state. At the same time, boundary conditions on these spatially decentralized decisions can be established and maintained through the control apparatus of general management and the primary decisions of the politico-managerial component of urban management.

Decentralization of the administration of social services merely to improve the effectiveness of the service delivery mechanisms, and without any relationship to a system of tightly coupled urban management, stands on a different footing. The former arrangement is intended to minimize the adverse incidence of area constraints on the utilization of services, but it also serves to provide employment opportunities for an area's indigenous personnel. Tightly coupled urban management, however, is interested in a decision structure, a method, and a strategy for area-focused socioeconomic planning, which is more than organizational arrangements to overcome dysfunctionalities in the administration of services. It is not indifferent to organizational redesign as a way of dealing with reperceived problems of urban services administration. Nevertheless, the larger concern of urban management is a system-designed capability to select and achieve a desired environmental state. Program implementation is no longer just functional management. At this level there has been both a dilution and a revision of once-accepted standards of professional competence in service fields through the increasing use of paraprofessional personnel.

URBAN MANAGEMENT MANPOWER

Design choices with respect to the manpower for urban management have been concerned with the relationship of professional manpower to political authority and to a style of urban management. A nonpoliticized, functionally rationalized style of urban management that would not subject professional manpower to nonmanagerial supervision in its performance of occupational tasks was the attempt of an earlier period to introduce the capability attributes of functionally rationalized management into the quality of urban management. The organization of urban management into functional enclaves and an emphasis on the systematizing of functional task operations meant that its quality was related to the capability attributes associated with the management of discrete functional enclaves of organization. The more simplified technology of management and the more simplified social structure for management in this earlier period provided the initial reference points for design choices with respect to the capability level of the manpower component of urban management.

The sustained federal investment in urban planning, renewal, and development since 1947 has introduced changes in both the technology and structure for urban management. The result of these changes has been to alter the requisite quality of the latter. In effect, another style of urban management has been encouraged, one that requires a different reference base for design choices with respect to its manpower. Through federally subsidized urban planning, urban management was induced to enter upon a variety of suprafunctional and functionally nonconstrained experiences in renewal-development planning. Not only did urban planning thereby break out of the organizational constraints of the functional enclave within which traditional physical planning for the city had been confined, but it was also brought into working contacts with the entrepreneurial, fiscal, allocative, investment, strategic, operational, coordinative, areal, representational, and decisional aspects of urban management.

The new linkages of urban planning with urban management were in actuality a striving for another style of the latter, one that could be characterized as tightly coupled (or tightly linked) management. It meant a revision in the quality of urban management through a change in the level of capability of its manpower sufficient to make such tightly coupled management practicable. The shift from functionally based to tightly coupled management (with a concomitant change in the technology of management) expresses a change in the strategy of urban management and foreshadows a revision of its social structure. These structural changes, as previously noted, represent a working out of the tighter coupling of urban management with urban planning. The adaptation of the manpower component to the redesign of the management structure can alter the quality of urban management through a meshing of the differential capability requirements of the managerial manpower to the differential organizational requirements of the structure for urban management. The level of the capability of urban management is the general objective.

The structure for a tighter coupling of the two should not be visualized in any specific organizational form. Rather, it should be thought of in terms of some primary components that are capable of taking on a variety of organizational forms. Each of these components will have its own social composition, its own information channels, and its own class of problems and decision processes. It is not a task specialization but a social specialization of function that will give pattern to the structure for tightly coupled urban management. What the coupling will require is a structure that is adapted to different orders of function, as represented by each of the component complexes.

The primary components (derived from a social specialization of function) are three in number: (a) the entrepreneurial (politico-managerials), (b) the urban general managers, and (c) the program operationalists. The urban management professionals will relate differently to each of these components, simply because the social utility and the technical contribution of the professionals will be different for each. The group of politico-managerials who will comprise the entrepreneurial component of urban management will represent a contingent and personalized social structure for urban management. They will be individuals of diversified motivations who will be playing a variety of career games and who will be the contingent by-products of political control of urban government. The sources and the modes of inclusion and utilization of technical competence within this group are critical for tightly coupled management. The inclusion of the professional within the social composition of the entrepreneurial component presents complex questions of team formation. More than just a desired level of technical sophistication is involved; for it is not only the level of technical competence that is relevant but—considering the diversified mix of political and professional types and of motivations and career orientations that enter into the entrepreneurial component of urban management—the social utilization of this competence. The manpower expertise will have to become more systematic about social utilization capabilities within the structure of the entrepreneurial component of urban management. The quality of the latter at this structural level concerns the capability of a conjoined group of politicals and professionals to deal with a class of decision choices precipitated by ill-defined problems and involving entrepreneurial options for the city.

The component of urban general management is structurally underdeveloped and lacks organizational forms in the existing pattern of city government. However, the link to urban planning will compel changes in the modes of urban management, since it introduces changes in its technology. The urban general management component is an alternative to the cabinet of departmental officers and to the staff of the office of the urban chief executive as the structural component for urban management. This component will compel a relatively greater investment by cities in plan-relevant management technologies and the structure for their utilization than is presently the case. These technologies have to do with program design, resource allocation, spatial or areal decentralization, environmental monitoring, fiscal and investment strategies, multiple program coordination, and management control. They are technologies that will be integrated into information systems for urban management decision. They are, moreover, technologies that are objects of science-oriented inquiry by the management science profession. The analytic base—or the logic—of these technologies cannot be learned on the job, but their organizational utilization and their application to the content of the urban management problems of a particular urban jurisdiction can be job learned. In all likelihood, the urban general management component will serve as the testing ground and supply the occupational paths for the urban professional seeking to move into the politico-managerial (entrepreneurial) field of urban decision. It is the structural link between the entrepreneurial and the program operations components, and within it are located the plan-relevant management technologies by which the transformation of the decisional output of the former into the expected activities of the latter can be done.

The program operations component will present a number of intractable problems for urban management. Program operations interlock area and clientele (or recipient) with the services content of the programs. For purposes of program operations, the urban jurisdiction will be disaggregated into territorial units and program differentiated clusters of spatially distributed recipients of services. Unstabilizing tendencies can be introduced by spatial factors and clientele motivations encountered in the management of service programs. Area has a representational as well as a managerial utility. At the same time, programs are differentiated responses to a range of need and aspirational levels of the service beneficiaries. Territorial control enables the service beneficiaries to impose their own aspirations on program management at the level of program organization and operations.

The extensive diversity of program fields (health, community, development, employment, family services, etc.), coupled with their interrelatedness, precipitates problems of coordination. Operational coordination is a managerial part of the planning implementation. In urban management, neither analytic coordination nor operational coordination are, as yet, highly developed competences. Knowledge of the appropriate combination of organizational forms and social processes with technical and sociopolitical skills for coordination should be included in the competence of the future manpower for urban management. In tightly coupled management, urban planning will expect urban management to have the capability of putting into effect area-focused, multiprogram, urban development strategies within some framework of comprehensive planning.

The urban professionals will separate out into (1) the nationals and (2) the locals. These terms describe categories of mobility patterns. The nationals are those who will have access to occupational paths which will enable them as urban professionals to move between urban jurisdictions on a national (or subnational) geographic scale. Their careers will lie in the urban industry—which is both an aggregate of jurisdictions and an aggregate of firms constituting an employment market for the urban professionals. From a motivational standpoint, these nationals will be searching for paths of rapid, upward mobility, both social and organizational. Their competences will be generalized (interjurisdictional and interareal), adaptive (responsive to the variances of different jurisdictions and areas), and social (effectively accommodated to differences in leadership types, leadership styles, and decision structures).

The locals among the urban professionals are those whose occupational paths will be constrained within an urban jurisdiction or within a local aggregate (metropolitan area) of such jurisdictions. Their careers will lie within an urban jurisdiction inside a locality, but modified by the possibilities of intralocal mobility provided for the urban professional. From a motivational standpoint, these locals will accommodate to queuing up in a "waiting line" of organizational opportunities for upward mobility. It is doubtful, however, whether the structural base for intralocal mobility will provide a significant alternative to the "waiting line" as a mobility path for them. Their career problems will be those associated with "waiting line" opportunities, including the equitable (seniority) and status (salary differentials) constraints or organization mobility through such a line. Their competences will be locally and specifically fitted to the social and environmental particularities of a specific jurisdiction or local area. The locals will also tend to become bureaucratized around position and task specialization.

In addition to accommodating to a "waiting line" of organizational opportunities, the locals of the urban professionals will find it necessary to accommodate themselves to an influx of the paraprofessionals. It is particularly in the urban services that the paraprofessional will be fed into the channels of urban management. The accommodation of the locals in this respect will take two forms: (a) competition for "waiting line" opportunities, and (b) a revision of the competences once accepted as the hallmark of the urban professional. It is doubtful whether mobility paths for the paraprofessionals will lead from program operations into general management; for between the structural components of urban management, there will be differentiated learning and performance requirements that will serve as mobility barriers.

REFERENCES

BARNARD, CHESTER I. (1947) The Function of the Executive. Cambridge: Harvard Univ. Press.

SAYRE, WALLACE S. and HERBERT KAUFMAN (1965) Governing New York City. New York: W. W. Norton.

VICKERS, GEOFFREY (1965) The Art of Judgment. London: Chapman and Hall.

Part IV

URBAN ORDER

Present and Potential

Introduction

THE preceding portions of this volume illustrate how the question of quality becomes more complex as one moves from the physical environment to the realm of the social. Standards can be determined for the potability of water and the cleanliness of the air—although even here disagreements arise, witness the controversy over nuclear fallout—but the question is infinitely more difficult when it relates to the welfare system, or the courts, or even to housing and the neighborhood environment. It is one side of the equation to ask how we operationalize such general goals as equal opportunity so that standards can be devised and the programs designed to achieve these ends evaluated. It is the other side to determine just what goals we really wish to attain and what programs we are willing to support in the endeavor.

The five chapters in this concluding section of the volume direct their attention to the institutional and normative foundations essential for the enhancement of the quality of urban life and on some of the broad avenues for moving in this direction. The first three deal, from different perspectives, with what might be called the "politics of order." Written by political scientists, each stresses the centrality of government, of the political process, in the maintenance and promotion of societal order. The quality of this order, as they demonstrate, is at base dependent upon the ability of the polity to create a structural framework which the overwhelming majority of citizens and groups in the society perceive as conducive to the satisfaction of their vital needs. The fourth chapter analyzes the role that private foundations can play in this process while the closing contribution, written in philosophic perspective, examines the kind of education urban man must have if the good life is to be achieved in the modern city.

In Chapter 16, Matthew Holden takes up the critical question of the quality of order as it relates to the security of urban residents. It is his position that criminal acts are in effect political acts since they usurp to themselves power which only government has the legitimate right to exercise: the power to take life and property by coercive force. Criminology, he argues, should therefore be regarded as a part of political science and should be treated as such with all the theory and knowledge relevant to that discipline. Viewed in this way, the question of order assumes new dimensions and becomes subject to new techniques, strategies, and processes. For, as Holden reiterates, political theory has long demonstrated that order, like a governmental system, cannot long endure on force alone. The goal is a system in which people perceive that they can serve their purposes within the institutional structures and rules of the game. When a substantial number believe otherwise, order is jeopardized and even destroyed. The policy objective, in short, cannot be the elimination of conflict (from which disorder stems) since this is an impossible task; it must be one of providing a conflict-management structure and pattern of procedure which all segments of the society regard as acceptable and conducive to the attainment of their ends.

Taking up the problem of racial conflict in the United States, Holden notes that blacks are widely sympathetic to violence, not because they prefer it or would themselves resort to its use, but because they believe it a justifiable way of breaking the long-standing order of white dominance which has consistently thwarted their goal aspirations. The same could be said of the rebelling students (and in another sense, of the police) who no longer perceive that their vital needs are served by the existing rules of the game.

Holden discerns several dangerous trends in our present preoccupation with "law and order." When the demands for public protection are high, the situation is conducive to a "politics of panic." Government, in such circumstances, tends to engage in a tension-reducing ritual by exercising its powers harshly upon the stigmatic victims ("the dangerous classes") despite the fact that such action does not change the social realities. Some elements in this direction have already emerged, such as the increasingly free use of the curfew at the slightest sign of "trouble," the growing public and official support for various forms of surveillance of "suspected" persons, and the mounting insistence on the part of the police—with latent civilian support—that they know best how to identify potential sources of disruption and therefore should be left free to deal with disorder as their judgment dictates. These factors assume greater importance in view of the fact that no explicit process exists for making decisions about order in any way comparable to the political process for determining other policy issues. As a result, the politics of order becomes a politics of panic in which the normal restraints on administrative power tend to diminish.

Assuming racial conflict to be the central issue of order in contemporary urban America, Holden asks what knowledge social science might bring to bear on this problem. His answer is not encouraging. He finds that urban literature has given virtually no attention to the conditions and processes of achieving domestic peace. He does, however, see the possible relevance of the experiences in international relations as a model for dealing with the issue of social peace. In this context, measures aimed at reducing the physical possibilities of danger, such as domestic arms control and disarmament, become important stabilizing conditions. Measures of this nature, he emphasizes, constitute only the first phase of a social truce. They must be accompanied by a new political process for defining the forms and objectives of order (the traditional values of police craftsmanship should not be permitted to define them) and the modification of institutional practices to assure equal treatment and opportunity for the black minority. Holden ends on an interesting note, observing that the question of order is ambiguous in American culture since whites are simultaneously prejudiced and fair-minded. Herein lies the hope. "The possibilities of defining new conditions for order, outside a framework of coercion, depend on this ambiguity, and the possibilities of political leadership in one direction or the other depend on a sophisticated interpretation of the latitudes which this ambiguity affords."

In Chapter 17, James Q. Wilson turns to the question of urban order as it is perceived by the average citizen. Noting the results of a poll which showed that a large percentage of the respondents—both blacks and whites—viewed crime, violence, and other improper forms of public behavior as the biggest problem facing the cities, he interprets this attitude as a rational concern for "community." As he employs the term, "community" does not refer to the need for belonging or to the ecological interdependence of people living in a common spatial area, but to the desire for the observance of standards of proper conduct in the public places in which one lives and moves—standards which are consistent with the values and life-styles of the particular individual.

This desire to protect one's immediate physical and social environment leads, the author notes, to a suspicion of heterogeneity and the support of measures which seek to insulate the neighborhood from the "threat" of those who are different. Wilson disputes the conception of the big city as the center of cosmopolitanism. Only a very few small fraction of its residents, he contends, want diversity; and that which is desired is of the "safe" variety, such as speciality shops, bookstores, cultural enterprises, and "ethnic" restaurants. Some readers may argue, however, that while only a minority of the people advocate extensive heterogeneity as essential for the "good life" of the city, a great many citizens have a substantial tolerance for a fair amount of "safe" diversity, defined in somewhat broader terms.

Wilson sees the central city as becoming increasingly made up of persons who face special disabilities in creating and maintaining a sense of community (as defined in his terms): the affluent whites who are often uninterested in participating in the affairs of the municipality; the poor whites, mostly elderly, who withdraw behind locked doors to protect themselves against perceived threats; and the blacks, who have little choice but to remain. Wilson argues that communal social controls tend to break down when people live in neighborhoods which are not territorially distinct from areas with different or threatening life-styles. In such cases, the demands for more formal or institutional controls are intensified.

Wilson believes that social class rather than race is the primary basis on which community-maintenance judgments are made. Behavior perceived as threatening may, if such an assumption is correct, be largely a reflection of inter-class differences. However, it is important to distinguish in this regard between attitudes which reflect the collision between differences in life-styles and those which transcend such variances. There is no group in our cities, for example, that positively values or indifferently accepts violence directed at their persons or destructive of such property as they may possess. It is also important to distinguish between two broad classes of disorder: one represented by criminal activity, the other by the deliberate use of disruptive action as a tactic for extracting changes from those so threatened. In the perceptions of most people, both of these forms of contemporary disorder may be the same, but the fusion may disguise important socioeconomic and cultural conflicts around which disruptive "politics of protest" have been generated.

Schmandt and Goldbach, in Chapter 18, approach the subject of urban order from a broad theoretical perspective. Like Wilson, they are acutely conscious of the spatial factor in determining social harmony. Specialization in the modern metropolis is manifested spatially as well as functionally. Just as sections of the urban community are identified with particular industrial and commercial uses, so are entire neighborhoods given over to groups socially differentiated by age, status, ethnicity, and color. The residents of these territorially demarcated subcommunities are concerned with space as a means of maintaining their individual life-styles. As the authors emphasize, life-style interests are basically place-related, but paradoxically the demands for greater neighborhood controls are being intensified at a time when nonplace-related interests are assuming greater importance in the life of most individuals. Technology, increasing interdependence, and growing affluence have expanded the geographical range of intercourse and multiplied the ties and interests which transcend space.

The problem, as the authors pose it, is to create a governmental framework within which the conflicting forces at the urban level can be accommodated without intensifying the place-related fears of the majoritarian society. They see the solution in a reordering of relations within the total political structure. Distinguishing between the distribution (allocation and equalization) and delivery (administration of goods and services) systems, they construct a matrix of city types to show the different mixes of the two systems. It is their argument that the redistribution function must be increasingly assumed by higher levels of public authority—the national and state governments—while the delivery function is left in local hands. Minority groups and the poor are less concerned with local government as the purveyor of services, important as this function is, than they are with it as a redistribution agent. And it is the latter rather than the former which causes the most conflict. The strategy, as Schmandt and Goldbach suggest, calls for raising the income levels of the disadvantaged directly through national—and to a lesser extent, state—policies rather than attempting redistribution through the service structure of the local polity. By minimizing the importance of the local delivery system as a redistributive mechanism, the tension which centers around it could conceivably be relieved.

Such a strategy would permit place-related interests—the primary concern of the middle class—to be protected by the local delivery system while the reconstituted redistribution system would enlarge the opportunity structure for the now deprived segments of the community. Thus in return for the assurance of continued control over their place-related concerns, established groups would likely be less resistant to policies which permit other groups to develop their own life styles. In this way, the latter could further their goals without having to challenge head on the place-related interests of the "haves." Such a possibility is not to be discarded as unrealistic in today's enlarged circumstances which allow needed gains without real losses to any group or segment of the society.

The concluding two chapters in this volume, as indicated earlier, take a broad and philosophically humanistic view of the need for remaking much of urban life, and of the obstacles which confront efforts at such a reconstruction. Chapter 19 suggests some of the innovations which might begin to flow rather immediately from a genuine commitment to urban leadership by the private foundations, whose substantial resources and relative autonomy uniquely qualify them to become major entrepreneurs of social change. Chapter 20, with a larger view, argues that the process of formal education, given an adequate recasting of the curriculum and instruction embodied in our schools, could do much to produce a new generation which would be the first capable of responding to urban problems as authentically urbane citizens. Both of these chapters thus repeat an insight implicit or explicit in many of the preceding contributions: that a relatively small number of individuals who occupy institutional positions with special leverage—in this case, foundation policy-makers and professional educators—have the possibility of contributing much to the creation of a society in which power and involvement would be broadly dispersed. The actualization of this possibility could result in the use of the immense resources generated by modern technology and organization to create an environment more fit for human habitation.

In Chapter 18, Paffrath takes the intersecting problems of poverty and racial conflict as paradigmatic for the ills of contemporary urban civilization. He argues, in effect, that eliminating the conditions from which impoverishment, social injustice, and consequent strife now spring would inevitably take us a long way toward the desired rebuilding of the whole urban order. He sees foundations contributing to the creation of "social inventions not known in the past," by (1) providing evaluations with a broader perspective than is usually available from institutions more directly involved in existing situations; (2) helping to establish some firm "fixes" for determining the direction of national policy and priorities for programming; (3) aiding in the development of regional and, even more, local competence and of communication between levels and groups in our society; and (4) providing an independent audit and critique of existing institutional arrangements.

A commitment by foundations to the urban crisis—one-fifth of their grants, directly or indirectly in 1968—is emerging but is still grossly underdeveloped. Foundations themselves, Paffrath argues, must have the will to shift their resources from traditional channels into new kinds of enterprises and to establish new relationships with both governmental agencies and organizations and groups in the private sector. The multitude of smaller foundations must consider new strategies of consorted action and of involvement with the impoverished and excluded in their own communities. All of this means, among other things, that foundations can no longer be judged "by the criteria proved valid for business operations"; rather, they must be judged by how well they convert their detachment and independence into experiments and innovations that visibly begin to cumulate into a new urbanism in which wide participation, authentic community, voluntarism, and individual worthiness have been translated from pieties into programs. As an illustration of the commitment to "see things whole," Paffrath asserts that an essential step in reconstructing our urban environment is to free our resources, psychological as well as fiscal, from the burden of militarized diplomacy: peace in our cities is thus inextricably linked to peace among nations.

The final chapter in this section approaches as near as one might hope to providing a "conclusion" for this volume's consideration, in diverse ways and from many perspectives, of the quality of urban life. Asserting that the "education of urban man must, perforce, be the education of Everyman," Johnson draws upon his own skills as social scientist, educator, and philosopher to consider the nature of the urban world in which Everyman now lives, the mode of education which would make that world comprehensible, the kind of schools which might provide such an education, and the type of society which might sustain such schools. He designs no blueprints; rather, his approach is often aphoristic, and prescriptions are frequently drawn only by implication.

Johnson argues that we cannot make sense of the complex and changing urban world so long as we know about it only in the fragmentary manner that is contemporary education's dubious achievement. Yet we must make sense of it if we are to be able to reconstruct it. To be an effective citizen and a free man, the resident of the city must have knowledge of the myths which give life meaning and the future hope. He must also have knowledge of the empirical world through which all aspirations must be made existential, and of the processes by which the best of what we dream and cherish may to a greater rather than a lesser degree become a part of the physical and institutional reality we call "the city"—poetics, physics, and politics. Given an education which would enable him to perceive, judge, and relate to his world in these terms, Everyman, who now merely resides within and is too often victim of the city, could become a new kind of "common man"—the active citizen who knows what he and his fellowmen have in common with each other and with a mythic Mankind whose existence is simultaneously past, present, and future. But Johnson does not belong to the ranks of those who make a panacea of "schooling." The kind of education he advocates can come into existence only through effective attempts to change educational institutions by an already aware minority, a "sacred remnant" that understands that the "educative school" can be created only as it is part of an emerging "educative society."

Like the other contributors, the author of this chapter leaves us with an apparent paradox: the city which would truly make one free seems a necessary environment for encouraging and sustaining the very innovations and reforms that alone can bring about the existence of such a reconstructed urban milieu in place of the one we now endure more than enjoy. He seems to say that a truly democratized urban citizenry can evolve only if there is a basic reconstruction of the education of the city's residents, but that such a reconstruction depends upon a willingness to support, or at least tolerate, it on the part of a substantial portion of the population. Their willingness to do this, however, would in turn suggest that such a citizenry had already come at least part of the way down the road toward the human-centered and humanistic view of things which are central to the education he advocates. Johnson does not say that this is the case; neither does he deny it.

W. B. Jr. and H. J. S.

16

The Quality
of Urban Order

MATTHEW HOLDEN, JR.

☐ CRIMINOLOGY IS NOT a part of political science, yet it surely ought to be. For to oversimplify, it deals with the five most critical political acts which any person (or group) could possibly commit: murder, robbery, kidnapping, theft, and rape. Could there possibly be a greater exercise of power than to deprive another person of his life, property, or physical freedom of movement, or to enforce upon a woman a possible pregnancy she may not want? The problem of order has many dimensions, yet it certainly must deal with precisely these acts and with the exercise or restriction of the capacity to apply coercive force toward them and their endless refinements and variations which human experience affords.

Order is created and maintained by a subtle joining of public law (at least in most societies), private law, various bodies of custom, informal agreement (of various degrees of rationality), and social prejudice. If effective, relatively few people will be willing to serve their purposes by going outside the understood rules; if ineffective, more will do so. And when there simply are no understood rules to which people may be predicted to adhere, one has then reached the Hobbesian anarchy. But disorder is always latent, and it is a constant challenge to political imagination and political knowledge to keep potential disruption at a minimum.

THE PROBLEM OF ORDER

As we usually think of it, the function of achieving and maintaining order is assigned to the government, chiefly through the mechanism of the criminal law. Perhaps the latter deserves closer attention from social scientists for, at minimum, it is the regularization of force and intimidation to serve the purposes of those who have the capacity to make their wills felt and respected in government. But it also has an additional feature: a tendency toward something we may call "quasi-totalitarianism." Although there may be a normative presumption of even-handedness, the criminal law system apparently could not operate well without the assumption that its targets are not anyone at random, but people of particular sorts (Silver, 1967). The invocation of the criminal law is thus a process of applying a form of social stigma not only to the individual in question but also to the group—however identified—of which he is a part. It tends to be a process of closed administrative politics in which decisions about the invocation of stigma and force are of very low visibility, and the party so stigmatized may thereafter be subject to virtually complete and arbitrary control by other administrative officials. (By "closed" politics I mean to adapt Snow's definition: a politics in which the deciding group feels little or no necessity to defend its decisions before any wider audience which can have any practical effect. See Snow, 1962.) To a great degree, the person under the impact of the criminal law becomes a subject dealing with his rulers rather than a citizen dealing with his government.

In its classic form, the prison is the epitome of the quasitotalitarian subsystem, within the constitutional democracy. As a power arrangement, it places some men under extreme control by others. It also deprives them of most of the normal channels for effective redress. The prisoner is, for instance, under much closer control than the enlisted man in the military service. Doubtless, this is the reason some men have chosen the army instead of jail when they had the chance. The prison is, of course, the extreme case. However, our recognition of private law and social prejudice points up a basic element in political theory: order cannot long rest upon force alone.

Two other forms of inducement to compliance are critical. The first is *genuine persuasion or the invocation of consensus.* Such persuasion holds in those cases where contending parties resolve their difficulties on the basis of shared norms, the validity of which each recognizes and accepts. These norms might be fidelity to Christian doctrine as enunciated by John Calvin, to the standards of scientific proof, or to the belief that the democratic process (however defined) stands above any specific substantive result. If there is a high degree of shared moral appreciation, so that parties regard each other as *fellow* participants whose presence is uniquely valuable, the system may be regarded as a "political community." The second form of inducing compliance is by some species of bargaining, trading, or exchange in which disagreement is real, but a calculation of the gains and losses leads fairly often to a peaceful settlement. This is essentially compliance on the basis of *purchase* and the system may be regarded as a "political market."

The achievement of order is merely a generic problem which constitutes the basic definition of "politics," but there is a special aspect about "urban" order. While we are still prone to consider it in the context of the *local* political cultures and social systems, (Downes, 1968; Wilson, 1968), the central fact is that urban order can be handled as "local" only under two conditions. One is when the larger political regime has so far collapsed that, de facto, there is no other possibility. The second is when the national political regime is so stable that local disorder constitutes no threat. Once a regime is functioning, but something short of fully stable, disorder in the cities is far more critical than disorder in the countryside. More governments (the new Congo and Burma are examples) have "held on" from an urban base until they were able to reassert control over the countryside than has been the case the other way around. The city is a center of command and control, from which organization is imposed upon the rest of the country. But the ecology and technology of the city make it subject to almost instant immobilization, a fact which causes the threat of urban guerrilla warfare to be so serious a prospect from the perspective of rulers. Europeans have been sensitive to this for a very long time, a sensitivity reflected in Elizabeth I's fear of the growth of London lest the city, as a place of "masterless men," also become a threat to the Crown (George, 1968). Americans have been less sensitive to this matter, probably because most nineteenth-century violence—though very significant in scale (Holden, 1969)—involved the confrontation of two or more minorities (e.g., Catholic Irish and Protestant Irish) isolable in parts of the city. However, when violence has threatened to engulf vital interests which could not be isolated (as in the Draft Riots of 1863, and anti-company agitation from the 1870's onward), the sense of involvement with the national system was immediately activated. This can be read, as early as fifty years ago when the Red Summer of 1919 brought a serious federal concern in several communities and actually made Omaha the sole city in American history to be under the *direct rule* of the United States Army (Waskow, 1967). We may see it even more clearly now, when—contrary to all its prior doctrine—the United States Army has come to be engaged in a constant readiness for action at home, no less than abroad. The problem, therefore, is not one merely of "law enforcement" or "maintaining" order, but of the creation of an order appropriate to the urban condition in the twentieth and twenty-first centuries.

VARIETIES OF DISORDER

The policy objective cannot be the elimination of conflict (whence disorder springs), for that is an impossible task. It is rather to provide a conflict management structure (pattern) which most participants will regard as dominant (at minimum) and (preferably) acceptable, desirable, or legitimate. The objective is to establish a "constitution," i.e., to constitute that structure by formulating "rules of the game" within which that action which evokes conflict may be managed. However, there will always be some participants whose perceived needs cannot be accommodated within those rules, with the consequent tendency for such participants to go beyond them in the pursuit of such needs. The degree to which they will do so is a question of "costs," both costs of violation and costs of accommodation. What material loss will the participant experience if he goes beyond? What material loss will he incur if he does not? What psychic pain will he suffer if he sees his opponent win, particularly if the opponent is defined as "enemy" (Finlay et al., 1967), while he himself loses? (Miller, 1967) To what extent will he, if he accedes to the rules (even where this will clearly mean that he loses), be "out" for good?

Internal order is something about which we tend to have, as Robert MacNamara suggested is the case in national security, ideas hardly more advanced than those of our predecessors several thousand years ago. Part of the problem in thinking about order and disorder in urban society is simply to be clearer about the varieties of disorder which occur. Criminal law speaks of "crimes against persons" and "crimes against property," but it may be more indicative of the problems to think of *moral offense* and *ordinary crime*. For these ways of thinking about the matter hint at the ways in which people who are not "disorderly" get excited about those who are. The question of other people's opinions is crucial, for the achievement and maintenance of order is not solely a matter of the official use of the public law.

The moral offense varies widely from culture to culture but nothing is more apparent than that all cultures contain some categories of behavior which are proscribed (or which some people want proscribed), regardless of whether they constitute any sort of threat to the collectivity. This type of offense is what Edelman (1964) has called "heretical" behavior. It may vary from an attempt to show a movie deemed sacrilegious by New York Catholics, to participation in proscribed behaviors such as having an abortion, consuming narcotics, or engaging in homosexuality (Waskow, 1966), to acting as attorney for a black tenant farmers' union in a Southern county dominated by white planters.

The second sort of disorderly action is "ordinary" crime. There is nothing "trivial" about it, but it is ordinary in the sense that the particular event by the particular actor does not evidently constitute a threat to the political system. Rather, it constitutes an injury to some specific person or group, in which other people take no serious interest unless they themselves identify extremely closely with the injured party or the one inflicting the injury. This type of crime may thus be treated as a spectator sport or a theatrical event. In consequence, if the injurer is apprehended, he is normally left to the more or less "automatic" workings of the administrative and judicial machinery without serious external intervention.

Both these forms of crime are politically relevant. Both impose pains upon some people (the victims) in cases where the positive law (which purportedly "allocates values authoritatively") asserts that they should be free from such pains. This is particularly true with respect to ordinary crime. When the inner-city woman who works as a maid returns home with her daily pay of $12 or so, she suffers a grievous privation if youngsters in the apartment building rob her. If government means, among other things, the protection of property, the taking of the money is also a political act performed outside the structure of the normal political process.

Behavior of this kind is substantially transformed when the number of such acts becomes somehow "too great." Common sense recognizes this fact in the assertion that "they shouldn't be permitted to get away with it, because then everybody will start doing it." It need not be correct that "everybody" will start doing it, but there is what Norton Long has more than once called the "critical mass" problem in disorder. He has estimated that (from the viewpoint of an existing system), not much more than four percent of the population can engage in serious deviation before entropy becomes apparent. In all likelihood, this is too high a figure (although the number will presumably vary in different situations). But some such level does exist. Problems of urban school administration bear out this point. In the last days of January, 1969, there was a rash of resignations by Detroit high school principals who found that their schools were out of control, that they could not reestablish control, and that they themselves could not stand the emotional pressure of working in these "out of control" places.

Granted that this estimate is correct, the offenders comprised only two and one-half percent of the student population. The question is why, in that situation, it was impossible to maintain control. Doubtless, there are several answers. One is that there was a visible external sympathy with the students, part of which is attributable to racial concern and racial hostility. The most active spokesmen of the external black community share the students' belief that the school system is simply all wrong and ill motivated. Moreover, it is probable that the principals—like the McCone Commission—have adopted a "riff-raff" or "incorrigible" hypothesis when, in fact, active student sympathy by nonparticipating students is similar to active black sympathy with those engaged in open disruption on the streets. (The insupportability of the "riff-raff" hypothesis is shown by virtually all the studies conducted after the various urban civil disruptions in the past five years. See, for example, Geschwender, 1968, and Tomlinson, 1968.) The process by which individual disruptions may turn into major public issues is indicated in another dimension of the high school situation: the much-vaunted athletic tournament. In the third week of January, the Detroit schools faced the prospect of cancelling the tournaments because referees announced they were afraid to come in at night. Possibly this was a bargaining tactic by which those who wanted the school administration to "get tough" could mobilize public support to that end. But even if it were a mere bargaining ploy, it could be made "realistic" by the spreading public concern about the scale of the individual acts reported—a concern in no sense abated by news stories of a public school teacher being set afire by students in New York City.

If we are not ordinarily prone to look to the high schools for verification of Hobbes and Machiavelli's concerns with order, we may have to change our minds. What we are dealing with in instances like the Detroit school situation is the transformation of individual acts into a form of *group conflict,* where the group desires to substitute one set of public values and policies for another. We have, however, lived with still another form of disorder beyond the critical mass level, namely the creation of private organization which we may call *crime as commerce,* and in which the commercial incentives lead to competition in the application of force. This system of exchange is carried on in an apparently regularized fashion, quite at variance with the commandments and injunctions of the positive law, including defiance of the rule that only the government has the right to put people to death. Although such action is defined as "criminal behavior," it is apparently often sanctioned by many people who are not the regular beneficiaries of the law. This attitude is the basis of the sociological-functionalist interpretation of the "underworld" as a system for providing law-abiding citizens with goods and services they desire, but in ways which are more convenient than the legal ways. There is apparently, also, some problem of disagreement on normative ends, for one has the impression that the Capone operations in Chicago were viewed with neutrality, at worst, by many people who were not directly involved.

Although the problems of order are, in one sense, generic, there is also a basis on which to argue that the most severe are the price for "social change." Disorder comes when qualitative changes in the relations of persons and groups are taking place and when a new "balance" has yet to be struck. It is here, above all, that people are willing to engage in disorderly behavior because they know (or believe) that if they "lose" at these points there will be no scope for future victories which make sense to them.

The class struggles involving the "rise of the trade union" represent such a break. The significant feature is not simply that workers are economically better off than they were before, but that a substantial set of relationships between the employed and the employer change. It is a decisive break when the formal organization of employed people (the union) ceases to be defined as a "criminal conspiracy" (on the early background see Morris, 1965) and becomes, first, an agency with which the employer *may* deal; then an agency with which he *must* deal as equal; then an organization with which he *prefers* to deal; and finally, the embodiment of various interests which are recognized symbolically in society by admitting "labor men" into the body of "responsible" *community and national* leaders whose approval makes the difference to many projects other than those defined by their "narrow" organizational interests. It was a slow and bloody process, however, to move from the state in which a worker (or nonpropertied man) was a "natural inferior" to the admission of the trade unionist as a figure "equal" to the potentates, powers, and principalities of industry.

Ethnic struggle in the American cities represents a different sort of break: the large-scale transplantation, within a relatively short time, of hundreds of thousands (then millions) of people from different cultures, and the placing of them side by side within the small space of the American city. This process—the sheerest case of densification—produced the religious riots and their modern equivalent, the inter-ethnic gang war, as people struggled to work out new relations to each other.

Not all breaks with history are consummated fully. In the case of black and white, the incidence of violence in the American South arises from the ending of slavery, the black effort to establish parity, and the white attempt to reassert a functional equivalent of the old order. If the latter could not be fully restored, the degree of restoration—by sheer force—was remarkable for the extent to which it created a new species of political "empire" to replace the former imperial relations of slavery. Interracial violence in the North is another part of the same break with history. The northward migration represented a break with the agricultural part, and for urban whites it marked a break with the city as a place in which they held sway. White immigrants from Europe, lower-middle- and lower-class native whites as well, found themselves in direct and persistent conflict with blacks for access to the public and private goods of the Northern centers (Grimshaw, 1959). If there was ever any doubt, it is now clear that the current forms of racial violence represent the sharpening of this break. It is not evident how much black violence is (or will be) deliberately calculated as a political strategy. But it is clear that blacks are widely sympathetic to violence, not in the sense of *preferring* it, but in the sense of believing that it is a justifiable way of behaving—for anyone who wants to behave that way—as a means of breaking the long-standing order of white dominance. In other words, the avoidance of violence on normative grounds is no longer a value so widely held as it once was, and the avoidance of violence on pragmatic grounds is something which more blacks are willing to leave other blacks to judge for themselves.

A different break with the past is reflected in the new attitudes of the student population and in the sympathy (as well as the hostility) which one can sense in their elders. This is a consequence, we believe, of the wide spread of higher education. Historically, education at a high level prepared an individual for an elite role and left him in a visibly superior position to most people. Near-universal education creates a new situation in which many more people are able to regard themselves as "every bit as good" as the elites and are able to "look down" on the public agents whom they regard (properly) as the instruments of such elites. Out of this break with the past, comes a reassertion of the past (although without reference to past symbolism). That is, there is the profound aversion to "authority," "control," "leadership," which is reflected, inter alia, in the specific disdain for the public officers of the law.

This brings us to a third break with history which, as I shall point out below, is critical. It is the emergence of the agents of public order as a self-interested and overtly active group, attempting to define as their own prerogative the determination of what the appropriate order ought to be and how it should be achieved. When a group so emerges (and defines itself as autonomous and "professional"), one of the behaviors to be expected is its own refusal to accept other definitions of the ends and means by which it ought to live. Police unionization, in short, reflects both the activation of the police as a group with social interests and the development of their capacity to enforce their will in the polity. This is a crucial process because it not only involves a certain rational strategy but also a *politics of collective psychiatry* in which the policeman's demonstration of "police power" (like the demonstration of "black power") is a necessity which the members of the group experience in order to make themselves feel better about themselves. This was revealingly stated by a Chicago representative at the founding convention of the National Police Association. People treat the police as "inferior citizens," he said, "but if we are going to be anything, we ought to be a superior citizen." The demonstration of power is critical (seen in that perspective) because it represents the form of action which will enforce upon the rest of society the necessity to show the symbols of respect, to treat the policeman as a *superior citizen*. But it also points to one of the severe problems in contemporary life: the increasing likelihood of what the Walker Report called "the police riot." From this perspective, we might well anticipate a new and more explicit problem: that the demand for superior citizenship will be reflected in police "direct action"–quite as averse to the conception of order within a constitutional democracy as any form of disorder experienced in recent years.

An appreciation of the varieties of disorder, as forms of behavior rather than as categories in the Uniform Crime Reports, leads us to a fuller understanding of why the problem of "urban" order is so critical. The problem arises out of a connection between ecology and politics. Among the multitudinous meanings of "urbanization," one aspect is extremely revelant here. Urban social systems are built upon density of human populations in territorial space. The city is, above all, a croweded place. But it is not the mere physical crowding which provokes disorder, as important a contributor as this may be; it is the crowding of social systems which necessarily occurs. For groups of people carry with them a great deal of their previous cultural experience and reestablish new lives organized around this experience. The greater the variety of such systems in contact (be they social classes, ethnic groups, or whatever), the greater the probability of conflict and the greater the probability that such conflict will move outside whatever is defined as the common rule structure.

The problems of governing cities have now assumed crucial importance because this crowding is intense and because the mechanisms of communication create a simulated crowding effect for the nation as city. Indeed, there is a sense in which one may now regard large stretches of the world as city. As a consequence, problems of urban order become at least as important to the whole American polity as London was to Elizabeth.

PROBLEMS IN THE "QUALITY"
OF URBAN ORDER

If our conception of "order" is accepted, then its "quality" may be assessed (if not "measured") by the extent to which people are induced to serve their purposes within the framework of understood rules. By this token, its quality today is relatively low. This conclusion must hold regardless of which crime indexes or indicators one accepts (Biderman, 1966). A high murder rate has a vital political meaning. It indicates that government has reached some limit in its capacity to protect human life, yet from a Hobbesian (or almost any other) viewpoint, no governmental function is more fundamental. However, we then have to ask what sorts of things would lead to a "better" and what sorts to a "worse" quality of order in urban society.

The plain answer is that, in many respects, we simply do not know. However, it appears that our present preoccupations in public policy ignore those aspects where some thought and action might be fruitful, while concentrating on those where any proposed line of action is questionable. It is very doubtful, for instance, that anyone knows much about how to bring down the murder rate in a short time (or even in the longer run) when almost half (48 percent) of killings occur within the family unit (President's Commission on Law Enforcement 1967). Similarly, it is doubtful that anyone knows enough to control crime waves (even if we reject Daniel Bell's [1960] argument that this is largely "myth"). After all, if we treat "business cycles" as analogous to "crime waves," it is only very recently that we had available a body of theory and a set of policy tools, even *apparently* effective for the business cycle which (by comparison to crime waves) involves a relatively simple set of variables. In the same way, if moral offenses reflect shifting social values—as changes in religious belief (which excited people just as much) once did—we neither know now to arrest these events nor even whether we should.

This knowledge gap involves a critical problem, if one views the matter from the perspectives of either effective or constitutional government. In the former sense, it means that there is a substantial measure of privation—of life, liberty and property—which no known governmental action could reduce. But it also means that public fears, in an era where the demand for protection is high, are conducive to a politics of panic. The fact that government is, in fact, unable to reduce perceived criminal threats significantly requires it (and the public) to engage in a tension-reducing ritual by identifying the group to be treated stigmatically. Thus, governmental power in such circumstances tends to be exercised harshly upon the stigmatic victims ("the dangerous classes") despite the fact that this does not change the social realities.

Moreover, in this process, it is very likely that a different dimension of disorder (organized crime) will be ignored. This type, which in principle is much more subject to control, may, for institutional reasons, be even less so than individual behavior. The problem of increasing the quality level of order by dealing with individual actions is difficult precisely because the variables explaining individual behavior are innumerable. In contrast, organized crime can probably be explained in relatively simple terms. We may regard the latter as the product of a set of economic and power strategies by persons who—though making no overt challenge to the norms of the system—persistently act outside those norms. The word "organized" is extremely important here, for it refers to the institutional character of the enterprise. It is a vital political phenomenon for, to take an example, the merchant who is subject to a "shakedown" or the businessman who declines to compete because he is physically afraid. Both, in such cases, have been deprived of an essential government protection. Without going into the argument here (see Holden, 1969), we may suggest that the essential reasons for the persistence of organized crime are twofold: (1) This form of enterprise does deliver some goods which *some* purchasers want, and which the positive law forbids. [We should be very careful, however, not to overemphasize this desire for illicit goods, as if all or most people were active consumers. Some students of crime have very much overdone this.] (2) The underworld enterprise functions with the collusion of police administrators and others very much as legal enterprises subject to government regulation often function in ways not fully contemplated by the statutes. Moreover, the reasons for official collusion are essentially the same: the necessity to work with a regulated constituency rather than to try to "eliminate it" (Edelman, 1964). The consequence, however, is that the underworld comes to be a sort of *imperium in imperio,* imposing its own rules, establishing its own procedures, and exacting its own penalties upon those who choose to operate within its boundaries (and upon many who do not so choose).

As in any such political situation, a multiplier effect is at work. Power begets power, in the sense that any group capable of acting cohesively has a far more decisive impact on the political process than its mere numbers would indicate. This fact is relevant, for example, to the problem of bribery and corruption. Data on this topic are virtually absent from the public record, at least in any form which social scientists have thus far been able to utilize. But it is a reasonable presumption (backed by some public testimony which may have at least probative value) that the underworld functions as a source of influence, income, and support for political figures who are sensitive to its interest. This situation, despite its sensational element, is not to be ignored. It is a principle of political science that some politico will be found responsive to virtually any interest in his constituency, and there is no a priori reason to rule out the prospect of bribery and corruption. What we cannot estimate is its precise effect on the political process.

THE "LAW AND ORDER" SYNDROME

The underworld poses still another intriguing intellectual problem. Social scientists who have conceived of private force and violence as a means by which groups secure sufficient leverage to have their interests legitimated have yet to show whether the underworld could be counted among the interests which subsequently "go respectable," abjuring their former techniques and methods (Nieburg, 1968). Is it possible to conceive, for instance, that a "community" group—energized to civic participation on the model of Saul Alinsky—could stand effectively against an underworld sector should their interests come into conflict? In what sense is the politics of enthusiasm a match for the politics of the gun? If this problem has been disregarded, what is even more neglected is an anticipation of the results to be expected from the present national preoccupation with "crime in the streets." Surely the image of "crime in the streets," taken in concert with the racial forms of urban disorder, cannot but lead to a reassertion and strengthening of the criminal law system. This is a predictable pattern, for the natural response to "threat" is "defense," but such action should be expected to (a) reinforce the quasi-totalitarian tendency of criminal law, and (b) contribute to a dynamics of mutual alarm in which the agents of order and their identified targets play off against each other to the detriment of social peace.

Some elements of the predictable pattern have already emerged: the increasingly free use of the curfew at the slightest sign of "trouble," the creation of new federal authority to regulate interstate movement ("the Stokely Carmichael law"), and the increased public and official support for various forms of surveillance of "suspected" persons. In context, such measures have a strong tendency to work against both the old-fashioned values of constitutionalism, let alone such newer criteria as opportunities for the enhancement of "human dignity" (Perloff, forthcoming).

There are at least two reasons for believing this will be the effect. First, no explicit process exists for making decisions about order in any way comparable to the political process for deciding other policy issues. The politics of order is a politics of panic in which the concerns of those most anxious about order become, in newspaper reportage and common discussion, a veritable tidal wave, and in which the "peril" to the community is magnified beyond all real proportion. Secondly, no explicit consideration of what is at stake takes place under such circumstances, but the decision-making process becomes increasingly dominated by a closed administrative politics in which the biases, self-interest, and doctrines of the administrative force stand as self-evident truth.

Closed administrative politics is enhanced by the legislative tendency to postpone major issues, a necessary part of the bargaining process, and by the judicial incapacity to know the factual details (and hence, major patterns) of administration. Both leave the critical decisions to be made through the accretion of small decisions within the administrative framework. This process, like other administrative processes, is not so much "public-spirited" as "self-interested," due to the significance of the "employee" interests as well as the craft values which are involved.

Those concerned with decisions about order, like other administrative personnel, have a natural desire to take care of their interests as employees. Policemen, for instance, have a strong desire to protect themselves against the possibility of physical and psychic harm. Some of the reaction attributed to Chicago policemen in the Democratic convention episode probably was due to their gross sense of calculated insult, the more activated when they believed they were being spattered with urine from the hotel windows. A man who thinks somebody is pouring urine on him is going to react angrily, even if later evidence should show that it was only clean water! In the same way, since policemen are likely to conceive of blacks as more dangerous and less likely to abide by the law (Skolnick, 1966), it is comprehensible that—however mistaken they may be—they will resist measures (such as "civilian review") which they believe will diminish their own scope and enhance that of their symbolic (black) assailants.

These employee interests, calculated to defend and protect the policemen in a narrow sense, interlock with the *craft* values which define for him what a good job is and how the good job is done. This is the basis of the well-developed practices by which law enforcement agencies purposefully, and as they see it, realistically, seek to evade or negate most of the procedural restraints which the federal courts impose upon them. Wire-tapping is practiced not because the officers in question are evil, nor because they do not know that it is illegal (except under specified conditions), but because their sense of craftsmanship demands that certain results be attained, and these results cannot be achieved within the limits of the formal law.

The significant feature of the problem of urban order, in this respect, is the virtual elimination of some prior restraints which were not normative but rooted in the ecology and technology of urban law enforcement. Until the last part of the nineteenth century, the police lacked the capacity to control more than the most minimal forms of disorderly urban behavior. Consider the simple problem of police pursuit of a man alleged to have taken money from an elderly woman. Prior to the rise of widespread basic literacy—sufficient for most people to read posters, if no more—the description of the man would have to depend on word of mouth, and the transmission of that description would inevitably have been garbled in short order. Prior to the invention of the telegraph and the telephone, oral communication could be conveyed only by the power of the lungs, which would have meant that the accuracy in transmission would soon be reduced still more. One need only recall to oneself how difficult it is for students to take lecture notes accurately on what has been said in a small classroom, or how difficult it is for a delegate in the back of the Democratic or Republican National Conventions to know what is happening on the podium. Prior to the invention of the electric light and its spread through most parts of the urban area, pursuit after dark would be hampered by the fact that a man might (in short order) find a hole, a doorway, or an alley in which to lose his pursuers and (with the garbled descriptions) be almost impossible to find and identify. Prior to the development of the hard-surfaced street and the internal combustion engine, rapid pursuit would have been limited to the physical capacity of the pursuing policeman. If one puts this combination together, one sees a certain realism—and not mere caricature—in the dismal Dickensian portraits of London, which must also have expressed a reality for the rapidly growing cities in the eastern part of the United States.

Because these constraints existed, the quasi-totalitarian character of the public mechanism did not have to be taken very seriously. As a practical matter, no one needed to worry about the values in the public mechanism, if that mechanism could not operate very efficiently. What we now have is a situation in which—with no appreciable increase in the constitutionalist character of the values—the capacity for intensive surveillance reduces the possibility of privacy virtually to a nullity (Westin, 1967).

The question is: how does one expect administrative (or other) decision-makers to behave when they are subject neither to compelling constitutionalist values nor to significant constraint on the possibilities of their violating such values? The answer derived from most experience is apparent. The probability estimate is that they will follow their preferences until they run into ineluctable constraints. What is still more significant is the tendency to seek means to evade the constraints, a tendency reinforced by the specific character of the break with history involved in current law enforcement problems. It is not merely that we have a technological pattern which reduces the empirical possibility of privacy. Nor is it merely that we have a new political pattern—the integral state—which makes it necessary for all decisions ultimately to go to the national center. It is that, together with both these factors, we also have a new condition in the severe status deprivation which policemen consider themselves to be undergoing. One of the best actual clues to police status is that, as Niederhoffer (1967) says, the policeman who used to be thought a fit marriage partner for the maid, may now aspire to marry the schoolteacher or the nurse. But, precisely as the policemen have begun to ascend this status ladder, they have become more explicit and concerned about the status issues. In this light, the constraints of the courts—about which they complain—and the adverse reflections in the media as people now suddenly come to debate whether the police are "adequate" or "appropriate" to the jobs they are doing, becomes profoundly insulting. As Dick Gregory purportedly said, there is a sense in which the police suddenly come to be (or see themselves as thought to be) America's "new niggers," the group that is convenient for many people to hold in contempt and disesteem. This means that until a new social arrangement is consolidated, one must expect the police to make severe and persistent demands that they be treated in a way which reflects their special desire for esteem. An administrative group which operates on such criteria, however, tends to become somewhat similar to the old

European (as against American) officer corps—a distinct body with its own political initiatives and incentives, frequently expounded and defended in the language of "nonpolitical" professionalism.

Professionalism, as it is coming to be used by the police, refers to the presumption that they "know best" what the techniques of order ought to be, that they know best how to identify potential sources of disruption, and that they should be left free to deal with such disorder according to their judgment of the appropriate methods. The police capacity to enforce this view upon society appears to be increasing, not only because there is a growing expression of latent "civilian" support (we cannot forget that, despite the public outcry via the media, most people who offered voluntary comment on the Chicago experience supported the police), but because the trend toward police unionization offers the working policemen the same kind of leverage which teacher unionization has offered in public education (Mayer, 1968). Let us, then, consider the broad outlines of a possible scenario. Assuming an increasing tendency for black "militants" to adopt tactics reminiscent of the Irish Republican Army or the Irgun (Holden, 1969; Avnery, 1968), and assuming a continuing acceleration in the scale of direct action by student groups and others, we would expect an increase in the pattern of confrontation with the armed establishments. If the latter has the view of itself which we offer above, then on those grounds alone—quite apart from any broader social policy interests—we could expect its actions to be equally direct, abrupt, and repressive. We might envisage a series of incidents of the sort described in the Walker Report on Chicago. In short, the politics of order would become more and more a dynamics of mutual alarm, in which the participants most prepared to resort to direct force would preempt those prepared to stop short of such forms of activity.

THE ACUTE QUALITY PROBLEM:
SOCIAL PEACE?

In this context, one can find little encouragement about the future of urban problems of order. Very simply, the question is how to achieve new forms of order-defining and order-maintenance which do not negate the constitutionalist values, how to establish viable rules of the game within which conflict may be managed on a reasoned basis, but which do not preclude multiple parties satisfying some of their most vital interests. This is apparent in the history of Roman politics and in the internal politics of some contemporary trade unions. It was critical in medieval and Renaissance Italy where the reputation for violence was crucial to the ability to participate in politics (Brucker, 1962), and in England and France, up to the nineteenth century, where riotous behavior was the poor people's de facto form of participation (Rude, 1964). Common American experience—although ill understood—makes the same point. Consider the nine great crises of American politics (from the Whiskey Rebellion to the Hayes-Tilden Settlement) before which the United States could not really be considered a stable nation-*state,* or the post-Reconstruction "civil war" between white and black which Charles Baer has aptly compared to the Nazi seizure of power, or the farmers' organized defiance of sheriffs and courts during Depression foreclosure proceedings (Allen, 1965; Shover, 1966).

In some of these cases, order was successfully achieved and in others destroyed; but all suggest that public force is an inadequate basis for order without a serious search for social peace.

Social peace is the condition in which contending parties find it increasingly "unthinkable" that they should seek to win by inflicting physical pain upon each other. It is the condition in which contending parties are *prepared to lose* in specific cases rather than violate the common rules of the system. If we suppose that racial conflict is the central issue of order in urban America (and hardly anyone will deny it), then we may consider both the options open to policy-makers and the character of the knowledge which scholars might bring to bear. Despite the pessimism which saturates scholarly writing today (and from which we are here not wholly exempt), it is possible that the new administration possesses (by its "conservatism") action-options which another administration might (by its "liberalism") find foreclosed.

The new administration begins with a political deficit-credit situation which *may* permit it to address the racial problem more realistically. Since it begins with a backlog of acute black distrust, it may perhaps be that any "realistic" action which it chooses to offer in the form of olive branches to blacks will have more credibility, precisely because such action was not expected. By the same token, its actual 1968 constituency—particularly including whites who are "hawks" on the law-and-order issues—may trust that administration enough to grant greater credence to its explanations as to why simple head-breaking coercion would be unproductive than they would grant to such an explanation from an administration believed to be "soft" on these issues. (In nineteenth-century England, where disorder and pressures for a social reform were somewhat associated, it was precisely the flexibility of a conservative elite which made adaptation possible [Silver, 1967: 1-24].) But the interesting problem is that, even were such an administration so disposed to act, it would have to invent its own rules of action, almost entirely without guidance from the professional students of urbanism. There is, in the urban literature, virtually no attention to the conditions and processes of achieving domestic peace, the necessary basis for reducing the elements of coercion in the maintenance of order. The riot studies, for instance, are virtually useless. For once they move from the data analysis which destroys the "riff-raff" hypothesis, they are bereft of suggestions about the conditions and processes that are appropriate to the serious social malady confronting the nation. In a sense, the studies may even be dangerous, for their emotional bent is toward the notion of changing "the system" by attacking "white racism" through "violence as a form of politics." Social scientists appear, as it were, to be developing a riot ideology of their own, but without the serious underlying analysis which would sustain a genuine conception of revolution.

Moreover, the intellectual disability extends beyond urban studies to the more inclusive literature of the social sciences. If one reads a publication such as *The Journal of Conflict Resolution* (or even more surprising, *The Journal of Peace Research*), he discovers very little comment on the appropriate theory of *peace*-making (or *peace*-achieving if "making" seems too deliberate and purposeful a term). Some theory may be found of *truce*-making or the process by which admittedly temporary cessations of open battle may be reached. There is also some comment on the theory of *treaty*-making, or the negotiating and bargaining process by which specific promises are exchanged, with the presumption that these will be adhered to for some time into the future, but with little or no supposition that they will really be "permanent."

To formulate the problem as the achievement of domestic peace, and to look at the sole available experience which may be at all relevant (international relations and industrial relations), is to indicate an historical process which may be out of phase with the current character of politics. The achievement of peace has, in the past, started with truce-making and treaty-making based upon the elites of the groups in conflict. By the process of dealing with each other, such elites have come to have (over very long times) a vested interest in the maintenance of the arrangments they have negotiated, and have eventually come to hold to those arrangements as things valuable in themselves. The process of institutionalization, by which wide popular support for these arrangements has been achieved, has taken still longer (Deutsch, 1957).

We do not *know* if this experience has any relevance to the current situation. But it does offer the best *available* model to deal with the central problem and the central objective: social peace instead of social war. If we look at what has been required in the past, we find that the process has necessitated: (1) an increase of compatible values and a sense of shared interest among the elites in question; (2) significant growth in the elites' capabilities for action, as indicated by improved political and administrative capabilities, significant economic development, the increase in the forms and varieties of communications among such elites and their constituencies, and the continuous broadening of the elites to include more and more persons of different sorts; (3) a certain freedom of mobility of persons across the boundaries of the units in question; and (4) the units in question come to exchange communications and rewards in many forms, symbolic and material (Deutsch, 1957: 43-52).

This listing points up both difficulties and possibilities in the current American context. The first major difficulty is that the present mood of thought actually impedes the responsible behavior of political elites, precisely because those who are currently in this position and those who soon will be (the student generation) lack the self-confidence and capacity to conceive what would be an appropriate and viable order. The response to this lack of self-confidence is a form of intellectual posturing in the romanticization of "participatory democracy" which means that no one is prepared to offer guidance and to be held accountable for the successes and failures of his guidance. The second major difficulty is that people, particularly white people, persistently tend to misunderstand the nature of the racial issues at stake. On one hand, they tend to define the issues as exclusively "economic," a definition which leads to a misunderstanding of the extent to which blacks who are not clients of the welfare programs share a certain degree of interest (although not a fully common interest) with those who are.

Had it not been for this misconception—which reflects itself in such enterprises as the Urban Coalition—little surprise would have been shown at the data which indicate that verbal support of "militancy" is about evenly distributed amongst all social classes in black America. On the other hand, they tend to define the issues as exclusively "symbolic," which leads to the misunderstanding that "separatist" orientations are as widely accepted as the common discussion might indicate. In reality, these two sorts of things are subtly and persistently entangled, and cannot be disentangled. The third major difficulty is that both white and black elites (including the "militant" activists who play elite roles, but refuse to accept full elite responsibility) increasingly tend to define the political values in exclusive terms, which is a form of escapism unless one accepts as an empirical proposition that somehow one population or the other will cease to live on the North American continent.

It would be more than one could attempt to indicate, in the absence of any appropriate theory, how logjams in elite perceptions can be broken, or even whether logjams in elite thinking can be broken. But it may not be inappropriate to indicate the sorts of measures which seem in order if one is to approach something like satisfaction of the Deutschian conditions for social peace:

1. People must increasingly, and rapidly, be relieved of the necessity to live in constant fear. The constant fear of being assaulted individually is relevant, but it is not the central problem. That problem is the constant fear that (a) one will individually be assaulted by someone of a specific disposition (a black robber or a white policeman, for example) and (b) the social situation in which one lives will be subject to large-scale disruption coming out of a particular part of the polity. (This is our contemporary version of the nineteenth-century Englishman's notion of the "dangerous classes.") The appropriate policy measure is a calculated effort at domestic arms control and disarmament, extending to virtually all official persons as well as to private individuals (see Holden, 1969: ch. 6). This seems a desideratum because the reduction of the arms situation would mean that people, however antagonistic, could have no capacity to assault each other, except by direct person-to-person contact, which qualitatively changes the character of the relationship. Reducing the *physical possibility* of danger is an important condition for stabilizing coexistence, thus leading to options for the enhancement of domestic peace.

2. But measures of this order should never be confused with the achievement of domestic peace. They are mere conditions, however essential. If these conditions, which amount to the first phases of a social truce, are to be transformed into something more, other pragmatic and symbolic acts are requisite. Given the nature of the presidency as a secular priesthood-kingship, much will depend on the dramatic style and example which emanates from that office (Gross, 1966). Consequently, by the choice of words, of persons, and of policies, the president must directly—and continuously—indicate that the achievement of social peace is a priority item equal, and sometimes superior, to the balance of payments question, the stability of the dollar, or the disposition of United States forces around the world. This objective must, if it is to be taken seriously, permeate every agency of the United States government in the same way that the cold war permeated every agency when that was defined as the prime problem.

3. There must be a new political process for defining the forms and objectives of order—in the absence of which the traditional values of police craftsmanship will define them—not only for the policemen but for public consumption. This process must extend to the entire range of social issues relevant to order—and must permit overt discussion of the psychic and pragmatic adjustments required of various segments of the population, and of the ways in which the costs of these adjustments may be reduced. If such a process is to develop, there must be serious attention to the modification of institutional processes—so long neglected in avant garde political science out of the mistaken belief that "institutionalism" and "behavioralism" were exclusive of each other—in order to permit the effective introduction and resolution of the issues which most concern black people.

Most proposals in this direction, even by such critics as Carmichael and Hamilton, have been remarkably "conservative" in their effects although rather "radical" in their supporting rhetoric. My own view is that this requires such newly possible measures as direct support for black private schools competitive with the public schools (a method of cultural affirmation), new forms of statehood for the large American cities, and new forms of representation in the central bureaucracies where—in the integral state—critical choice *must* lie (whether we like it or not).

The question may properly be raised whether the strength of racial prejudice (the only useful definition of "white racism") even permits action in these directions. There is no clear basis for an answer. The negative response (by no means self-evidently wrong), must depend on the presumption that the dynamics of mutual alarm have already pushed us beyond the point of no return. The reason for doubting that answer, however, is rooted in history. St. Clair Drake has observed that, around 1910, the United States and South Africa were remarkably alike in racial politics. Yet the paths of the two countries have unquestionably diverged since then. How could one account for the divergence? As far as I can tell, the answer lies in the fact that "order" (and the control of blacks as a part of the problem of order) is in American culture not a clear, but an ambiguous, symbol. White Americans, as the dominant population, are *simultaneously "prejudiced" and "fair-minded."* The possibilities of defining new conditions for order, outside a framework of raw coercion, depend on that ambiguity, and the possibilities of political leadership in one direction or the other depend on a sophisticated interpretation of the latitudes which that ambiguity affords.

REFERENCES

ALLEN, WILLIAM S. (1965) The Nazi Seizure of Power. Chicago: Quadrangle Books.

AVNERY, URI (1968) Israel without Zionists. New York: Macmillan.

BELL, DANIEL (1960) The End of Ideology. New York Free Press.

BIDERMAN, ALBERT (1966) "Social indicators and goals." Pp. 111-129 in Raymond A. Bauer (ed.) Social Indicators. Cambridge: MIT Press.

BRUCKER, GENE A. (1962) Florentine Politics and Society. Princeton: Princeton Univ. Press.

DEUTSCH, KARL W. et al. (1957) Political Community and the North Atlantic Area. Princeton: Princeton Univ. Press.

DOWNES, BRYAN T. (1968) "The social characteristics of riot cities." Social Science Quarterly 49 (December): 504-520.

EDELMAN, MURRAY (1964) The Symbolic Uses of Politics. Urbana: Univ. of Illinois Press.

FINLAY, DAVID J. et al. (1967) Enemies in Politics. Chicago: Rand McNally.

GEORGE, M. DOROTHY (1968) London Life in the 18th Century. New York: Harper & Row.

GESCHWENDER, JAMES (1968) "Civil rights protests and riots." Social Science Quarterly 49 (December): 474-484.

GRIMSHAW, ALLEN (1959) "Lawlessness and violence in America and their special manifestations in changing Negro-white relationships." Journal of Negro History 44 (January): 52-72.

GROSS, BERTRAM (1968) "Some questions for presidents." Pp. 308-350 in Bertram Gross (ed.) A Great Society? New York: Basic Books.

HOLDEN, MATTHEW, JR. (1969) The Republic in Crisis. San Francisco: Chandler Publishing (forthcoming).

---(forthcoming a) "Achieving order and stability." In Harvey S. Perloff (ed.) volume on problems of American government, under the auspices of the Commision on the Year 2000.

---(forthcoming b) "Politics, public order, and pluralism." In James R. Klonoski and Robert I. Mendelsohn (eds.) The Allocation of Justice. Boston: Little, Brown.

MAYER, MARTIN (1969) "The full and sometimes very surprising story of Ocean Hill, the teacher's union and the teacher's strikes of 1968." New York Times Magazine (February 2): 18.

MILLER, WALTER B. (1967) "Violent crimes in city gangs." Pp. 127-141 in Thomas R. Dye and Brett W. Hawkins (eds.) Politics in the Metropolis. Columbus: Charles E. Merrill.

MORRIS, RICHARD B. (1965) Government and Labor in Early America. New York: Octagon Books.

NIEBURG, H. L. (1968) "Violence, law and the social process." American Behavioral Scientist 2 (March-April): 17-19.

NIEDERHOFFER, ARTHUR (1967) Behind the Shield. Garden City: Doubleday.

PERLOFF, HARVEY S. (forthcoming) "Frameworks for thinking about U.S. government in the year 2000." In Harvey S. Perloff (ed.) volume on problems of American government, under the auspices of the Commission on the Year 2000.

President's Commission on Law Enforcement and the Administration of Justice (1967) The Challenge of Crime in a Free Society. Washington, D.C.: Government Printing Office.

RUDE, GEORGE (1964) The Crowd in History. New York: John Wiley.

SCHUR, EDWIN M. (1965) Crimes Without Victims. Englewood Cliffs: Prentice-Hall.

SHOVER, JAMES L. (1966) Cornbelt Rebellion. Urbana: Univ. of Illinois Press.

SILVER, ALLAN (1967) "The demand for order in civil society." Pp. 1-24 in David J. Bordua (ed.) The Police: Six Sociological Essays. New York: John Wiley.

SKOLNICK, JEROME (1966) Justice Without Trial. New York: John Wiley.

SNOW, C. P. (1962) Science and Government. New York: New American Library.

TOMLINSON, T. M. (1968) "Riot ideology in Los Angeles: a study of Negro attitudes." Social Science Quarterly 49 (December): 485-503.

WASKOW, ARTHUR (1966) From Race Riot to Sit-In. Garden City: Doubleday.

WESTIN, ALAN F. (1967) Privacy and Freedom. New York: Atheneum.

WILSON, JAMES Q. (1968) Varieties of Police Behavior. Cambridge: Harvard Univ. Press.

17

The Urban Unease:
Community vs. City

JAMES Q. WILSON

☐ ONE OF THE BENEFITS (if that is the word) of the mounting
concern over "the urban crisis" has been the emergence, for perhaps
the first time since the subject became popular, of a conception of
what this crisis really means, from the point of view of the urban
citizen. After a decade or more of being told by various leaders that
what's wrong with our large cities is inadequate transportation, or
declining retail sales, or poor housing, the resident of the big city,
black and white alike, is beginning to assert his own definition of
that problem—and this definition has very little relationship to the
conventional wisdom on the urban crisis.

This common man's view of "the urban problem," as opposed to
the elite view, has several interesting properties. Whereas scholars are
interested in poverty, this is a national rather than a specifically
urban problem; the common man's concern is with what is unique to
cities and especially to large cities. Racial discrimination deeply
concerns blacks, but only peripherally concerns whites; the problem
that is the subject of this article concerns blacks and whites alike,
and intensely so. And unlike tax inequities or air pollution, for which
government solutions are in principle available, it is far from clear
just what, if anything, government can do about the problem that
actually concerns the ordinary citizen.

EDITORS' NOTE: *This chapter is reprinted, with references and source commentary added, from* The Public Interest, *No. 12 (Summer, 1968), pp. 25-39, by
permission of the author and the publisher. ©by National Affairs, Inc.*

This concern has been indicated in a number of public opinion surveys, but, thus far at least, the larger implications of the findings have been ignored. In a poll of over one thousand Boston home-owners that I recently conducted in conjunction with a colleague, we asked what the respondent thought was the biggest problem facing the city. The "conventional" urban problems—housing, transporta-tion, pollution, urban renewal, and the like—were a major concern of only 18 per cent of those questioned, and these were expressed disproportionately by the wealthier, better-educated respondents. Only 9 per cent mentioned jobs and employment, even though many of those interviewed had incomes at or even below what is often regarded as the poverty level. [The survey was sponsored by the Joint Center for Urban Studies of MIT and Harvard under a grant from the National Science Foundation.] *The issue which concerned more respondents than any other was variously stated—crime, vio-lence, rebellious youth, racial tension, public immorality, delin-quency. However stated, the common theme seemed to be a concern for improper behavior in public places.*

For some white respondents this was no doubt a covert way of indicating anti-Negro feelings. But it was not primarily that, for these same forms of impropriety were mentioned more often than other problems by Negro respondents as well. And among the whites, those who indicated, in answer to another question, that they felt the government ought to do *more* to help Negroes were just as likely to mention impropriety as those who felt the government had already done too much.

Nor is this pattern peculiar to Boston. A survey done for *Fortune* magazine in which over three hundred Negro males were questioned in thirteen major cities showed similar results. [Beardwood, 1968]. In this study, people were not asked what was the biggest problem of their city, but rather what was the biggest problem they faced as individuals. When stated this generally, it was not surprising that the jobs and education were given the highest priority. What is striking is that close behind came the same "urban" problems found in Boston —a concern for crime, violence, the need for more police protection, and the like. Indeed, these issues ranked *ahead* of the expressed desire for a higher income. Surveys reported by the President's Commission on Law Enforcement and Administration of Justice showed crime and violence ranking high as major problems among both Negro and white respondents. [Commission on Law Enforce-ment and Justice, 1967: 85-89].

THE FAILURE OF COMMUNITY

In reading the responses to the Boston survey, I was struck by how various and general were the ways of expressing public concern in this area. "Crime in the streets" was *not* the stock answer, though that came up often enough. Indeed, many of the forms of impropriety mentioned involved little that was criminal in any serious sense—rowdy teenagers, for example, or various indecencies (lurid advertisements in front of neighborhood movies and racy paperbacks in the local drugstore).

What these concerns have in common, and thus what constitutes the "urban problem" for a large percentage (perhaps a majority) of urban citizens, is *a sense of the failure of community*. By "community" I do not mean, as some do, a metaphysical entity or abstract collectivity with which people "need" to affiliate. There may be an "instinct" for "togetherness" arising out of ancient or tribal longings for identification, but different people gratify it in different ways, and for most the gratification has little to do with neighborhood or urban conditions. When I speak of the concern for "community," I refer to a desire for the observance of standards of right and seemly conduct in the public places in which one lives and moves, those standards to be consistent with—and supportive of—the values and life styles of the particular individual. [My use of the term "community" refers to the *function* of a community in regulating behavior in public places through face-to-face contact.] Around one's home, the places where one shops, and the corridors through which one walks there is for each of us a public space wherein our sense of security, self-esteem, and propriety is either reassured or jeopardized by the people and events we encounter. Viewed this way, the concern for community is less the "need" for "belonging" (or in equally vague language, the "need" to overcome feelings of "alienation" or "anomie") than the concerns of any rationally self-interested person with a normal but not compulsive interest in the environment of himself and his family. [For a treatment of the psychological needs met by a community, see Nisbet, 1953.]

A rationally self-interested person would, I argue, take seriously those things which affect him most directly and importantly and over which he feels he can exercise the greatest influence. Next to one's immediate and particular needs for shelter, income, education, and the like, one's social and physical surroundings have perhaps the greatest consequence for oneself and one's family. Furthermore, unlike those city-wide or national forces which influence a person, what happens to him at the neighborhood level is most easily affected by his own actions. The way he behaves will, ideally, alter the behavior of others; the remarks he makes, and the way he presents himself and his home will shape, at least marginally, the common expectations by which the appropriate standards of public conduct in that area are determined. How he dresses, how loudly or politely he speaks, how well he trims his lawn or paints his house, the liberties he permits his children to enjoy—all these not only express what the individual thinks is appropriate conduct, but in some degree influence what his neighbors take to be appropriate conduct.

These relationships at the neighborhood level are to be contrasted with other ways in which a person might perform the duties of an urban citizen. Voting, as a Harvard University colleague delights in pointing out, is strictly speaking an irrational act for anyone who does not derive any personal benefit from it. Lacking any inducement of money or esteem, and ignoring for a moment the sense of duty, there is for most voters no rational reason for casting a ballot. The only way such an act would be reasonable to such a voter is if he can affect (or has a good chance to affect) the outcome of the election—that is, to make or break a tie. Such a possibility is so remote as to be almost nonexistent. Of course, most of us do vote, but primarily out of a sense of duty, or because it is fun or makes us feel good. As a way of influencing those forces which in turn influence us, however, voting is of practically no value.

Similarly with the membership one might have in a civic or voluntary association; unless one happens to command important resources of wealth, power, or status, joining such an organization (provided it is reasonably large) is not likely to affect the ability of that organization to achieve its objectives. And if the organization *does* achieve its objectives (if, for example, it succeeds in getting taxes lowered or an open occupancy law passed or a nuisance abated), nonmembers will benefit equally with members. This problem has been carefully analyzed by Mancur Olson [1965] in *The Logic of Collective Action* in a way that calls into serious question the ability of any organization to enlist a mass following when it acts for the common good and gives to its members no individual rewards. Some people will join, but because they will get some personal benefit (the status or influence that goes with being an officer, for example), or out of a sense of duty or, again, because it is "fun." As a way of shaping the urban citizen's environment, however, joining a large civic association is not much more rational than voting.

CONTROLLING THE IMMEDIATE ENVIRONMENT

It is primarily at the neighborhood level that meaningful (i.e., potentially rewarding) opportunities for the exercise of urban citizenship exist. And it is the breakdown of neighborhood controls (neighborhood self-government, if you will) that accounts for the principal concerns of urban citizens. When they can neither take for granted nor influence by their actions and those of their neighbors the standards of conduct within their own neighborhood community, they experience what to them are "urban problems"—problems that arise directly out of the unmanageable consequences of living in close proximity.

I suspect that it is this concern for the maintenance of the neighborhood community that explains in part the overwhelming preference Americans have for small cities and towns. According to a Gallup Poll taken in 1963, only 22 per cent of those interviewed wanted to live in cities, 49 per cent preferred small towns, and 28 per cent preferred suburbs. (Only among Negroes, interestingly enough, did a majority prefer large cities—perhaps because the costs of rural or small town life, in terms of poverty and discrimination, are greater for the Negro than the costs, in terms of disorder and insecurity, of big-city life.) Small towns and suburbs, because they are socially more homogeneous than large cities and because local self-government can be used to reinforce informal neighborhood sanctions, apparently make the creation and maintenance of a proper sense of community easier. At any rate, Americans are acting on this preference for small places, whatever its basis. As Daniel Elazar [1968] has pointed out, the smaller cities are those which are claiming a growing share of the population; the largest cities are not increasing in size at all and some, indeed, are getting smaller.

A rational concern for community implies a tendency to behave in certain ways which some popular writers have mistakenly thought to be the result of conformity, prejudice, or an excessive concern for appearances. No doubt all of these factors play some role in the behavior of many people and a dominant role in the behavior of a few, but one need not make any such assumptions to explain the nature of most neighborhood conduct. In dealing with one's immediate environment under circumstances that make individual actions efficacious in constraining the actions of others, one will develop a range of sanctions to employ against others and one will, in turn, respond to the sanctions that others use. Such sanctions are typically informal, even casual, and may consist of little more than a gesture, word, or expression. Occasionally direct action is taken—a complaint, or even making a scene, but resort to these measures is rare because they invite counterattacks ("If that's the way he feels about it, I'll just show him!") and because if used frequently they lose their effectiveness. The purpose of the sanctions is to regulate the external consequences of private behavior—to handle, in the language of economists, "third-party effects," "externalities," and "the production of collective goods." I may wish to let my lawn go to pot, but one ugly lawn affects the appearance of the whole neighborhood, just as one sooty incinerator smudges clothes that others have hung out to dry. Rowdy children raise the noise level and tramp down the flowers for everyone, not just for their parents.

Because the sanctions employed are subtle, informal, and delicate, not everyone is equally vulnerable to everyone else's discipline. Furthermore, if there is not a generally shared agreement as to appropriate standards of conduct, these sanctions will be inadequate to correct such deviations as occur. A slight departure from a norm is set right by a casual remark; a commitment to a different norm is very hard to alter, unless of course the deviant party is "eager to fit in," in which case he is not commited to the different norm at all but simply looking for signs as to what the preferred norms may be. Because of these considerations, the members of a community have a general preference for social homogeneity and a suspicion of heterogeneity—a person different in one respect (e.g., income, or race, or speech) may be different in other respects as well (e.g., how much noise or trash he is likely to produce).

PREJUDICE AND DIVERSITY

This reasoning sometimes leads to error—people observed to be outwardly different may not in fact behave differently, or such differences in behavior as exist may be irrelevant to the interests of the community. Viewed one way, these errors are exceptions to rule-of-thumb guides or empirical generalizations; viewed another way, they are manifestations of prejudice. And in fact one of the unhappiest complexities of the logic of neighborhood is that it can so often lead one wrongly to impute to another person some behavioral problem on the basis of the latter's membership in a racial or economic group. Even worse, under cover of acting in the interests of the neighborhood, some people may give vent to the most unjustified and neurotic prejudices.

However much we may regret such expressions of prejudice, it does little good to imagine that the occasion for their expression can be wished away. We may even pass laws (as I think we should) making it illegal to use certain outward characteristics (like race) as grounds for excluding people from a neighborhood. But the core problem will remain—owing to the importance of community to most people, and given the process whereby new arrivals are inducted into and constrained by the sanctions of the neighborhood, the suspicion of heterogeneity will remain and will only be overcome when a person proves by his actions that his distinctive characteristic is not a sign of any disposition to violate the community's norms.

Such a view seems to be at odds with the notion that the big city is the center of cosmopolitanism—by which is meant, among other things, diversity. And so it is. A small fraction of the population (in my judgment, a *very* small fraction) may want diversity so much that it will seek out the most cosmopolitan sections of the cities as places to live. Some of these people are intellectuals, others are young, unmarried persons with a taste for excitement before assuming the responsibilities of a family, and still others are "misfits" who have dropped out of society for a variety of reasons. Since one element of this group—the intellectuals—writes the books which define the "urban problem," we are likely to be confused by their preferences and assume that the problem is in part to maintain the heterogeneity and cosmopolitanism of the central city—to attract and hold a neat balance among middle-class families, young culture-lovers, lower-income Negroes, "colorful" Italians, and big businessmen. *To assume this is to mistake the preferences of the few for the needs of the many.* And even the few probably exaggerate just how much diversity they wish. Manhattan intellectuals are often as worried about crime in the streets as their cousins in Queens. The desired diversity is "safe" diversity—a harmless variety of specialty stores, esoteric bookshops, "ethnic" restaurants, and highbrow cultural enterprises.

[I suspect that the tolerance for social diversity, especially "safe diversity," increases with education and decreases with age. This tolerance, however, does not extend to "unsafe diversity"—street crime, for example.]

ON "MIDDLE-CLASS VALUES"

At this point I had better take up explicitly the dark thoughts forming in the minds of some readers that this analysis is little more than an elaborate justification for prejudice, philistinism, conformity, and (worst of all) "middle-class values." The number of satirical books on suburbs seem to suggest that the creation of a sense of community is at best little more than enforcing the lowest common denominator of social behavior by means of *kaffee klatsches* and the exchange of garden tools; at worst, it is the end of privacy and individuality and the beginning of discrimination in its uglier forms.

I have tried to deal with the prejudice argument above, though no doubt inadequately. Prejudice exists; so does the desire for community; both often overlap. There is no "solution" to the problem, though stigmatizing certain kinds of prejudgments (such as those based on race) is helpful. Since (in my opinion) social class is the primary basis (with age and religion not far behind) on which community-maintaining judgments are made, and since social class (again, in my opinion) is a much better predictor of behavior than race, I foresee the time when racial distinctions will be much less salient (though never absent) in handling community problems. Indeed, much of what passes for "race prejudice" today may be little more than class prejudice with race used as a rough indicator of approximate social class.

With respect to the charge of defending "middle-class values," let me stress that the analysis of "neighborhood" offered here makes no assumptions about the substantive values enforced by the communal process. On the contrary, the emphasis is on the process itself; in principle, it could be used to enforce any set of values. To be sure, we most often observe it enforcing the injunctions against noisy children and lawns infested with crabgrass, but I suppose it could also be used to enforce injunctions against turning children into "sissies" and being enslaved by lawn-maintenance chores. In fact, if we turn our attention to the city and end our preoccupation with suburbia, we will find many kinds of neighborhoods with a great variety of substantive values being enforced. Jane Jacobs described how and to what ends informal community controls operate in working-class Italian sections of New York and elsewhere. Middle-class Negro neighborhoods tend also to develop a distinctive code. And Bohemian or "hippie" sections (despite their loud disclaimers of any interest in either restraint or constraint) establish and sustain a characteristic ethos.

PEOPLE WITHOUT COMMUNITIES

Viewed historically, the process whereby neighborhoods, in the sense intended in this article, have been formed in the large cities might be thought of as one in which order arose out of chaos to return in time to a new form of disorder.

Immigrants, thrust together in squalid central-city ghettos, gradually worked their way out to establish, first with the aid of streetcar lines and then with the aid of automobiles, more or less homogeneous and ethnically distinct neighborhoods of single-family and two-family houses. In the Boston survey, the *average* respondent had lived in his present neighborhood for *about twenty years.* When asked what his neighborhood had been like when he was growing up, the vast majority of those questioned said that it was "composed of people pretty much like myself"—similar, that is, in income, ethnicity, religion, and so forth. In time, of course, families—especially those of childrearing age—began spilling out over the city limits into the suburbs, and were replaced in the central city by persons lower in income than themselves.

Increasingly, the central city is coming to be made up of persons who face special disabilities in creating and maintaining a sense of community. There are several such groups, each with a particular problem and each with varying degrees of ability to cope with that problem. One is composed of affluent whites without children (young couples, single persons, elderly couples whose children have left home) who either (as with the "young swingers") lack an interest in community or (as with the elderly couples) lack the ability to participate meaningfully in the maintenance of community. But for such persons, there are alternatives to community—principally, the occupancy of a special physical environment that in effect insulates the occupant from such threats as it is the function of community to control. They move into high-rise buildings in which their apartment is connected by an elevator to either a basement garage (where they can step directly into their car) or to a lobby guarded by a doorman and perhaps even a private police force. Thick walls and high fences protect such open spaces as exist from the intrusion of outsiders. The apartments may even be airconditioned, so that the windows need never be opened to admit street noises. Interestingly, a common complaint of such apartment-dwellers is that, in the newer buildings at least, the walls are too thin to ensure privacy—in short, the one failure of the physical substitute for community occasions the major community-oriented complaint.

A second group of noncommunal city residents are the poor whites, often elderly, who financially or for other reasons are unable to leave the old central-city neighborhood when it changes character. For many, that change is the result of the entry of Negroes or Puerto Ricans into the block, and this gives rise to the number of anti-Negro or anti-Puerto Rican remarks which an interviewer encounters. But sometimes the neighborhood is taken over by young college students, or by artists, or by derelicts; then the remarks are anti-youth, anti-student, anti-artist, or anti-drunk. The fact that the change has instituted a new (and to the older resident) less seemly standard of conduct is more important than the attributes of the persons responsible for the change. Elderly persons, because they lack physical vigor and the access to neighbors which having children facilitates, are especially vulnerable to neighborhood changes and find it especially difficult to develop substitutes for community—except, of course, to withdraw behind locked doors and drawn curtains. They cannot afford the high-rise buildings and private security guards that for the wealthier city-dweller are the functional equivalent of communal sanctions.

In the Boston survey, the fear of impropriety and violence was highest for those respondents who were the oldest and the poorest. Preoccupation with such issues as the major urban problem was greater among women than among men, among those over sixty-five years of age than among those under, among Catholics more than among Jews, and among those earning less than $5,000 a year more than among those earning higher incomes. (Incidentally, these were *not* the same persons most explicitly concerned about and hostile to Negroes—anti-Negro sentiment was more common among middle-aged married couples who had children and modestly good incomes.)

The third group of persons afflicted by the perceived breakdown of community are the Negroes. For them, residential segregation as well as other factors have led to a condition in which there is relatively little spatial differentiation among Negroes of various class levels. Lower-class, working-class, and middle-class Negroes are squeezed into close proximity, one on top of the other, in such a way as to inhibit or prevent the territorial separation necessary for the creation and maintenance of different communal life styles. Segregation in the housing market may be (I suspect it is) much more intense with respect to lower-cost housing than with middle-cost housing, suggesting that middle-class Negroes may find it easier to move into previously all-white neighborhoods. But the constricted supply of low-cost housing means that a successful invasion of a new area by middle-class Negroes often leads to that break being followed rather quickly by working- and lower-class Negroes. As a result, unless middle-class Negroes can leapfrog out to distant white (or new) communities, they will find themselves struggling to assert hegemony over a territory threatened on several sides by Negroes with quite different life styles.

This weakness of community in black areas may be the most serious price we will pay for residential segregation. It is often said that the greatest price is the perpetuation of a divided society, one black and the other white. While there is some merit in this view, it overlooks the fact that most ethnic groups, when reasonably free to choose a place to live, have chosen to live among people similar to themselves. (I am thinking especially of the predominantly Jewish suburbs.) *The real price of segregation, in my opinion, is not that it forces blacks and whites apart but that it forces blacks of different class positions together.*

WHAT CITY GOVERNMENT CANNOT DO

Communal social controls tend to break down either when persons with an interest in, and the competence for, maintaining a community no longer live in the area or when they remain but their neighborhood is not sufficiently distinct, territorially, from areas with different or threatening life styles. In the latter case especially, the collapse of informal social controls leads to demands for the imposition of formal or institutional controls—demands for "more police protection," for more or better public services, and the like. The difficulty, however, is that there is relatively little government can do directly to maintain a neighborhood community. It can, of course, assign more police officers to it, but there are real limits to the value of this response. For one thing, a city only has so many officers and those assigned to one neighborhood must often be taken away from another. And perhaps more important, the police can rarely manage all relevant aspects of conduct in public places whatever may be their success in handling serious crime (such as muggings or the like). Juvenile rowdiness, quarrels among neighbors, landlord-tenant disputes, the unpleasant side effects of a well-patronized tavern—all these are matters which may be annoying enough to warrant police intervention but not to warrant arrests. Managing these kinds of public disorder is a common task for the police, but one that they can rarely manage to everyone's satisfaction—precisely because the disorder arises out of a dispute among residents over what *ought* be the standard of proper conduct.

In any case, city governments have, over the last few decades, become increasingly remote from neighborhood concerns. Partly this has been the consequence of the growing centralization of local government—mayors are getting stronger at the expense of city councils, city-wide organizations (such as the newspapers and civic associations) are getting stronger at the expense of neighborhood-based political parties, and new "superagencies" are being created in city hall to handle such matters as urban renewal, public welfare, and anti-poverty programs. Mayors and citizens alike in many cities have begun to react against this trend and to search for ways of reinvolving government in neighborhood concerns: mayors are setting up "little city halls," going on walking tours of their cities, and meeting with neighborhood and block clubs. But there is a limit to how effective such responses can be, because whatever the institutional structure, the issues that most concern a neighborhood are typically those about which politicians can do relatively little.

For one thing, the issues involve disputes among the residents of a neighborhood, or between the residents of two adjoining neighborhoods, and the mayor takes sides in these matters only at his peril. For another, many of the issues involve no tangible stake—they concern more the *quality* of life and competing standards of propriety and less the dollars-and-cents value of particular services or programs. Officials with experience in organizing little city halls or police-community relations programs often report that a substantial portion (perhaps a majority) of the complaints they receive concern people who "don't keep up their houses," or who "let their children run wild," or who "have noisy parties." Sometimes the city can do something (by, for example, sending around the building inspectors to look over a house that appears to be a firetrap or by having the health department require someone to clean up a lot he has littered), but just as often the city can do little except offer its sympathy.

POVERTY AND COMMUNITY

Indirectly, and especially over the long run, government can do much more. First and foremost, it can help persons enter into those social classes wherein the creation and maintenance of community is easiest. Lower-class persons are (by definition, I would argue) those who attach little importance to the opinions of others, are preoccupied with the daily struggle for survival and the immediate gratifications that may be attendant on survival, and inclined to uninhibited, expressive conduct. (A lower-*income* person, of course, is not necessarily lower *class;* the former condition reflects how much money he has, while the latter indicates the attitudes he possesses.) Programs designed to increase prosperity and end poverty (defined as having too little money) will enable lower-income persons who do care about the opinions of others to leave areas populated by lower-income persons who don't care (that is, areas populated by lower-class persons).

Whether efforts to eliminate poverty by raising incomes will substantially reduce the size of the lower class is a difficult question. The progress we make will be much slower than is implied by those who are currently demanding an "immediate" and "massive" "commitment" to "end poverty." I favor many of these programs, but I am skeptical that we really know as much about how to end our social problems as those persons who blame our failure simply on a lack of "will" seem to think. I suspect that know-how is in as short supply as will power. But what is clear to me is that *programs that seek to eliminate poverty in the cities will surely fail,* for every improvement in the income and employment situation in the large cities will induce an increased migration of more poor people from rural and small-town areas to those cities. The gains are likely to be wiped out as fast as they are registered. To end urban poverty it is necessary to end rural poverty; thus, programs aimed specifically at the big cities will not succeed, while programs aimed at the nation as a whole may.

The need to consider poverty as a national rather than an urban problem, which has been stated most persuasively by John Kain and others, is directly relevant to the problem of community [Kain and Persky, 1967]. *Programs that try to end poverty in the cities, to the extent they succeed, will probably worsen, in the short run, the problems of maintaining a sense of community in those cities—and these communal problems are, for most persons, the fundamental urban problems.* People migrate to the cities now because cities are, on the whole, more prosperous than other places. Increasing the advantage the city now enjoys, without simultaneously improving matters elsewhere, will increase the magnitude of that advantage, increase the flow of poor migrants, and thus make more difficult the creation and maintenance of communal order, especially in those working-class areas most vulnerable to an influx of lower-income newcomers. This will be true whether the migrants are white or black, though it will be especially serious for blacks because of the compression effects of segregation in the housing market.

THE DIFFERENCES AMONG PEOPLE

It is, of course, rather misleading to speak in global terms of "classes" as if all middle-class (or all working-class) persons were alike. Nothing could be further from the truth; indeed, the failure to recognize intraclass differences in life style has been a major defect of those social commentaries on "middle-class values" and "conformity." The book by Herbert J. Gans [1967] on Levittown is a refreshing exception to this pattern, in that it calls attention to fundamental cleavages in life style in what to the outside observer appears to be an entirely homogeneous, "middle-class" suburb. Partly the confusion arises out of mistaking economic position with life style—some persons may be economically working-class but expressively middle-class, or vice versa.

To what extent can persons with low incomes display and act upon middle-class values? To what extent is there a substitute for affluence as a resource permitting the creation and maintenance of a strong neighborhood community? Apparently some Italian neighborhoods with relatively low incomes nonetheless develop strong communal controls. The North End of Boston comes to mind. Though economically disadvantaged, and though the conventional signs of "middle-class values" (neat lawns, quiet streets, single-family homes) are almost wholly absent, the regulation of conduct in public places is nonetheless quite strong. The incidence of street crime is low, "outsiders" are carefully watched, and an agreed-on standard of conduct seems to prevail.

Perhaps a strong and stable family structure (as among Italians) permits even persons of limited incomes to maintain a sense of community. If so, taking seriously the reported weakness in the Negro family structure becomes important, not simply because of its connection with employment and other individual problems, but because of its implications for communal order. Indeed, substantial gains in income in areas with weak family and communal systems may produce little or no comparable gain in public order (and I mean here order as judged by the residents of the affected area, not order as judged by some outside observer). What most individuals may want in their public places they may not be able to obtain owing to an inability to take collective action or to make effective their informal sanctions.

"BLACK POWER" AND COMMUNITY

It is possible that "Black Power" will contribute to the ability of some neighborhoods to achieve communal order. I say "possible"—it is far from certain, because I am far from certain as to what Black Power implies or as to how dominant an ethos it will become. As I understand it, Black Power is not a set of substantive objectives, much less a clearly worked-out ideology, but rather an attitude, a posture, a communal code that attaches high value to pride, self-respect, and the desire for autonomy. Though it has programmatic implications ("neighborhood self-control," "elect black mayors," and so forth), the attitude is (to me) more significant than the program. Or stated another way, the cultural implications of Black Power may in the long run prove to be more important than its political implications.

In the short run, of course, Black Power—like any movement among persons who are becoming politically self-conscious, whether here or in "developing" nations—will produce its full measure of confusion, disorder, and demagoguery. Indeed, it sometimes appears to be little more than a license to shout slogans, insult "whitey," and make ever more extravagant bids for power and leadership in black organizations. But these may be only the short-term consequences, and I for one am inclined to discount them somewhat. The long-term implications seem to be a growing pride in self and in the community, and these are prerequisites for the creation and maintenance of communal order.

Historians may someday conclude that while Negroes were given emancipation in the nineteenth century, they had to win it in the twentieth. The most important legacy of slavery and segregation was less, perhaps, the inferior economic position that Negroes enjoyed than the inferior cultural position that was inflicted on them. To the extent it is possible for a group to assert communal values even though economically disadvantaged, Negroes were denied that opportunity because the prerequisite of self-improvement—self-respect—was not generally available to them. The present assertion of self-respect is an event of the greatest significance and, in my view, contributes more to explaining the civil disorders and riots of our larger cities than all the theories of "relative deprivation," "economic disadvantage," and the like. The riots, from this perspective, are expressive acts of self-assertion, not instrumental acts designed to achieve particular objectives. And programs of economic improvement and laws to guarantee civil rights, while desirable in themselves, are not likely to end the disorder.

The fact that these forms of self-expression cause such damage to the black areas of a city may in itself contribute to the development of communal order; the people who are paying the price are the Negroes themselves. The destruction they have suffered may lead to an increased sense of stake in the community and a more intense concern about the maintenance of community self-control. Of course, no amount of either self-respect or commitment to community can overcome a serious lack of resources—money, jobs, and business establishments.

NO INSIDE WITHOUT AN OUTSIDE

Because the disorders are partly the result of growing pride and assertiveness does not mean, as some have suggested, that we "let them riot" because it is "therapeutic." For one thing, whites who control the police and military forces have no right to ignore the interests of the nonrioting black majority in favor of the instincts of the rioting black minority. Most Negroes want *more* protection and security, not less, regardless of what certain white radicals might say. Furthermore, the cultural value of Black Power or race pride *depends in part on it being resisted by whites.* The existence of a "white enemy" may be as necessary for the growth of Negro self-respect as the presumed existence of the "capitalist encirclement" was for the growth of socialism in the Soviet Union. As James Stephens once said, there cannot be an inside without an outside.

Nor does Black Power require that control over all political and economic institutions be turned over forthwith to any black organization that happens to demand it. Neighborhoods, black or white, should have control over some functions and not over others, the decision in each case requiring a rather careful analysis of the likely outcomes of alternative distributions of authority. Cultures may be invigorated and even changed by slogans and expressive acts, but constitutions ought to be the result of deliberation and careful choices. The reassertion of neighborhood values, by blacks and whites alike, strikes me as a wholly desirable reaction against the drift to overly bureaucratized central city governments, but there are no simple formulas or rhetorical "principles" on the basis of which some general and all-embracing reallocation of power can take place. Those who find this reservation too timid should bear in mind that functions given to black neighborhoods will also have to be given to white neighborhoods—it is not politically feasible (or perhaps even legally possible) to decentralize power over black communities but centralize it over white ones. Are those radicals eager to have Negro neighborhoods control their own police force equally eager to have adjoining working-class Polish or Italian neighborhoods control theirs?

In any case, no one should be optimistic that progress in creating meaningful communities within central cities will be rapid or easy. The fundamental urban problems, though partly economic and political, are at root questions of values, and these change or assert themselves only slowly, if at all. And whatever gains might accrue from the social functions of Black Power might easily be outweighed by a strong white reaction against it and thus against blacks. The competing demands for territory within our cities is intense and not easily managed, and for some time to come the situation will remain desperately precarious.

REFERENCES

BEARDWOOD, ROGER (1968) "The new Negro mood." Fortune 77 (January): 146-151.

Commission on Law Enforcement and Justice (1967) Task Force Report: Assessment of Crime. Washington, D.C.: Government Printing Office.

ELAZAR, DANIEL J. (1968) "Are we a nation of cities?" Pp. 89-97 in Robert A. Goldwin (ed.) A Nation of Cities. Chicago: Rand McNally.

GANS, HERBERT J. (1967) The Levittowners. New York: Pantheon Books.

KAIN, JOHN F. and JOSEPH J. PERSKY (1967) "The North's stake in southern rural poverty." Harvard Program in Regional and Urban Economics, Discussion Paper No. 18.

NISBET, ROBERT (1953) The Quest for Community. New York: Oxford Univ. Press.

18

The Urban Paradox

HENRY J. SCHMANDT
and JOHN C. GOLDBACH

☐ "MEN COME TOGETHER in cities in order to live. They remain
together to live the good life." Since Aristotle wrote these words
some 2,400 years ago, the pursuit of the good life in the urban
communities of the world has had a varied course. Cities have
enjoyed "golden ages:" those of medieval Italy reached high pinnacles in the development of the arts, and those of the Dutch
burghers achieved a nobility of domestic living seldom duplicated in
the annals of urbanism. But chaos and destruction, tension and
violence, poverty and disease have also been the common lot of cities
everywhere. The "black plague" in the fourteenth century took more
than half the population of London, and allied bombs in World War
II virtually destroyed the industrial cities of the Ruhr valley. The
story is the same in the United States. The British burned the
nation's capital during the War of 1812, and a half century later
General Sherman wreaked total destruction on town and farm alike
in his march from Atlanta to the sea. Even the civil disorders which
have struck American cities today find their parallel in the collective
violence of the past: rioting in Boston in 1773 against British taxes,
the draft protests in New York during the Civil War, and the clashes
between strikers and police in Chicago in 1937 are but a few
examples.

Today the cry of "law and order" is a familiar strain heard in American cities across the land. It is the favorite theme of candidates of all parties, eliciting the loudest popular applause and generating the most hysteria among the citizenry. It dominates the debate over national domestic goals and overshadows the discussions about the quality of our urban environment. Never was it more important, however, that this theme be put in proper perspective. The common problem facing all societies, at whatever stage of their development, has been to devise and maintain a satisfactory system of order—not in the pejorative sense of force and suppression as currently implied by the term "law and order," but in the more fundamental sense of organizing and integrating human activities among individuals and groups that share a common spatial locality. This task is basically one of politics, of government. Classical urban theory explicitly recognized this responsibility by its emphasis on the role of the city or polity as an ordering and integrating agent of social control. Modern theory, however, rejected this notion by viewing local government as a service-providing bureaucracy and formulating its basic problems in terms of administrative efficiency. The result was a distortion of the role of the local polity and its significance for contemporary urban life.

Those who call American society "sick" point to the demonstrations and riots, the rising crime rate, the use of drugs, the disaffection of the youth, and the heresy of the intellectuals. But as Arthur Schlesinger argues, the "sickness" of American society does not reside in the existence of problems; it rests in the nation's incapacity to deal effectively with them. This incapacity, moreover, is less a question of economic resources than one of political and governmental responsiveness. As John Gardner, former HEW secretary and now head of the Urban Coalition, observed: "Beneath all the city's physical problems and its social ills, we are faced with problems of social organization, of governance, of politics in the Aristotelian sense of the word."

Contemporary developments in the political life of the nation have brought with them new demands on the public instrumentalities of social control, from the policeman patrolling a beat to the justices of the Supreme Court ruling on critical issues of human rights. The massive problems of poverty and discrimination, together with the numerous other civic maladies resulting from rapid urbanization and industrialization, call for high degrees of expertise and sympathetic understanding by all who are engaged in the formulation and execution of public policy. They also call for institutional responsiveness commensurate with the size of the crisis and sensitive to the needs and aspirations of all segments of the society. No strategy or action is likely to succeed if applied in small doses or a niggardly fashion. Challenge and response must be of the same magnitude.

There are those who say that the local polity is no longer capable of responding adequately to the urban challenge, that it has been drained of the capacity to act effectively as power has gravitated upward to higher levels of government, industry, and organizational groupings. Hans Blumenfeld is correct when he says that it is important to distinguish between problems which are urban in character, such as traffic congestion and air pollution, and those which happen to have their principal situs in the urban community, such as unemployment, poverty, and racial discrimination. The latter affect all sections of the country, rural as well as urban, so that action to eliminate poverty or reduce unemployment and discrimination within the cities cannot succeed without corresponding action to correct these deficiencies outside of them. Problems of this nature, simply put, are national problems which require national policy and national action for their solution.

True as the distinction is, we cannot ignore the fact that the urban community is the major battlefield for the most crucial domestic issues of national concern. One need not accept the proposition, advanced by some, that local governments are little more than a delivery system for national products—"the local community is the place where the national community administers things locally"—to appreciate the dependence of national urban policy upon effective implementation at the local level. The instances are numerous in which such policy has been denied by neglect, misinterpretation, or outright defiance by local authorities and locally based institutions. In a sense, the principle of self-help is as applicable here as it is to individuals. Unless we are willing to discard our federal system of government and further centralize power—with all the dysfunctional consequences this implies—national policies and action can only put local polities in a position to help themselves. For unless there is vigorous response to current needs and demands at the community level, problem-solving is severely circumscribed.

In some ways, the urban community is a set of paradoxes. It is the center of the nation's finest intellectual, cultural, and productive achievements, but it is also the scene of its greatest social failures and its greatest disorders. It is a highly interdependent area where the aggregation of varied skills, talents, and capital has brought about a high level of prosperity, but it is also an area where central city is set off against suburb, where racial separatism predominates, and where political power is fragmented and dispersed. It is at one and the same time beautiful and ugly, planned and unplanned, friendly and hostile, a playground and a battlefield, a magnet of hope and a symbol of despair. It generates activism and involvement on the one hand, and alienation and withdrawal on the other.

These paradoxes underlie the "plight" of the modern city, finding expression in the counter-forces and tendencies which characterize its functioning and development. Specialization begets both power fragmentation and greater interdependence; centrifugal forces pull against centralization; technology and bureaucratization militate against citizen involvement, and at the same time give impetus to it; the rising expectations of the deprived are matched by the increasing resistance of the "haves;" the geopolitics of social and racial separatism conflict with functionalist policies of integration; the psychological needs for local territorial identification run counter to the continuous expansion of human interests into national space. The critical structural problem is to devise a political framework at the local or regional level within which these conflicting forces might best be mediated and accommodated. As we shall attempt to show, the question is less one of restructuring local government than of restructuring relationships between levels of government and between the public and private sectors.

THE ATTITUDINAL CLIMATE

The vision of local communities composed of citizens working together in peace and harmony, dedicated to the general good, and closely attached to their city, as the ancient Greek to his polis and Calvin to his Geneva, has long intrigued man. Pre-industrial society saw this ideal in the city as a place of religious assembly, in which worship was the common bond, and the altar, not the central business district, the focal point. Jefferson saw it in the small rural village, the ward republic, "the wisest invention ever devised by man for the perfect exercise of self government." Jane Jacobs and others see it in the revitalized neighborhood of the large city.

THE URBAN PARADOX [477]

Americans are peculiarly disadvantaged in their search for the ideal community. Short in history, devoid of an urban tradition, and captivated by a "rural romanticism," they retain, like Jefferson, a fear and distrust of the city and its concentration of people and activities. To many of them, the city is where one encounters different and threatening styles of life. For some, it is Gargantua; for others, a monster which man has created and now does not know how to control. Plutarch's "city as great teacher" gives way to Steffens' "shame of the cities." This anti city bias is little conducive to fostering sentiments of attachment for, and allegiance to, the commune on the part of the urban dweller. It negates any possibility of endowing the city with a civic personality in which its residents can share and identify themselves.

But the cord also pulls in the other direction. The sprawling industrial giant may blur the civic image and destroy its symbolic value, yet it holds the spotlight steadily on its economic opportunities, its aggregation of wealth, its material rewards, its modern culture, and the conveniences of its advanced technology. Like Jason's golden fleece, these features continue to attract an aspiring populace and hold it within the urban confines. Although the Secretary of Agriculture pleads for a return to the countryside and the Economic Development Administration channels funds into the declining rural areas to make them more attractive, the tide refuses to be reversed. The bright lights entice even while they repulse.

This attraction-repulsion dichotomy leads to an ambivalence in man's posture toward the city. He wants both Gargantua and the rural village, the advantages of urbanism and the freedom from its costs. To obtain both, he endeavors to compartmentalize his private world spatially, entering the city proper for his economic pursuits (and at times for his entertainment) and leaving it for the virtues of the "less urban" sanctuary. The difficulty is that the city keeps catching up with him. Even his efforts to bar sidewalks in his residential world so as to preserve "the semi-rural character" of his sub-community eventually fail. Thus he pushes for more expressways to take him farther out, for greenbelts to hold back the human tide, and for stricter zoning controls to keep out the "undesirables."

Unfortunately for his peace of mind, his economic well-being is too closely linked to the urban core for him to disregard completely its fate. (Even the lower-middle-class suburban white has some awareness of this interdependence.) He lends vocal support to slum clearance and renewal efforts by the central city, speaks vaguely of city-suburban cooperation, participates in United Fund drives for the area, and contributes to the cultural activities of the inner city. Aside from these token acts, however, he remains unwilling to face up to the realities of contemporary urban life. Although recent events have made him well aware of the serious nature of the social problems faced by the center city, he conveniently relegates their solution to local public functionaries and higher levels of government. "Support your local police force," becomes the symbolic expression for avoiding personal involvement and responsibility. Place the task in the hands of the "experts" and let them satisfy the minimal needs of the underprivileged, remove some of the gross disparities which trigger off trouble, and above all, maintain order. As Scott Greer and others have referred to the modern urbanite, he is indeed a man of limited civic commitment.

PLACE AND NONPLACE

The conflicting forces or counterpulls at work in human settlements have at least one dimension in common, that of space. Specialization, for example, a major distinguishing characteristic of the modern urban community, is manifested spatially as well as functionally. The ecological map of every sizeable city or metropolitan area invariably shows both a spatial aggregation of economic activities and a spatial distribution of social types. Just as sections of an urban community are identified with particular industrial and commercial uses, so also are entire neighborhoods given over to groups socially differentiated by age, socioeconomic status, ethnicity, and color.

The idea of community, moreover, is closely linked to the traditional notion of place or space. Aside from the question of shared values, the traditional mark of community, people and groups in urban settlements are united by the interdependence which arises among them as they pursue their diverse interests in a common locality. The sheer fact of one's dependency on another, the trademark of a specialized world, in turn gives rise to a mutual concern among urbanites in maintaining the operation of their place-based systems.

For the average individual unversed in the mysteries of regional science, the macro-community and its interrelationships are too complex for him to comprehend and understand. Much closer and more meaningful to him—and therefore the subject of more intense feelings on his part—is his spatially defined neighborhood or sub-community. Here the rhetoric of shared values acquires substance even though the values may lack the nobility of the utopian dream. The print is simple in pattern and sharply etched: people who live in close proximity to each other share a common interest in what occurs near and about their place of residence. This concern, unlike that for the larger settlement, does not relate to the economic or ecological interdependence of the area but to the quality of those public services and regulatory devices which affect the social and physical surroundings of the residential castle. As James Wilson has observed elsewhere in this volume, community is less the need for identification or belonging than the concern for the immediate environment in which the individual and his family live and move.

The members of these spatially demarcated subcommunities, particularly the large majority in the population who are family- or home-centered, are concerned essentially with the notion of place as a means for maintaining their style of life (Williams *et al*, 1965; Williams, 1966). Hence, they are particularly interested in govern-mental activities which they perceive as contributing to this end. These include direct services such as the control of traffic flow, protection from criminals and miscreant neighbors, and the provision of child-rearing facilities. They also encompass what is sometimes referred to as gate-keeping functions, the exclusion of people (and enterprises) with conflicting values: those who do not eat the same kind of food, share the same beliefs, dress in the same fashion—the minority groups spilling out of their place, the flower children, the social activists, be they clergymen or "kooks," and other deviant individuals with "strange" manners. Each family-oriented group, in short, demands that the others conform to its values and accept its priorities. In the process, as Herbert Gans (1967) points out, each seeks power to prevent others from shaping the institutions that must be shared, for otherwise the family and its culture are not safe.

Men shy away from the "threat" of diversity, from the task of living with others of different values and ways of life who share the world with them. Unable to empathize with diversity, they cherish the myth that the one way to avoid conflict is to put enough space between themselves and these alien spirits, a buffer zone that will isolate them from "undesirables" and minimize the threat believed implicit in close contact. When the neighborhood of the large city no

longer provides this protective shield, they move to the outer peri-
meters where space is less at a premium and where compatriots can
be found. Thus, the exurbanite who holds dear his home, with hedge
and fence is, in a sense, "a linear descendant of the legendary pioneer
who moved when he could see a neighbor's smoke." The rural tradition of the family as a self-sufficient unit still survives even though
those who cherish it most are conscious of its unreality.

Life-style interests are basically place-related, a function of
location. Unlike work, friendships, religion, professional contacts,
and other spatially undefined relations, they are territorially
bounded by the individual's residential environs. Paradoxically, the
demand for greater control over life-style functions is being intensified (even in the black neighborhoods) at a time when place-related
interests represent a decreasing proportion of the total bundle each
individual holds (Webber, 1964). The widening sweep of interests
unrelated to place is manifested in numerous ways. With mounting
affluence and more leisure time, the urban dwelling loses something
of its primacy as recreation, sports, travel, and other attractions draw
an increasing number of families away from the hearth during week-
ends and holidays—to a second home on the lake or to one of the
numerous public and private recreational areas that span the nation.
In both the social and work worlds, also, wealth and technology have
permitted the geographical range of intercourse to expand phenome-
nally, freeing interactions of many types from the requirement of
physical proximity and multiplying the ties and interests which
transcend space. Moreover, many concerns vital to urban dwellers—
among them poverty, civil rights, and unemployment—are not
territorially defined and are beyond the scope of effective neighborhood or local action.

The notion of neighborhoods of urbanites vigorously place-
oriented and rigidly dedicated to the maintenance of the status quo
is subject to qualification. There are many such subareas where the
defense of hearth and home is intensely pursued. But there are also
others with residents little committed to place, their major interests
oriented outward. Between these two extremes the intensity
continuum ranges widely. Neighborhoods in which the concern of
the residents revolves around the preservation of life-style values are
likely to be more highly cognizant of space-related functions and
extremely resistant to changes which might weaken or jeopardize
their control over these activities. Conversely, those whose predominant interests are undefined by space are less likely to evince a high

degree of concern over place-related functions or to feel as threatened by change.

The stereotype of upper-income suburbs—the parkland of the organizational and professional elite—jealously guarding the walls of their domestic citadel in contrast to the "openness" of city neighborhoods and lower-income satellites, has been effectively punctured by empirical evidence. In fact, spatially related interests have in many ways become more important to the lower-middle class than to those higher on the socioeconomic scale. The blue-collar worker, taxicab driver, filling station attendant, and grocery store clerk are more intimately bound to their neighborhoods by an equity in their house, which often represents their life's savings, by lower mobility opportunities, by feelings of insecurity, and by fewer ties to the outside world. This territorial circumspection leads to such manifestations as the fierce and at times violent resistance to the entry of Negroes by lower-middle-income white enclaves, both in suburbia and in the central city.

At the other extreme, a neighborhood dominated by apartment dwellers in the professional and managerial classes is less likely to manifest frantic concern over spatially related activities. Outside of basic security to their person and property—a concern held equally by all classes and all races—such activities are not perceived by them as vital to the maintenance of their style of life. Not only do their prime interests and personal contacts transcend their residential environment, but they also enjoy a large measure of mobility, geographical as well as social and occupational, and greater opportunity of choice. Cosmopolitan in outlook and well-equipped by education and resources for the competitive life of modern urban society, they have neither the insecurity of the lower-middle-class white nor the localist bias of the suburban tract dweller. Their investment in the neighborhood is negligible, their interest in it as a potential community minimal.

The neighborhoods of the poor and the ghetto-dwellers present still another case. How meaningful to the residents of these areas are place-related activities which perpetuate poor housing, provide school curricula irrelevant to their needs, and represent a standard of public services well below that of the more affluent sections of the urban complex? Black separatists, as we know, argue for programs of ghetto enrichment instead of "piecemeal integration" which they see thwarted by demographic trends toward more segregation. They also argue against the "bureaucratic paternalism" of big-city administrations from which suburbanites have already escaped. Yet, how much do ghetto residents want to preserve the life-style of their neighborhoods, a style developed largely as a defensive mechanism against the hostility and unresponsiveness of the majoritarian society? Adequate housing, better services, and environmental improvements are clearly called for. However, the ultimate solution to the problem of poverty and race lies less in the betterment of these space-related factors than in enhancing the social and economic mobility of the ghetto prisoners and providing them with an improved set of options.

The dilemma here is obvious. On the one hand, the nation must provide a satisfactory living environment for its underprivileged citizens, a task that involves essentially place-related activities and interests. On the other hand, it must solve the basic problem of incorporating this minority into the mainstream of American life, a task that is spatially undefined. The two objectives are not, of course, incompatible, and each can be pursued simultaneously with the other. The difficulty, however, is that they keep coming in conflict with each other in their execution. Educational policies which stress the neighborhood school (place-related) clash with those which call for the establishment of educational parks and the mingling of children of all races and socioeconomic status (nonplace-related). Similarly, the demands of black separatists for neighborhood autonomy (place-related) run counter to the integrationist programs (nonplace-related) long advocated by civil rights proponents. These counter-pulls lead to confusion in the definition of objectives and, in turn, to public policies ambiguous in their formulation and execution.

TOWARD A PLACE-NONPLACE THEORY

The paradoxes conspicuous in urban affairs arise out of the greater number of social differences in the city, the more opportunities it offers and the more contests it generates, and the correspondingly greater concern for additional controls over the urban aggregate. But the changes and reforms that most concern cities in turmoil also challenge the underlying beliefs and attitudes of cultural systems. The resultant controversy over these challenges is, in turn, proportionate to the amounts of control required for the public regulatory systems to implement needed changes or new opportunities. Social differences that were contested in city settings of the past shifted forms of control, from hereditary status to proscriptive contracts in Europe, and from community power structures to pluralistic politics in urban America. These historic shifts were realized through the cumulative power of the spatially undefined interests of newcomers and at the expense of the place-related interests of oldtimers. However, both the old American ambivalence over public power—the less government the better—and the cumulative developments of rapid urbanization have diffused the controls necessary for the satisfactory coexistence of spatial and aspatial interests. This ambivalence, as noted earlier, is reflected in the often confused regulatory policies of government.

The economic developments and institutional reforms, which challenge older place-related interests by augmenting new functions, affect policy proposals and the quality of urban life in many ways. The challenges to cultural systems traceable to inventions or economic expansion effectively do away with the previous scarcities which so often restricted the field of political systems identified with local cultures. However, the durability of local politics against economic and technological breakthroughs suggests that place-related beliefs and attitudes cover more than negative feelings of restriction or fear. Thus the ascriptions more positively invoked by man's primordial and everlasting dependence on land are traced all the way from his need for privacy and identity to ethnic culture and even Western civilization.

Place-related interests in the American urban experience more specifically connote: (1) belief in localism as patriotism toward one's home ground, and suspicion of governments in proportion to their remoteness (Key, 1949); (2) belief in property rights as the key to the good life, expressed in residential zoning as a protective device for culture, and in economic controls ranging from the freeman's "forty acres and a mule" of a hundred years ago to the subdivision techniques of the present (Dietze, 1963; McKissick, 1966); (3) the "bottoms up" attitude in political practices which are sufficiently attentive to certain constituents and promotive of a district loyalty that immobilizes geographic representatives against a total-community view (Fesler, 1949); and (4) the growing emphasis given to the physical environment by planners and conservationists as the essential determinant of everything from child-rearing to the overall ecosystem (Park, 1952; Wilhelm, 1962).

For a long time, such place-centered commitments allowed little or no public observance or control over the growth and direction of spatially undefined interests. As a result, the development of the latter was left to the free play of economics and technology, a policy that drained powers from place-centered controls. The general belief persisted that private enterprise would foster more balanced and responsive cities. But the breakthroughs of sophisticated economies and the "technostructure" of modern society shifted powers to such nonplace intangibles as anonymous management in lieu of local ownership, and to income that could flow out of the area. Each functional innovation released centrifugal forces that successfully challenged the physical confines of local governments, if not their political control altogether.

Entrepreneurs evaded the space limitations of local boundaries by gaining corporate power with state and national encouragement. Then economic notables themselves withdrew from active participation in the civic affairs of the local polity. (Schulze, 1961). Meanwhile, automobiles and freeways effectively destroyed spatial immobility and instigated the movement and settlement of more urban residents to more distant points. Concomitantly, the worst cases of poverty, though originating outside the city, migrated to its inner core. Here the minority groups began to center their demands for earning power and human rights against the primacy of property rights and the limitations of the ghetto environment. And warriors in battles against poverty increasingly emphasized income strategies at the national level rather than service strategies which are part of local jurisdictions (Gross, 1968).

In addition to marking successive shifts in urbanization, each of the above contests increased the mobility and therefore the power of aspatial challengers at the expense of controls possessed by spatial interests. The new specializations and additional services accompanying each shift not only advanced against old shortages and over place restrictions, but also broadcast new opportunities about still other options to still more challengers. Paradoxically, the gains secured by the nonplace developments also reformulated place-centered beliefs suitable to the successful challengers: property rights and tax havens for industrial plants, localism founded on home ownership and its cares for suburbanized workers, participatory politics and neighborhood activism for core-city minorities. Instead of resolving the American ambivalence over power, these gains and developments served to accentuate differences, actual and perceived, between urban publics and their governments. The more widespread the sensitivity to the good life of affluence, the more pressure is put upon governments for the redress of perceived wrongs and the delivery of actual goods and services (Gurr, 1968). The whole fission of events hardly marks the end of place-relations in favor of an aspatial utopia, a redundant mirage. But no longer can it be said that the local political enterprise is circumscribed by territorial limits.

Viewed in this perspective, the essential aims of today's urban minorities are not much different from challenges to the polity made by entrepreneurs and workers in earlier periods. The net effect of each contest is an assault on older place-related interests ranging from the flight of industries and elites to service "spillovers" which burden long-time residents and their city administrations.

A "chain reaction" inventory of the leading economic developments, social dissents, and groups mobilized against established systems of control, reveals not only more demands beyond the capacity of place-related systems (local governments), but also new forms of politics altogether apart from place loyalties. American political practices have implicated all kinds of governments in the challenges and direction of urban life, designating in the process two major roles: the public sector as nonplace master and as spatial servant.

In nonplace respects, governments—predominantly the national—mastermind distributive policies in which programs for the economy overcome environmental restraints or scarcities that deter equalization: everything from freeways, eminent domain, gas taxes, and rapid transit subsidies, to fair employment, minimum wage and income strategies, and the "soft justice" decisions of supreme courts which so often reject local police practices in seeking to enforce a modicum of equality of treatment. Execution of these programs goes beyond the initial enforcement, and oftentimes against the norms, of local officials imbued with the ideology of the indigenous polis and its folkways. The extent to which national rules in such areas as open housing, welfare, and reapportionment are public-regarding, is probably of less concern to localists than what these actions portend for the administrative integrity and jurisdiction of the place-based polity. By now also, economic surpluses, the refinements of cost-benefit analysis, and the spread of the mass media have sensitized not only the affluent but other Americans as well to public controls against closed organizations and for more "fair shares" commitments (Bredemeir, 1968).

The enlarged functional choices accompanying economic enterprise have broadened opportunities and enhanced life-styles, and these by-products have in turn resulted in more urban governments serving integrational place-related ends. The epochal shifts from production to consumption economies, and from piecemeal physical to regionalized social planning, have made local subsystems more responsive to the financial and psychological resources which extraterritorial organizations, both public and private, can invest in block grants, urban coalitions, and income transfers. But this latest series of shifts also suggests that the process model of pluralistic politics based on "who governs" is being adapted to content models of intervention more concerned with what these governments "should do."

A growing number of subsystems made possible by the investment and intervention strategies of supralocal agencies are place-related. These strategies include everything from suburban police grants, model cities, tax deductions for new property-owners, and owner-occupancy loans, to community action projects and "advocate planning" for inner core neighborhoods (Marris and Rein, 1967). Paradoxically or not, the increased mobility that drained local government capacities and led to specialized and outside services has also widened the range of choice for many groups opting for the creation of new subareas for themselves without having to challenge, head on, the place-related interests of the "haves." Instead of a clear trend for aspatial utopias, the movement this time may be toward new decentralization schemes and representative systems for urban newcomers and suburban consumers alike. This, in a sense, is what Gans meant when he concluded from his Levittown study that the best approach to change "is to give up the single solution that compromises between the wants of different groups, and to experiment with new solutions for dissatisfied groups and cultures in the total population" (Gans, 1967: 412).

PLACE-RELATEDNESS AND PROFESSIONALISM

The distaste for open politics commonly evidenced by civic notables and even middle-class whites reflects the simplistic notion that contests or conflicts, at the local level at least, are the work of "self-seekers" or "outsiders," and threats to efficient management and services. This bias towards apolitical public administration is receptive to the criteria and routines important to business and technology but it is not attuned to the realities of modern urban communities. The material and ideological surpluses that accumulate in urban settings not only permit, but cannot prevent, a loosening up of attitud s and allocations toward more choices, which also means more politics. But more politics, today, does not mean more of the same kind of public controls as in the past. The new political forms are promoted at the outset by dissenters, as was true with the Progressives; and then justified against the "critic notables" in terms of social investment.

As stated by Roger Starr, "The city seems less promising to its critics because more people are getting more of its services" (Starr, 1966: 32). The old politics may have reflected a "bottoms up" attitude but it became overcommitted and therefore restricted to bossism and selective bargaining. Even so, this process model of politics—"who gets what, when, and how"—simulated the production systems of its day in its styles of membership and position, and concerned itself with the immediate problems of maintaining group cohesion against rival competitors. The political tradition focused on basically internal and quantitative standards for the allocation of public goods. The new politics goes beyond such process models of governance and inclines to a content model cued to consumer expectations and federal and state aids. It focuses on "public interest" as dedicated to quality controls improving the level and equitable distribution of goods and services. Not only quantitative standards, but also norms promoted by outside organizations, are involved in this model.

Standards have not only risen vertically with incomes and services; they have also been opened horizontally to new demands from new sources in the city. The corresponding increase in contests—an increase that spells turmoil to many civic notables and local administrations—has less to do with challenging American values than with applying them to publics formerly excluded from full participation in the reward structure. In the old system, "have-nots" were maintained as cheap labor pools and selectively admitted to group and party membership for mutual benefits in jobs, votes, and access to the distributive system. Today, commercialization and antipoverty campaigns extend choices to urban newcomers in more compelling ways than membership in an interest group or victory through elections (Davis, 1965; Dentler, 1968). In the older system, for example, all kinds of associations mediate between urban publics and high-level administrators. The newer systems of manufacturing, sales, and media, on the contrary, by bringing more choices to the attention of a mass audience, make these publics less resigned to a circumscribed status and less respectful of the mediating processes of interest groups (Smelser, 1962; Bell, 1968). In the older system, to cite another example, police departments helped make possible, through recruitment and selective enforcement, the coexistence of newcomers and notables. Today's police, on the other hand, face dissenters and social sanctions beyond their "station-house culture" and place-related norms.

Production and communication revolutions lead to the loss of insulating space, which enhanced the status of old elites; the obsolescence of mediating processes by previously more influential groups; and the increased pressures on governments to respond directly and immediately to large social issues. Urban pluralism that ordinarily describes the "dispersed inequalities" of intervening groups, is now seen as the saving grace of a national system that extends the range of qualitative varieties and political choices (Thompson, 1968). New contests thus challenge the old political and spatial systems of urban control. Many of the programs of city officials are criticized as holding actions. Many of the notables and their heirs, dispersed in suburbs, are criticized for noninvolvement and scapegoating. And ghetto antagonists are criticized most of all for rising crime rates and "aggressive demands." Although more choices and shares are mandated for more publics in the new systems, these criticisms—and the content of new programs for that matter—shed little light on whether "politics has faded amidst general plenty," or if "new forms of participation can produce bureaucratic responsiveness" (Rein and Miller, 1966: 48).

With the increase of their scale and cost, services become less local and political, the important factors in the older process model. As a consequence, the apparent difference between bureaucratic professionalism and citizen involvement is still presented as one between "administrative leadership and political dabbling" (Sommers, 1958). This bipolar distinction has a direct bearing on the process and content of policies in the metropolis. As put by William Riker, only political leaders must stay in one place to rise in the world while bureaucrats must move about the country for career development (Riker, 1964: 107).

Although the new politics, like modern economic power, has become fluid and national in many regards, the civic controls of older local politics still prevent, as noted by Neil McDonald, the more professional, nonplace environment from "taking over completely" (McDonald, 1955: 84-85). The new administrators also realize that many service problems involve differences in social values more than they do technical facts. Their bureaucratic response to these values can still be realized in the normative terms of the content model of "public interest" as much or more, and sometimes in spite of, the disposition of local representatives (Kristol, 1968).

Confrontations between the professional bureaucrat and the involved citizen are complicated by cross-factored claims from both sides. The democratic ambivalence in which power is both feared and required now has a metropolitan corollary where power is increasingly dispersed and brought in by two kinds of outsiders: the nonlocal official and the noncivic antagonist. As cities became larger or leaner, the politics of bossism and bargaining gave way to bigger issues of alienation and "group mobilization." The importance of the councilman or district representative interceding on behalf of interested citizens, appears to yield to the more functional representations of pressure groups and protest movements. This usually does not occur until after local officials have abdicated their responsibility to a citywide public interest by failing to provide grievance machinery and effective access to city hall for the less affluent newcomers. The larger the city, measured in terms of more organizations, different roles, and intervening authorities, the more critical do rules, communication, and overhead planning become. But such long-range and expert considerations undeniably relegate citizen participation to a seat far back.

The process model of city politics is criticized today, even where operating, as unable to help those most in need of outputs designed to enhance human dignity and stimulate civic involvement. The upward-mobiles among the disadvantaged point out that representation of minorities in council chambers no longer provides an effective means of influencing or controlling the administration of such functions as police and planning (Young, 1964). Some maintain that elections of Negroes, even to the mayor's office, will not pacify "have-nots" so much as accelerate their pace of expectations and further aggravate their disillusionment (Hadden, 1967). Finally, black power leaders argue that separatism is more conducive to solidarity than integration policies (Handlin, 1966). The choice of ghetto enrichment implicit in this argument is not unlike the long practiced separatism of suburban whites. Striking at politics and administration alike, these objections question the tidy efficacy of political negotiating skills and professional expertise. The critical issues that have spilled beyond local systems of control, and which demand normative standards and new kinds of action, seem also to be beyond scientific, and therefore professional, determination.

The leading concepts of participatory politics and bureaucratic responsiveness involve both information and operation gaps. The conflicts of value arising from militant black demands and stubborn white resistence will be experienced for some time in metropolitan areas. These conflicts will perpetuate the "communication gap" between dispersed and diverse publics and the "responsibility gaps" increasingly charged to local governments. But because of the critical nature of the social issues and their threat to the well-being of the nation as a whole, urban and metropolitan areas will become the staging grounds for increasing amounts of long-range professional assistance. The likely result is that more citizens, disaffected by the gap between promise and fulfillment, will charge high-level administrators with all kinds of "credibility gaps" (Gladwin, 1967). Forecasts of process strategies—in redevelopment projects for core cities, relocation of low-cost housing for core residents, and the various regional development programs for metropolitan areas—redefine political issues and power in terms of the functional expertise of outsiders. But a growing set of content strategies developed in the War on Poverty, with choices in community amenities, job investment in "human capital," and decentralized lay advisory boards acknowledge a revolution of rising involvement with nonprofessional information from new local sources.

RESTRUCTURING URBAN GOVERNMENTAL RELATIONSHIPS

It is against the background of trends and issues we have described that the local polity's potential for improving the quality of the urban environment must be assessed. Dissatisfaction with the performance of urban government, and its apparent inability to respond energetically to contemporary needs, has led to innumerable efforts to reshape local political structures. These attempts have ranged from simple tinkerings with internal administrative mechanisms, to campaigns for the creation of unified metropolitan governments. Major impetus for the reform movement has come from those whose interests are less place-related than spatially undefined: businessmen who serve a regional or national clientele; technocrats who view local governmental fragmentation as an impediment to efficient adminis- tration and the effective handling of urban problems; political scientists who see no adequate structure in the present system through which the demands for integrated services and greater equity can be arbitrated. The efforts to bring about change, however, have met with general public indifference and active resistance from the entrenched localists who view a decentralized or suburbanized government pattern as critical to the protection of place-related life- styles. The record of accomplishment, as a result, is not impressive; to most critics, it is dismally poor.

What has been missing in much of the debate over reorganization is an assessment of the social costs of structural or organizational change. The essential question that has been consistently skirted is whether and in what way the translation of social demands into public policy, and the responsiveness of local governmental systems to class, racial, and other community differences, vary with the kinds of political structures adopted. Descriptions of the process model of politics have tended to measure policy outcomes in terms of local influence, costs, and compromises: the larger the city, the harder it is to get agreement (Presthus, 1964); informal agreements are needed to overcome formal decentralization (Banfield, 1961); decisions are made at the lowest common denominator of voluntary assent (Wood, 1963); and local leaders are most likely to manage conflicts by deferring or deflecting action (Wheaton, 1964). These propositions, however, undervalue the grievances of those who have been excluded from sharing local political power and who consequently have little respect for the city hall or court house game. The newly found activism of groups formerly "powerless" and their resort to tactics of "crisis precipitation" and "creative disorder" are increasingly chal- lenging the older processes and structures. In so doing, they have cast the issue of urban political reorganization in a new context with new dimensions.

The concerns of the "new activists," in short, overstep structural mechanisms for new forms of problem-solving. Similarly, the issues which have reached crisis proportions in many cities, call for new approaches to the dilemma of urban governance. These develop- ments, in turn, suggest that the focus be shifted from the question of

reorganization to a reordering of relations within the total governmental structure as it presently exists. Such a reordering would be
based on the present and emergent forces in modern society and be
designed to take advantage of the pressures generated by the growing
disabilities and fears of the urban populace and by their rising
expectations of the good life.

Two forces, originating at opposite poles, are already converging
to alter the urban political structure. The first, referred to above, is
the activism of racial minorities and the poor who are attempting to
change the control structure of the local polity in order to have a
greater voice in the allocation of its rewards. The second is federal
action in the form of conditional grants-in-aid, poverty and model-
cities programs, open-housing legislation, and similar measures. These
interventions in local systems are seeking to redistribute values and
resources so as to increase the opportunity and worth of the deprived
segments of the urban community. As both of these forces have
grown stronger, they have met with increased resistance from the
majoritarian society.

A CONCEPTUAL MODEL

To conceptualize the present situation, it is helpful to distinguish
analytically between the distribution and delivery systems of government. The first involves essentially the allocation of public goods and
resources. It is the principal means whereby the inequities which a
capitalistic society inevitably generates among its members are
tempered, through income redistribution (welfare, rent subsidies,
guaranteed annual incomes, etc.) and through public services. The
second consists of the organizational mechanisms through which
public goods and services are delivered to the citizen consumers. The
latter system also serves as a redistributive agent to the extent that its
output is not paid for equally by all the recipients. Its main function,
however, is service, not income redistribution, a distinction that is of
importance to governmental restructuring.

The matrix in Figure I is designed to indicate in broad outline the
relationship between the local delivery and distribution systems and
those of higher governmental echelons. Four "ideal-type" cities or
urban communities are represented in the schema. The first, the
"political city" occurs where both the distribution and delivery
systems are largely local and where city politics in the process sense

(competition over resource allocations) plays a predominant role in community life. This type of polity existed in the United States until the early decades of the present century, but it has rapidly disappeared in more recent times as the national and state governments have assumed an increasing share of the distributive functions.

DELIVERY SYSTEM

DISTRIBUTIVE SYSTEM	Local	Extralocal
Local	Political City	Contract City
Extralocal	Administrative City	Dependent City

Figure 1. **A SCHEMA OF CITY TYPES**

The second, or contract city, is a special type of limited significance. It is best illustrated by the "Lakewood Plan" in the Los Angeles area which permits a city to contract with the county government for virtually all of its municipal services—"functional consolidation without political consolidation"—as the slogan goes. In such cases, that portion of the distributive system which today remains in local hands continues to be controlled by the city government while the delivery system is contracted away to another level of government.

The third category, the administrative city, represents the direction in which local government in the United States appears to be evolving. Under this form, the distribution function—including a large share of the financing of urban services—is assumed by the national and state governments. Local government becomes essentially a delivery system which must operate within the general framework cast by the larger distribution system. This arrangement does not necessarily mean diminution of local governmental powers, although it is regarded as such. Considerable administrative discretion in policy interpretation and application is likely to remain with those who have charge of the delivery system. In addition, the redistributive policies of the upper administrative levels can offer new opportunities for local governments by making available to them resources which they themselves could not possibly muster because of political realities or lack of economic means.

The fourth and final types shown in the matrix, the "dependent city," represents a further devolution of local powers to higher levels. This type is common in cities of the developing nations where not only the distribution system but also the actual administration of many key local services are vested in extralocal governments. This latter development can be expected to occur in the United States with increasing frequency (creating, for example, special authorities or endowing councils of governments with operating powers) in the field of such nonplace related functions as air and water pollution control, the development and management of large-scale recreational facilities, and transportation.

THE DIRECTION OF CHANGE

Local polities are not effective redistribution agencies, either of income or nonmaterial values, for at least two reasons: (1) their relative lack of resources, and (2) the intense opposition that is generated locally by efforts to change the distribution system. At the same time, local authorities are less concerned with who has control over societal distribution policy than they are with maintaining their autonomy over the delivery system. The attitudes of the urban populace in this regard take two forms. On the one hand, middle-class whites view the local delivery system with its social controls as vital to the preservation of their place-related life-styles. On the other hand, minority groups and the poor are concerned with this system primarily for the distributive function it performs. This concern is heightened by the fact that the redistributive policies of the national and state governments are channeled largely through the local delivery system. Both segments of the community, the "haves" and the "have-nots," seek control over this structure, but for basically different ends.

This analysis suggests that reorganization efforts might be directed with more success at the distribution system than, as is commonly done, at the delivery system. In this context, the arguments of those who call for a strategy of raising the income levels of the disadvantaged directly through national policies, rather than subsidizing this portion of the population indirectly through the service structure, make eminent sense. Such a strategy, by minimizing the importance of the local delivery system as a redistributive mechanism, and by enlarging the opportunity structure of the deprived through increased income, would help ease the tensions which now threaten to make armed camps out of many of the nation's urban communities. The adoption of income strategies would also allow the greater employment of fees and "factor prices" in local services without causing undue hardship on the less affluent. This practice would enable local governments to concentrate on the more urgent programs by reducing the burden of "free" public services across the board. (Thompson, 1968). The inevitable result would be a strengthening of the local polity's delivery system and an enhancement of its viability.

With the assumption of a larger role by the national government in income reallocation, place-related interests would continue to be protected by the local delivery system, as they are presently. At the same time, the reconstituted distribution system would permit the now deprived members of the society to establish, should they so desire, their own neighborhoods or suburbs according to their own taste. For the urban community as a whole, this would mean a diversity of housing, environmental arrangements, and institutions keyed to the diversity of background, cultures, and aspirations of its various segments.

Given the existing urban crisis, the demands of the separatists, the uneasiness of the "haves," and the rising level of prosperity which is constantly enlarging the potential distribution pool, the possibility of such a settlement is less remote than one might assume. There are gains in it to be realized by all participants. The local struggle for equality and life-style protection need not be a zero sum game in which there are only winners and losers; it can, if properly structured and articulated, be converted into an "economic-type" game in which all or most of the parties profit. By opening up new opportunities for the new contestants on the urban scene, less of a threat will be posed to the now resistant majority. In return for the assurance of continued control over their place-related concerns, established groups are likely to be less resistant to policies which permit other groups to develop their own life styles.

A settlement of this nature can occur only if control over the distribution system is shifted predominantly to the national and state levels, and if private industry, with its nonspatial orientation, is given sufficient governmental encouragement to assist on a large scale in broadening the opportunity structure for the now deprived elements of the population. The thrust must come from extralocal sources, since long experience has demonstrated the virtual impossibility of modifying the distribution system at the local level. The odds against such modification have, moreover, multiplied with the proliferation of local governments and taxing units in urban areas. With each new municipal incorporation or special district, the redistributive abilities of local units, like the chances for metropolitan government, become more remote. But with each new demand and expectation of public service, decision-making also becomes less municipal. The instrumental difference between local politics and professional administration no longer centers around which system is *the* policy-making branch, or which is "closest" or "on top." The growing irrelevance of place-related rivalries to the revolution of rising expectations, leads to attitudes more disposed to accept higher-level resources. The outcome, as dramatized in the urban issues over order and justice, amounts to a dispersal of power with a range of controls sufficient to bridge the gaps between overcommitted local abilities and undersubscribed regional needs (Clark, 1967; Martin, 1965).

The principal factors underlying the attraction-repulsion paradox can be sorted out as an inventory of forces affecting the qualitative perceptions and quantitative changes in urban settings. The paradigm

THE QUALITY OF URBAN LIFE

Structural Factors	Local Concerns	National Commitments
Life-style Interests:	place-related norms (involvement and environment)	aspatial standards (mobility and professionalism)
Styles of Political Control:	the process model (selective bargaining)	the content model (policy-oriented)
Attitudinal Climate:	separatism (the gate-keeping functions of subareas)	functionalism (the goals of regional planning and integration)
Stages of Resource Control:	delivery systems (local services to citizen consumers)	redistributive policies (state and federal allocation of public goods and opportunities)

shown presents some complementary sets of distinctions for local incentives and national inducements which illustrate the stages and dispersal of power accommodated in urban life.

In today's enlarged circumstances, which allow needed gains without real losses to any group or segment of the society, the dispersal of power introduces aspects of decision-making, conventionally missed by city hall, in new forms ranging from community action to national goals. Such performance standards modify the spatial intervention of local players. As much as crises draw attention to new proposals, the urban paradox may be settled by extending services and at the same time enhancing community values. Charting these developments as complementary agents, rather than as aberrant conflicts, extends the scope of urban life beyond the well-documented patterns of social stratification and brokerage politics to the wider complex of issues and choices more representative of today's metropolis.

REFERENCES

BANFIELD, EDWARD C. (1961) Political Influence. New York: Free Press.

BELL, DANIEL (1968) "The adequacy of our concepts." Pp. 127-161 in Bertram M. Gross (ed.) A Great Society? New York: Basic Books.

BREDEMEIER, H. C. (1968) "The politics of the poverty cold war." Urban Affairs Quarterly 3 (June): 3-35.

CLARK, JOSEPH (1967) "The new urbanism." Pp. 38-51 in B. Berry and J. Meltzer (eds.) Goals for Urban America. Englewood Cliffs, N.J.: Prentice-Hall.

DAVIS, KINGSLEY (1965) "Some demographic aspects of poverty in the United States." Pp. 299-319 in Margaret S. Gordon (ed.) Poverty in America. San Francisco: Chandler.

DENTLER, ROBERT A. (1968) American Community Problems. New York: McGraw-Hill.

DIETZE, GOTTFRIED (1963) In Defense of Property. Chicago: Henry Regnery.

FESLER, JAMES (1949) Area and Administration. University, Ala.: Univ. of Alabama Press.

GANS, HERBERT (1967) The Levittowners. New York: Pantheon Books.

GLADWIN, THOMAS (1967) Poverty U.S.A. Boston: Little, Brown.

GROSS, BERTRAM [ed.] (1968) A Great Society? New York: Basic Books.

GURR, T. (1968) "Urban disorder: perspectives from the comparative study of civil strife." American Behavioral Scientist 11 (March-April): 51-54.

HADDEN, J. K. et al. (1968) "The making of Negro mayors, 1967." Trans-Action 5 (January): 29-31.

HANDLIN, OSCAR (1966) "Negro American." Daedalus 95 (Winter): 268-283.

KEY, V. O. (1949) Southern Politics in State and Nation. New York: Knopf.

KRISTOL, IRVING (1968) "Decentralization, for what?" Public Interest 2 (Spring): 17-25.

MCDONALD, NEIL A. (1955) Study of Political Parties. Garden City, N.Y.: Doubleday.

MCKISSICK, FLOYD (1966) Hearings Before Subcommittee on Executive Reorganization of Committee on Government Operations, 90th Congress. Pp. 2289-2331 in Federal Role in Urban Affairs, Part 2. Washington, D.C.: Government Printing Office.

MARRIS, PETER and MARTIN REIN (1967) Dilemmas of Social Reform. New York: Atherton Press.

MARTIN, ROSCOE C. (1965) Cities and the Federal System. New York: Atherton Press.

PARK, ROBERT E. (1952) Human Communities: The City and Human Ecology. New York: Free Press.

PRESTHUS, ROBERT (1964) Men at the Top. New York: Oxford Univ. Press.

REIN, MARTIN and S. M. MILLER. (1966) "Poverty, policy, and purpose: the dilemmas of choice." Pp. 20-64 in L. H. Goodman (ed.) Economic Progress and Social Welfare. New York: Columbia Univ. Press.

RIKER, WILLIAM (1964) Federalism: Origins, Operations, Significance. Boston: Little, Brown.

SCHULZE, ROBERT O. (1961) "The bifurcation of power in a satellite community." Pp. 19-80 in Morris Janowitz (ed.) Community Political Systems. New York: Free Press.

SOMMERS, WILLIAM (1958) "Council-manager government, a review." Western Political Quarterly 11 (March): 145-155.

STARR, ROGER (1966) The Living End: The City and Its Critics. New York: Coward-McCann.

THOMPSON, WILBUR (1968) "The city as distorted price system." Psychology Today 23 (August): 28-33.

WEBBER, MELVIN M. (1964) "The urban place and the nonplace urban realm." Pp. 79-153 in Melvin M. Webber et al. Explorations Into Urban Structure. Philadelphia: Univ. of Pennsylvania Press.

WHEATON, WILLIAM (1964) "Public and private agents of change in urban expansion." Pp. 154-191 in Melvin M. Webber et al. Explorations Into Urban Structure. Philadelphia: Univ. of Pennsylvania Press.

WILHELM, SIDNEY M. (1962) Urban Zoning and Land Use Theory. New York: Free Press.

WILLBERN, YORK (1964) The Withering Away of the City. University, Ala.: Univ. of Alabama Press.

WILLIAMS, OLIVER et al. (1965) Suburban Differences and Metropolitan Policies. Philadelphia: Univ. of Philadelphia Press.

WOOD, ROBERT C. (1963) "The contributions of political science to urban form." Pp. 99-128 in Werner Z. Hirsch (ed.) Urban Life and Form. New York: Holt, Rinehart & Winston.

YOUNG, WHITNEY (1964) To Be Equal. New York: McGraw-Hill.

19

Foundations as Urban Reformers
Between Vision and Reality

LESLIE PAFFRATH

☐ WHEN I PASSED through Harlem a few days ago on the New York–Chicago train, a giant yellow billboard attached to a slum building advertised a current Broadway show—"Promises, Promises." I lived for years on the edge of Harlem and began walking its streets as an observer more than thirty years ago. I have taken the same train route through Harlem for that length of time also. With the exception of high-rise public housing projects which have replaced some of the slum blocks, progress must seem to Harlem families largely another case of "Promises, Promises"—still unfulfilled. The same decaying tenements are there, but less habitable and safe because older. The gap between reality and new prospects is wider than a decade ago because the human environment is at such variance with what technology can provide and the nation can afford. The tenement children of school age when I first knew Harlem now have their own children, many of them being brought up in the same physical conditions. New generations have begun life in the same neighborhoods. Years ago, I knew homes so poor that I wondered what kept more of the "breadwinners" from stealing. I recall the children of a Puerto Rican family, and have never discharged the feeling that malnutrition lowered one of the children's resistance to the illness that was fatal. The rise in the national crime rate demonstrates a number of things, including a rampant attitude that what cannot be earned can be taken, if necessary by violence. The temper of the slums has shifted from frustration, to anger, to violence.

While this volume and chapter focus on urban affairs, the perspective would be fogged if we did not recognize that the story of promises unfulfilled is not limited to urban needs. An almost identical profile can be drawn for the urban poor and the rural poor. The equation—poverty equals handicap—applies with similar force to those bereft through isolation and economic stagnation in rural areas as it does to the dwellers in choked urban centers. The hollows of Appalachia underscore this fact. The proud, stoical inhabitants of this area, descendents of early nineteenth century settlers, have been given promise after promise. They have been assured roads which would bring services in and carry men and products out to the market place. Many of the roads they travel are still too primitive to bring in a physician or, ironically, to carry out the dead during closed seasons. They have been promised a higher level of education for rural schools, including transportation from the remote hollows, more experienced teachers, modern buildings. In most areas of Appalachia the people are still waiting, recalling past promises.

This chapter is purposely preoccupied with poverty; it does not endeavor to sum up all of the other circumstances in cities which call for the involvement of foundations. Literature on that subject is extensive, fast growing, and available. What it does attempt to do is to set down thought and experience which may find a place in some communities. As an author comes to appreciate that all he writes will not appeal equally to all, so the value of this chapter to readers will vary. For the author, one impulse transmitted and then converted to a working program will be compensation.

THE FUNCTIONS OF FOUNDATIONS

Confronting this landscape of human need, and aware of the forces of change which have produced it in the historical process, foundations in the United States are a special tool to be used with skill, courage, and imagination to help alert the nation. Beyond this role, foundations can be a potent force—as a refined expression of society's concern for the individual—in reshaping the human environment for man's benefit. The task will require foundations, as well as the private sector at large, to create social inventions not known in the past. As part of this task, foundations, if they are to fulfill their potential, must be willing to engage in some extremely difficult and sometimes risky undertakings and to confront some literally awesome questions.

Foundations can assist man in evaluating how he can live on earth. It is Buckminster Fuller who reminds us that certain patterns are invisible to man; that because his motion sense is limited, he is relatively insensitive to the pattern of change. What man can feel, touch, see, and hear is less than one millionth of reality. There is under way at all times a strong tide of change which will affect him, but which remains invisible. If we can see only a portion of the earth, as the astronauts viewed it, should it surprise us that we cannot "see" more than a portion of the social scene? It must continue to be a role of foundations to make us aware of this limitation, an awareness which can move us toward greater understanding. This understanding, in turn, should not lead us to abdicate but to place goals in a perspective which is realizable.

In considering policy planning, a comparison of perhaps some value may be made. The United States, in the field of foreign affairs, has found it difficult to set and adhere to a long-term trajectory of policy. Such a projection is, in practice, virtually impossible, due to the variables involved in a world of nation-states of differing cultures and the shifting tactics of our own national policy, including changing administrations. Our goals in foreign policy are said to suffer, under these circumstances, from a lack of constancy. In dealing with unidentical cultures within the nation (North-South, rural-urban) we face a similar dilemma. Can we achieve some constancy in carrying out national goals to renew or reconstruct our inner cities? Can we learn which goals must be retained for stability and which amended because of changing conditions? Can private institutions, including foundations, help fix the polar star which may keep government on course in those policies essential to orderly progress, and also distinguish those goals which must be recast?

Foundations clearly can be useful instruments in helping to set priorities for programs relating to urban affairs, as a part of essential policy planning. In a statement criticizing foundations for rigid positions on risk-ventures, Bayard Rustin, speaking at the 1967 annual conference of the Council on Foundations, said, "Basically, the problem is not money. The problem is that the United States does not establish the proper priorities in this society which would make it psychologically, financially and economically possible to get rid of poverty" (Rustin, 1967). Policy planning in meeting urban needs to raise the quality of urban existence is essential. To be effective it should be structured with some degree of continuity, but flexible enough to yield to contingencies not apparent in advance. The planning process must take regional factors into account, with communication established to insure flow in and flow out—neighborhood to community to region, and the reverse.

Because men work chiefly through the institutions they have created, foundations, as spokesmen for revision and renewal, have the responsibility to examine these institutions. In any society, the latter takes on a patina of antiquity. Their structures calcify, and positions become too rigid to adapt to change. This aging process is an impartial one. It does not exempt even those institutions which by their mandates are meant to be instruments of change. As age takes hold, the competitive process diminishes. Material survival is assured and job tenure guaranteed. Or by unspoken agreement, tenure may be the reward for conformity. In this context, where are the institutions which will revitalize? Where are those which will reorient? Where are those which will call the roll on the institutions which have become resisters when they were meant to be catalysts? The American foundation is a candidate for this function, at least until other instruments are devised for this purpose. As an "ombudsman" of sorts, foundations can become instruments for the review of goals and programs carried out by institutions which have been caught in the aging process. Foundations have the capacity to enlist talent from other institutions in order to diagnose. Hopefully, they will prescribe on the basis of findings, and also show an awareness that the heart action of an institution will be healthy to the degree that its people are involved in free and creative ways.

The relationships, actual and proposed, of foundations to the problems of poverty and majority-minority-group conflict, which are described in the following pages, may be taken as a paradigm for that more general concern for the role foundations can and should play in improving the quality of urban life in the United States.

FOUNDATIONS AND MINORITY-
GROUP PROBLEMS

The United States has no more serious question than what it is going to do about the festering areas inhabited at present by minority-group citizens—areas not redeemable for human use in their present form. These are the habitations of the abandoned people. They have been left behind by economic and social change, including a shift in the center of the economy. The urban hub was once the source of livelihood for the minority groups. They served and were served by the many facets of the marketplace. They came where the jobs were, but their coming proved to be a factor in driving jobs away as earlier residents of the city could afford to seek the relative solace of the outer limits of the urbs. It is of course true that demands for aptitudes and job skills have changed, that these citizens are not in tune with new requirements. But the marketplace has also shifted, with many industries moved to the metropolitan rim. Hence the stronger economic base which provides jobs is no longer where the minorities reside.

A combination of government and private sources will be required to put into effect changes which will utilize minority-group citizens. It is an open question whether, and through what means, the aggregate resources of the American culture can bring this shift about. Can we create new centers to absorb in productive ways the people who make up the tight clusters of our inner cities? Reconstruction during the postwar period in Europe, the Soviet Union, and Japan proves that man can replace his physical environment. In those countries, redevelopment was accomplished despite the heavy loss in human life occasioned by the war. We in the United States approach our task of reconstruction with industrial and manpower resources at a summit of strength. Current examples of needed ingenuity are found in such actions as the passage of a bill in New Jersey opening 18,000 acres of swampland for long-range development, in a sense as a separate city; or in Buckminster Fuller's plan to develop "floating cities," communities engineered by principles of shipbuilding. These units would occupy now unusable waterfront or shallow-water areas adjacent to metropolitan centers and would accommodate several thousand persons each. It is this kind of innovation which might be supported by foundations in the planning stage, and later find its capital in private investment and/or government sources.

In our evaluation, if we decide that we can maintain parks and golf courses, zoos, and Playboy Clubs but cannot support a proper level of nutrition, housing, preventive medicine, and education for citizens, we will then have reached the point where it is no longer moral to maintain parks, zoos and other chosen forms of recreation. Should we ignore the moral factor, we will discover that parks and golf courses are off limits, because unsafe. This is not an extreme forecast. We have only to look at the change taking place on our once pastoral campuses—those retreats set apart from social reality—which have now become centers of discord and pitched battle between youth and authority. Forms of recreation are dividends of a prospering, balanced society and are not to be deplored. What is deplorable is the resort to anarchy by those who live in need and hopelessness alongside luxury. When we ignore the possibility of anarchy, we help to invite it. If it comes, we will have lost a chance to ameliorate hopelessness through the use of reason.

FOUNDATIONS AND POVERTY

James Tobin (1968), writing in the Brookings Institution's volume, *Agenda for the Nation,* poses several important questions:

> In the coming years the issue of distribution may be squarely joined: Will the affluent majority explicitly tax itself to improve the lot of the poorest fifth of the population? Is the majority wise enough or frightened enough to do so? How can the poor, with so little national electoral force, bring effective political pressure for redistributive measures? If their disaffections cannot find political outlet, will it pour forth into the streets?

The enemy which must be conquered, with foundations helping to use old and new weapons in the attack, is maldistribution, a condition which results in a lack of life's essentials for some and in guilt and fear for others. Whatever we attempt, it must be more than derring-do; it must work under the stress of social testing.

If the nation approached urban revival as it would a disaster, the scope of the relief operation would dwarf any other peacetime undertaking. It is, in fact, questionable whether or not meaningful relief and remedy can be effected, unless the country comes to grips with the problem in the same manner it would an acute human disaster. The introduction to *Agenda for the Nation,* published prior to the time President Nixon began his term of office, puts it thus:

> Before it can formulate its domestic program, the new administration will have to think through, and find its own key to, the central paradox of American society in 1969: on the one hand, we are a nation which sees itself as wracked and divided over problems of poverty, riots, race, slums, unemployment, and crime; on the other hand, we are a nation which is clearly enjoying high prosperity, rapid economic growth, and a steady diffusion of affluence at a rate almost unimaginable a decade ago [Gordon, 1968].

If the American nation—meaning its citizenry and not just the administration currently in office—does not relate its economic growth and expanded affluence to achieving equal opportunity and combating the disaster conditions which are so visible, it will have to face the grim prospect that it is living on borrowed time. For it is clear—and there has already been ample forewarning—that the economic and social fabric will be ripped apart by these conditions unless major steps are taken to remedy them.

In surveying the poverty field, it is apparent that little examination of the conditions of the older indigent citizens has taken place. Social security and recently enacted federal programs providing medical aid have offset the extremes, but they do not cope with the central emotional problems spawned by a blighted environment. Prominently, this includes loss of dignity and personhood as a member of the active community. Simple loneliness is high among the emotional problems. It is a condition which haunts the lives of those whose families have moved to other areas, or those who have survived their contemporaries. We know also that these circumstances are not limited to the elderly poor. The nation has only begun to concern itself with the nature of problems in this realm of human needs and with remedies to cope with them. Important field work can begin on the community level, with foundation support—even on a modest scale. It is in the urban areas, due to size and congestion, that the anonymity of the old is seen in its extreme. The results of work relating to the aged poor in our cities—both research and experimental programs—will undoubtedly have broad application to the problem of improved care for the elderly.

Segments of American society remain untested with respect to the upper limits of their capacity and willingness to support projects for social improvement. A valuable contribution by foundations would be an estimate of the giving potential of individuals, industry, service corporations, retailers, the professions, and others. Programs requiring funding over an extended period of years must look to public financing, increasingly from federal sources. We know these costs will be great. However, we must not underestimate the dimension of funds needed from private sources for projects not eligible for government support, and for experimental efforts which can lead the way. The threshold of giving power among foundations themselves, directed to eliminating urban ills, is untested. Less than a decade ago, many foundations would have claimed it was impossible to transfer funds to urban programs from fields customarily supported by them. Events in the nation, sometimes of a violent character, have helped to change this. We now find a substantial commitment of funds and staff to urban needs. As Harvey B. Mathews, vice-president of the Foundation Library Center reports:

> In an ever broadening thrust into the "poverty area," U.S. philanthropic foundations gave one dollar in every five in 1968 directly or indirectly, for action on urban blight, the ghetto, racial relations, and minority group problems.
>
> A recent survey by The Foundation Center on grants of $10,000 or more reported in the bi-monthly *Foundation News* revealed that grants totaled $753 million, of which $138 million, or 18 percent, were for grants specifically for attacks on "poverty" or for activities in six fields in which such attacks were part of the purpose of the allocated funds. The fields were: welfare, education, sciences, health, religion, and humanities. The interest and concern of foundations in urban affairs has been accelerating over the past five years and appears to be continuing on an upswing.

Foundations and other private organizations have provided the starter funds for projects which have been later used as models for government programs. Some church leaders credit foundations for effective leadership in bringing about ecumenical action on urban problems by focusing attention on urban affairs in ways which have brought church groups together. Among these efforts, the Inter-religious Foundation for Community Work in New York City is an example. And among denominations, special arrangements have been reached to tackle urban problems. The $3 million fund established by the Protestant Episcopal Church for urban work in New York City was done with counsel from foundations. Thus, as the message of the need has become clear, religious denominations have begun to structure and finance themselves for the job. On the trend of foundation involvement in helping to raise the level of urban life, David F. Freeman (1969), president of the Council on Foundations, reports:

> While any generalization about foundation programs is dangerous, many general-purpose private, corporate and community foundations have responded to the urban crisis in the past few years. Early grants served as models for O.E.O. programs. In the past three years grants have provided matching funds for various job-training programs, helped to launch community housing and planning efforts, and supported experiments in school decentralization, and police-community relations. Urban specialists have been added to staffs or employed as consultants. Currently "program-related" investments of foundation capital funds are being made available for minority-group business and low-cost housing projects in urban areas.

Trustees of smaller foundations sometimes feel overwhelmed by the sight of the massive requirements related to urban affairs, and naturally wonder whether grants of a few hundred or a few thousand dollars can make a difference. Evidence shows that small grants allocated to important social pressure points can be important. There is something about our society which looks for object lessons, and thrives on those provided by "pilot projects." The human factors—citizen participation, sensitivity, patience—are indispensable factors in putting even the smallest grant to work. An alert, informed staff person in a neighborhood can spell out the difference between public apathy and programs of implosive force, and can wring change out of stagnation. The financial outlay may be small measured against the gains. Trustees of smaller foundations located in a community or region sometimes wisely arrange for central staff, combining resources and boards of directors. It is a method to be considered by small trusts and foundations that feel the pinch of size and absence of executive staff.

FOUNDATIONS AND THE
PRIVATE SECTOR

All institutions, especially those enjoying tax exemption, must be ready to act in ways which will extend their influence beyond yesterday. It is a commonplace to say that a central ingredient of their strength is a freedom to perform, bearing the brunt of criticism for running unusual risks. Many foundations are bridled by traditional definitions of discretion. The "prudent man" approach has led them to reject in the past proposals which might have given society model solutions, on which present social programs could be based. The origin of this attitude appears to be rooted in the approach to problems usually taken in the field of commerce, and is understandable, for the latter, given considerations which relate to corporate responsibility, limited capital, competition factors, and the profit motive. It is much less applicable to organizations whose special strength is in being venturesome.

At least a part of the ambivalence of foundation decision-making lies in the gap between what a foundation believes it should do and the fact that funds are derived from the business community, with policies often directed by boards composed of business leaders. Therefore, the work of foundations is frequently judged by the criteria proven valid for business operations. The urgency and complexity of problems rooted in economic and social disparity in the United States should notify governing boards of foundations that they must distinguish between the criteria which pertain to business decisions, and those which apply to the decision-making process for foundations.

The involvement of industry in programs geared to help the economically handicapped has increased rapidly and wholesomely during the past three years. To a degree not predicted a few years ago, large and small corporations have thrown their hats into the ring of social change. Their corporate contributions include grants to special projects aiding the poor, work-study programs, support for new or renewed housing, capital and/or counsel for enterprises owned by nonwhite citizens, support of the Urban Coalition, and public education programs, of which the Xerox Corporation's television network series, "Of Black America" is an example. Of all contributions, lending leadership and "know-how" has been as important as other forms of aid.

The traditional and economic relationship between corporations and foundations makes them natural partners for much of the civic and human betterment which must be carried out on the American scene. Indeed, an increasingly broad approach in corporate giving is moving many profit-making organizations closer to the role of private, philanthropic foundations. Always recognizing inherent differences betwen the two, communication between the spokesmen of each should be frequent and full. They have much to learn from one another, including knowledge of each other's programs in order to set their respective priorities. Combining judgment and swapping experience, corporations and foundations can be strong but independent colleagues. This can be achieved only if there is sustained and frank communication about goals and current projects. The nation's foundations have a professional obligation to take the initiative in this area of cooperation.

The accumulated total of urban needs is now so great that the nation cannot expect to cope with them through the use of anything less than the resources of government. Nevertheless, what can be contributed by nongovernmental sources is important, even indispensable, because of the freedom with which independent organizations can act. Examples of this strength-in-freedom can be cited in many parts of the nation. The creation of the Urban Training Center for clergymen and laymen in Chicago is an example. Under the combined auspices of church denominations, with foundation financial support, the center was established on terms probably not possible under government financing and administration. This thriving project is a model for similar centers in other cities, principally because its sponsors gave it great latitude in setting the curriculum and selecting those who would train. Free of the accountability attached to public spending, its staff was also free to change combinations as a result of continuing experience. In short, they were free to innovate and to take chances. A recent report that a new school for training rabbis has been located in the midst of a north Philadelphia nonwhite area demonstrates further the move by private institutions toward experimental action.

The strength-in-freedom attribute of the private organization, including foundations, is rooted in the right and the obligation of the independent man or institution to experiment. This is not as easily done when the conditions of public law and fiscal rule must be met. This fact accounts for the trend by government in recent years to contract with the private sector for substantial parts of important research in the sciences, arts, health, education, the humanities, and public educational broadcasting. A strength of the nongovernmental effort is its capacity to speculate—thoughtfully and responsibly, but freely—without an eye on the voting machine or challenge from legislative committees.

Foundations and other private institutions, including those of higher education, can, moreover, be a resource in guiding special programs to their unusual objectives, which sometimes must be achieved by uncommon methods. Foundations can contribute, by reason of their detachment and their independence, an overview of the larger social scene. Foundations can help citizens to move outside and above their daily station to view the battleground from a distant vantage point; and with detachment leading to better strategy. Community task forces of leaders and workers assigned to observe conditions and programs elsewhere can inject citizens with new ideas and vitality for the return home.

THE NEIGHBORHOOD LEVEL

Much of what has been described calls for involvement of many segments of the community. This may seem an unwieldy method, a tortuous path to a goal. Some will recall how the "power-elite" supposedly worked things out in local affairs—a few people getting together to impose a solution. We still need these leaders, but much as the benevolent rational approach has done in the past, the decision-making pattern of the present must involve citizens in a comprehensive way. The only apt comparison which comes to mind is the way in which citizens have been swept into action during the nation's past war efforts. In the modern urban community, the special skills and temperaments of a wide range of citizens are needed. The psychological need for involvement is also present, and it cannot be met unless the individual grips the tool. The diverse requirements of neighborhoods demand this degree of participation. The poor have views to express and this can probably be accomplished best through actual involvement in problem-solving. A situation to be avoided, because it will be fruitless, is dialogue between those not close enough to hear one another. Work relationships will insure having citizens close enough to hear, and hopefully to understand.

A few sentences are in order on the value of independent, small-scale efforts—sometimes important *because* they are very small. The formal, committee process of decision-making tends to dilute originality. Legal and corporate criteria have their proper place in the setting of their origin, but these criteria applied rigidly in locales where communication is weak and action is the watchword may sterilize the best idea. The more effective role of foundations in many instances may be the freedom of initiative, with financial support, given to independent groups on a community or neighborhood level. For many foundations it will mean a deliberate departure from customary ways of monitoring, and sometimes managing, what they support. The foundation need not be cast in the role of "give-and-forget." Rather, it is a case of following supported efforts with interest and assessment. Foundations with ample resources to do so could bring into existence satellite foundations, free to operate in their own orbits, impulse having been supplied by the parent organization. Encouraging and developing the spirit and fact of individual participation is paramount, from idea stage to performance.

Policy-making carried out in some distant headquarters cannot be sensitive to the daily throb of the neighborhood pulse. The regional planning approach relies on the expert and too often he has become detached from local conditions. There is a need for local planning units, made up of citizens having a personal or family stake in the outcome; men and women (including youth) who have pressing reasons to improve the local street scene. Enthusiastic participation in block and neighborhood projects will not come about by legislation; it must come literally from the sidewalk. The strength of Peace Corps activities overeseas has been derived from these principles—trained, energetic leadership granted broad authority in the field, and voluntary participation by those who stand to benefit most by local improvement. Funding need not be in large amounts for the projects to be important catalysts. Foundations can project themselves effectively in this role. We are past the time when the foundation, as a vehicle of education and human betterment, should be thought of exclusively as a grand-scale operation. The growth of the community foundation concept is proof of this. In 1967 there were 214 separate community foundations in existence in the United States, with total assets placed at more than $708 million (market value). The larger, older foundations can become parents and advisors to smaller foundations which apply their resources to problems on block, neighborhood, and community levels.

Industrial and business corporations supporting neighborhood projects to combat urban ills also must outgrow the strait-jacket of corporate caution. The factors which influence a corporate decision are based on long experience, sometimes on near-scientific method.

These same factors may not apply to projects intended to meet urban conditions which have grown out of long-standing neglect—conditions where the very language may be different and situations which cannot "show a profit" in the accepted sense. Initiative in business enterprise on a neighborhood level, including help with skills and capital, are ingredients which the private sector can most readily provide. The stimulus which a new, small industry or service organization, formed for profit, can give a neighborhood may be significant. An owner-managed small industry can be a working symbol of progress. In terms of jobs, income, and identification with the outside world of commerce it can be a counterforce against the often dominant sense of hopelessness in urban areas.

As the nation seeks and finds new forms for preventing or handling international crises—the "hot line" between Washington and Moscow, United Nations peacekeeping operations, and the like—the foundations can work to develop new forms for communication, programming, and financing domestic experimentation. What new forms, or combination of forms, should be put together is a matter which foundations and other private institutions must address themselves to. In some situations it may be the use of task forces combining skills and experience drawn from different categories of citizens; an amalgam of experience from industry, small business, the ranks of consultants, academic institutions, and government. For the purpose of analyzing needs of a neighborhood, results might be obtained with a jury-like composite: an executive, an accountant, a shop foreman, a mother, professional men—a mix of differing religious, ethnic, and work backgrounds combined to advise on civic decision-making.

THE VOLUNTEER IN
COMMUNITY WORK

The United States prides itself on the role of the volunteer in community service. Voluntarism in this country is developed to a higher degree than in any other nation. However, in relation to the expanded role required of the volunteer, the procedure—including recruitment, training, scheduling—is unbelievably casual. The volunteer is needed to provide the human element, preferably on a one-to-one basis or the nearest possible approach to it. He should, however, be part of a structure which lends discipline to his work and which raises the level of his expertise to that of the professional. The problem is how to create this structure of organization within which the volunteer will work without depersonalizing his efforts. A professional approach is required in order for the volunteer to understand better the alien and sometimes hostile circumstances he will encounter. Many professionals, despite their training, find themselves at a loss, enmeshed in human cross-fire alien to their backgrounds. A recent Rockefeller Foundation report (1968) recorded this episode:

> A youngster with a record of defiance in the classroom, thrown out of his junior high English class by an exasperated teacher, became the poet of . . . [a] mimeographed student newspaper. Turned loose to express his thoughts, he wrote:
>
> > People laugh and talk,
> > There are walls between our thoughts.
> > About them, I'm lost.
> >
> > Close your eyes, go to sleep
> > Close your eyes, do not peek
> > Go to sleep, horror dreams
> > It is not what it seems.

In our efforts to muster manpower for the needed tasks we are leaning too much on the people who are already overburdened with obligations that require their services most nights of the week. There is a need to find and give confidence and responsibility to those *not* now participating. The categories and numbers of these citizens are large, including men, women, young people never before used. They include the bridge-club set, senior citizens, members of fraternal orders and service clubs, mothers with children old enough to have released them from full-time home duties. This more extensive involvement will energize communities. It is the kind of activity that can link the nation's needs with its most precious resources: human concern and skill. A useful private venture would be the creation of registries for skilled volunteers. If maintained with high standards, and made available to the several project directors whose programs concentrate on urban affairs, a current inventory of available volunteers could prove to be the difference between project success or failure. The salutary influence of broad citizen participation could change the social conscience and outlook of a community. Citizens who had been onlookers would gain the experience and commitment growing from an active role. It could be a living example of Thomas Merton's (1956) counsel:

> A purely mental life may be destructive if it leads us to substitute thought for light and ideas for action. The activity proper to man is not purely mental because man is not just a disembodied mind. Our destiny is to live out what we think, because unless we live what we know, we do not even know it. It is only by making our knowledge part of ourselves, through action, that we enter into the reality that is signified by our concepts.

McGeorge Bundy, former White House advisor and now President of the Ford Foundation, estimated recently a budget requirement of $30 billion annually for "as far ahead as we can see" to meet urban needs and move the nation toward equal opportunity. Others share his view that the nation is only beginning to comprehend the volume of public money which will be required. To the estimates of financial need must be added manpower requirements. While national surveys have value, it is the communities which will draw up the estimates which have meaning for themselves. Community foundations are natural supporters for a service of this kind. Their efforts can help overcome the loss of unused talent through the placing of individuals in positions of strategic importance to the community.

CULTURAL AND ARTISTIC UNDERSTANDINGS

Cultural and artistic endeavors in the United States are supported substantially by private sources (including foundations) and have a potential for use in meeting the problems of the cities. We are witnessing prodigious expansion in this field. It can be seen in schools, community projects, and the establishment of major art centers in cities like New York, Washington, D.C., Milwaukee, Minneapolis, Los Angeles, Dallas, Atlanta. These are undertakings which provide versatile settings for the expression of art forms. They are appropriate, and they will enrich a society determined to refine the use of its resources. How do we relate this cultural affluence to our efforts to raise the level of education and life in general for those who have no sense of belonging to these centers? Private wealth has largely created the new art centers and will be necessary for their future support—by no means assured in all cities. These same private sources, among them philanthropic foundations, must find ways to extend the influence of art centers to citizens whose lives have not yet been touched by artistic effort. It is part of the contradiction of wealth that it provides cultural enrichment to those who already possess a good share of it, some even a surfeit. The cultural "have-nots" are usually not on the sharing end. It is not a case of being closed out by deliberation, but through passive acceptance of the cultural pattern of nonparticipation.

As it has been necessary in the expansion of education to invent new ways to involve individuals in programs of adult education, cultural centers must also experiment with and install their own new devices to extend programs beyond their usual audiences. In doing this, they should examine for its value in this connection the most advanced electronic communication available. Foundations can be instrumental in providing "starter funds" for these purposes. The augmenting of staff attached to cultural centers of a neighborhood character, for example, might bring to these localities many of the resources available at the giant art centers. As experiment takes place it will become apparent that rules which have worked for many years are inapplicable to many of the new human contact situations. New rules will therefore have to be worked out. The community and national reputations of major art centers will in the future have a direct relationship to their efforts in the ghettos, including the uncovering of new talent there and the awakening of new spirits among the poor.

Among "object lessons," there are few more dramatic than Budd Schulberg's writers' workshop in Watts (Los Angeles), which had to devise its own working format—including the rule of patience. His report (1967) of a poignant episode during his experience is particularly enlightening:

> He [Schulberg] was interrupted by a teen-ager who had taught himself to play moving jazz on the clarinet and flute: "What's the use of writing what we want? We've been trying to say what we want for years but who listens to us? We're not people. If you really thought we were human beings you wouldn't allow us to live like this. Just look up and down this street. The rubble hasn't even been cleared away. It's full of rats. All of us have been raised with rats. Uptown you're sleeping two in a king-sized bed and we're sleeping four in a single bed. A game of checkers or setting up Teen Posts won't solve this. If we were some foreign country like the

Congo, you'd be worried that we might go Communist and you'd send us millions of dollars to keep us on your side, but here at home you just take us for granted. You think you've got us on the end of your string like a yo-yo. Well, we're not going to hang on that string anymore. . . . We're ready to take our stand here and to die for our freedom in the streets of Watts."

Schulberg asks: "Do these words frighten and shake you? I heard them week after week. Many evenings I walked out of the Coffee House, into the oppressive darkness shaken and frightened by the depth and intensity of the accumulative anger."

Young Negroes, as Schulberg noted, came to the workshop at first with reluctance and hostility, hardly the way to approach creative endeavor. The parents and children of poor, uneducated families may have a similar reaction at being offered "opportunities" with the arts. It will require tact, human understanding and patience to keep from withdrawing when pupil reactions do not rise to middle-class, white expectations. The task may fall to foundations to assist major urban art centers in creating satellite centers which function as an integral part of neighborhoods, directed by some of those for whom the help is intended. Major art centers would then be the wholesale suppliers of accomplished performers and teachers.

Private industry is learning to adopt new approaches in creating interest and skills among citizens who have been told, directly or indirectly, that they were not equipped for a place in industrial society. Educators are now convinced (even though practice lags far behind philosophy) that the special conditions spawned by urban slum culture must be met by radical changes in school curriculum. Similar and persistent approaches will have to be tried by institutions devoted to the arts. Where other efforts may fail to break the "bad spiral" of ignorance, apathy, and hostility, the arts might be effective. If we admit that the arts are important instruments for advancing understanding between the people of different nations, we ought to recognize them as a useful vehicle for understanding—or at least communication—between classes in the United States.

COMMUNICATION

In dealing with urban needs, the community-based foundation can encourage and support lively communication between those who represent the several interests of the community. The more sensitive its ear is to human tremors, the more useful its decisions are likely to prove. It is common to hear questions raised about the values of meetings and conferences, of whatever type and size. Some foundations have ruled out grants for conferences, explaining that they want to put their money where the action is. In this dispute, when the talk about talk subsides, it may become clear that communication on a person-to-person basis is a helpful process. Planned conversation is as basic between members of a community as it is essential to harmony and growth with a family. The "youth worker" in his relationship to organized "gangs" is an example of this at work in the struggle against human tension created by urban conditions. He serves both as a confidant for the youth group and as a live telegraph line to convey (and often literally, to interpret) the thoughts and actions of the group to the "outside." When necessary and possible, often on a casual basis, he brings the group together with those who might be in an acceptable position to guide them.

In bringing about effective communication, we have to recognize again the handicap of a short supply of trained leadership. There is a wheel-within-wheel situation in the fact that communication is required for the training of leaders and we are handicapped in developing leaders because of insufficient communication. There are no shortcuts in expanding the ranks of leadership. The cure cannot be implanted from outside; only the methods can be borrowed. The steady buildup must be based on a confidence in people's ability to acquire both a sense of responsibility and an active set of skills. In relation to the nation's shortage of trained leaders, private institutions—and prominently, private foundations—face one of their most complicated challenges because of the freedom they have to invest in experimental methods for locating and activating leadership.

THE FOREIGN POLICY FACTOR

There is a direct relationship between defense appropriations and what the nation can allocate for programs combating urban decay. Those foundations which carry out programs aimed at arms reduction on a parity basis among great powers are in line to make a contribution to social progress in the United States. The threadbare slogan of the 1940's—"Scholarships not Battleships"—might have its present-day equivalent in "Stocktaking not Stockpiling." A clear correlation exists between the burden of national military expenditures and, for example, the financing of an expanded preschool Head Start program. This correlation cannot be pushed aside by the argument of those who say we can afford both. In fact, we are not presently making it national policy to afford both. Private institutions, foundations among them, can assist public opinion—and public policy—in making the choice between an increase in the stockpile of nuclear weapons or creating and expanding programs which sometimes mean continued life for children. A privately supported program (a study, publication, conference, exchange) aimed at agreement to reduce military expenditures by mutual agreement can be regarded as a blow struck in favor of reducing some pocket of ignorance and poverty. For a foundation, it is highly consistent to share its resources between projects affecting life in the United States and those dealing with foreign policy. A balance of this kind is the equivalent of the kind of diversification viewed by private industry as sensible, and even essential to assured growth.

In our need to conserve on military expenditures in order to allocate more substantially to social programs, we must study the Soviet Union's situation, which may be more comparable than commonly thought. There are portents of serious domestic problems there, perhaps no less virulent than in the United States. In the inescapable great power confrontation, how do we balance and reconcile the use of resources to help bring our own house in order? The Soviet Union has given privately expressed signs that it wishes to get out of the present scale of military burden for at least two reasons. First, they are tense, as we are, living under the ominous cloud of a possible nuclear attack, which both nations recognize could come by miscalculation. Secondly, the Soviet economy groans under the dual burden of armaments and internal development. Socialism has its own dreams for expanding education, medicine, leisure, and the other ingredients which lead to a desired way of life. Soviet leaders foresee that domestic programs will advance more readily if a greater slice of the economic pie can be shifted from armaments to the needs of the home front. In the United States, for reasons consistent with our principles, we have the same need to devote a larger share of human and fiscal resources to the accumulated problems which have come in the wake of the industrial and agricultural revolutions.

In addition to public education on foreign policy issues at home, private foundations perform a valuable service when they assist institutions with informal exchanges involving citizens from abroad. The interlocking nature of the United States' relations overseas makes it imperative that we reach out continually to sustain and expand private contacts wherever possible. Meetings between citizens of the Soviet Union and the United States, supported by foundation funds, are an example of the substantive contribution to be made through private initiative. In the next few years, foundations should be alert for similar openings for unofficial exchange with persons from Mainland China. As the cultural revolution there stabilizes and economic growth accelerates, the United States may find itself facing a situation similar to that long in existence with the Soviet Union. It would be witless for us to achieve agreement with the latter while ignoring the possibility of a new stage in the arms race with Mainland China. Foundations and other nongovernmental organizations can be valuable peace-makers and nation-builders at the same time. They can serve as physicians to the socially injured; and as agents of prevention on behalf of those who might be injured but for their efforts. The nation's policies at home and abroad are entwined in the same braid, woven of one hemp. The foundations, in direct action and in supporting other institutions which help the nation make prudent use of its resources, can contribute to the wise construction of domestic and foreign policy.

THE VIEW AHEAD: A PERSPECTIVE

All human activity takes place within a context of time, place, and prevailing attitudes. This context may at times induce friction, often for reasons so invisible or complex that one is at a loss to know why such a condition dominates. It is clear that certain combinations of motivation, human intercourse, and emotion lead to friction. There are also combinations which lead to harmony, understanding, and constructive acts. One of the tasks of foundations is to help individuals understand and distinguish between these combinations and turn them to human advantage. In contributing to conflict resolution, foundations have an obligation to seek the hidden keys to new solutions to be found only in new forms. Can we find new answers to old equations; is it possible to have ten plus ten equal more than twenty, using ingenuity in place of traditional arithmetic?

A noteworthy example of success in this approach is found in the dispute between India and Pakistan over the distribution and use of the waters of the Indus River. Earlier approaches to a solution were based on the quantity of water known to be available, but insufficient for a satisfactory settlement. Through the intervention of the International Bank, a project was developed which increased substantially the available quantity. In place of acceptance of the traditional status, technological change helped bring about a solution. In meeting our own urban and rural poverty problems, can we not change the past order of things to facilitate solutions? A few generations ago a man could not expect to travel more than 30,000 miles during his life. Today, millions of individuals have the means of travelling several million miles during a lifetime. Our parents recalled the wonder of their childhood at the news of people moving 60 miles an hour on a train; the serious question was whether people could withstand the force of such extreme speed. The Apollo team, in December 1968, survived safely and comfortably at a speed of approximately 25,000 miles an hour. The lesson of acceleration of man's knowledge of the physical sciences has many pages—underwater exploration, deep-earth probing, computer technique, study in the life sciences. It is neither logical nor beneficial to separate these areas of human achievement from the potential in other areas, especially the application of knowledge to human existence. If we have the skill to discover how to use energy to transmit the human voice from earth to moon areas, we can surely work out organizational techniques to help ourselves lead fuller lives. For many millions, in the United States and elsewhere, the present organization of society is ill structured for human profit, sometimes even for endurance. Can there be any discovery more important to man than the new uses of his mind?

The universe has been described as "an aggregate of nonsimultaneous and only partially overlapping events," a description sufficient to give astronomer, physicist, or philosopher pause. Use of this cosmic diagram for an understanding of earth-bound social forces can be helpful. The events affecting the lives of the poor—or any of us—are often nonlocal in origin. Yet their pattern is consistent with the principle of cause and effect, even though the cause and effect actions are widely separated. For example, competition in trade from another region, or a nation overseas, can influence local conditions. Because the underlying causes of social problems are "nonsimultaneous" and "only partially overlapping," remedial measures by planners and workers can lead them into the forest of frustration, especially if they seek cures based on the assumption that all or most conditions are of local origin, simultaneous and not overlapping. It is therefore not immature or shoddy to approach complex problems with the readiness to accept solutions which are only partial. If we can harness man's emotional drives, conceding that a gallop is not always possible in the short run, and if we can accept this as an act of integrity and not compromise, we will have advanced. Notwithstanding the magnitude of the needs, the assignment must be viewed as realizable for human minds and hands.

One of the astronauts, viewing the earth in its entirety, described it as an oasis in the desert of the universe—green, moist, and life-sustaining. His perspective, shared with earth-bound man, has given new meaning to the question: "What are we doing with planet Earth as a place of human habitation?" We begin in the United States with an asset of great strength—the freedom to examine and diagnose our human failures. It is an attribute not condoned or shared by all societies; among them are systems where human frailty cannot be admitted lest it be regarded as a weakness in the political-social scheme. We are free to conclude that certain conditions in the United States are unthinkable, even though man-made, and then to think the impossible as we work for change.

Although the web of social conditions spread over society seems incredibly stubborn and perverse, we can bolster our efforts with the reminder that man alone is this planet's adaptable creature, having the capability to alter the design of his life environment. If social innovation seems long in coming and the human price intolerably high, we can recall that man existed for many centuries without realizing the power and leverage he might release from the earth he inhabited. The metals which were extracted and refined for the Apollo spacecraft—able to tolerate intense heat and stress—were literally under man's bare foot when he was shaping a wooden spear to hunt and survive. What he then lacked, and later discovered, was the tool of discovery within himself. We must now ask, figuratively, what is under our feet, ready to be converted and used to human advantage. In the realm of social invention, what implements remain to be discovered for man's use in shaping his environment? We have only begun to use man's full power. With the richer loaf yet out of reach, man has lived on the crumbs of his intellectual and moral resources. In seeking a significant encounter with the tragic conditions of our cities, it is helpful to recall that the history of man is marked by progress. Evolution in a physically hostile environment has made it clear that man is designed for self-refinement and fulfillment, if fear does not turn him back.

REFERENCES

FREEMAN, HARVEY B. (1969) Statement prepared at request of author (January).

GORDON, KERMIT (1968) "Introduction." P. 4 in Agenda for the Nation. Washington: Brookings Institution.

MATTHEWS, HARVEY B. (1969) Statement prepared at the request of author (January).

MERTON, THOMAS (1956) Thoughts in Solitude. New York: Dell Publishing.

RUSTIN, BAYARD (1967) Address before Annual Conference, Council on Foundations. P. 19 in report of 18th Annual Conference. New York.

Rockefeller Foundation (1968) Rockefeller Foundation Quarterly 3: 6.

SCHULBERG, BUDD (1967) Address before Annual Conference, Council on Foundations. P. 35 in report of 18th Annual Conference. New York.

TOBIN, JAMES (1968) "Raising the incomes of the poor." P. 80 in Agenda for the Nation. Washington: Brookings Institution.

20

Toward the Education
of Urban Man

EARL S. JOHNSON

☐ THE EDUCATION OF urban man must, perforce, be the
education of Everyman. Thanks to the evolution in communication
and the impact of a complex technology on the human community,
the former distinction between city and country has given way, even
in education. Thus we need an education whose emphasis is on
process rather than on *place*. That process is an urban one. It is the
process, or rather the interlacing processes, by which the urban
community and the urban character are formed.

An image of that process and of the community and character
which it shapes was conceived by Charles H. Cooley at the opening
of the twentieth century in highly personalized terms: "In order to
have society it is ... necessary that persons should get together
somewhere, and they get together only as personal ideas in the mind.
Where else?" His question, being rhetorical, gave its own answer. A
third of a century later it fell to Louis Wirth to report that the play
of greatly more complicated social forces now makes the pictures we
have of our society "blurred and incongruent." It is due, he wrote, to
"this period of minute division of labor, of extreme heterogeneity
and profound conflict of interests." Because of them "the world has

been splintered into countless fragments of atomized individuals and groups." In such a confused pattern man finds it difficult to know himself because such self-knowledge is premised on knowledge of his relations to his fellows, his community.

The social world in which Everyman now lives, increasingly urban in its nature, has resolved the issue of its integrity much better in symbiotic than in social terms. Technologically it is well integrated but sociologically and morally, which is to say in terms of a common and congruent set of values and social images, it is, as Wirth observed, "blurred and incongruous." The same general diagnosis which Wirth gave is shared by other scholarly critics. For Alfred N. Whitehead our problem is one of "the feebleness of coordination," social rather than technological. For Karl Mannheim it manifests a "lack of awareness" from which follows a "lack of concern."

The diagnosis of our time given by Graham Wallas, the distinguished British social philosopher in his book *The Great Society*, written on the eve of World War I, while given with greater particularity than that which we owe to the scholars already cited, is in substance the same. "We are forced now," he writes,

> to recognize that a society whose intellectual direction consists only (or chiefly) of unrelated specialisms, must drift and that we dare drift no longer. We stand, as the Greek thinkers stood, in a new world. And because that world is new, we feel that neither the ever-accumulating records of the past, nor the narrow experience of the practical man can suffice. We must let our minds play freely over all the conditions of life, till we can either justify our civilization or change it [Wallas, 1916: 3].

The theory of education which I now propose takes its general nature from what is eloquent in the insights of Wirth, Whitehead, Mannheim and Wallas. It is premised on the view that if urban man's conception of himself in the society of his fellows is to make sense, knowledge about society must make sense. That it does not now make sense is due, I believe, to its fragmented state—the "unrelated specialisms" of Wallas' image. Such a view does not affirm that knowledge alone can change the nature of the social realities. That task lies perhaps chiefly with the need for a comprehensive revision of our culture's basic social, economic, and political institutions. But if the fragmented state of knowledge, across its full spectrum, can be repaired, man will at least be in a position to know with greater reliability and greater clarity the structure of the society of his time, i.e., the kind and quality of its integrity. From such a stance he will be better prepared, at least intellectually, to mend the flaws in its structure. Certainly, only a *will* to do so, however dedicated it be, will not suffice. It is my view, further, that a relevant will depends on a clearer set of perceptions than presently characterizes urban man's image of himself and his social world. If my view is in error in these matters one must ask how "the will to virtue" can come to be and be put to work unless it be nourished by more reliable perceptions. Even if better virtues of the mind and spirit are not guaranteed by more reliable knowledge, I believe they are at least made more probable by it.

I wish now to offer a caution. It is that Matthew Arnold's wish that life be seen "steadily and whole" does not dispose us to see it either more or less "steadily and whole" than it is, given our ability to see it as objectively as possible. This is no small order, and if it can be honestly served we shall be in a secure mental and spiritual position to see it as "steadily and whole" as orderly change will allow.

Whitehead's judgment is that contemporary man's perception of the world reveals a "feebleness of coordination" due to excessive specialization and by that circumstance to a fragmentation of knowledge about it. Such relevant passages as the following from the chapter "Requisites for Social Progress" in his book, *Science and the Modern World* justify his diagnosis: specialization "produces minds in a groove"; "there is no groove of abstractions which is adequate for the comprehension of human life"; "the directive force of reason is weakened" as a consequence of which "the leading intellects lack balance"; and "the specialized functions of the community are performed better and hence more progressively, but the generalized direction of knowledge lacks vision" (1925: 275, 276, 277).

If to Whitehead's observations be added Wallas' prophetic vision that because a society whose "intellectual direction consists only (or chiefly) of unrelated specialisms, must drift" the need for a comprehensive repair of such a state of knowledge is readily justified. The nature of that repair now invites my concern.

My major premise is that urban man's experience with his social world will make little sense unless his knowledge about it makes sense. This is true because, as never before in his history, man's confrontation with reality is indirect and symbolic. While such confrontations extend the range of his experiences, their reliability depends heavily on the degree to which these indirect and symbolic representations are patterned. In the degree to which they are unpatterned his world is much the same as that in which William James' hypothetical baby lived, a "booming, buzzing confusion." A world, so known, cannot and does not meet the criterion of relatedness.

So it is that contemporary man needs related or patterned knowledge. Only it can provide the basis for the wisdom he needs to act with intelligence and purpose in it; for wisdom is, by definition, an interdisciplinary product. It is not gained by merely adding up units of knowledge gained from specialists, however true to fact they may be.

Such a requirement does not do away with specialized knowledge; it only insists that it be related in patterned ways, for those are the ways in which experiences are had, held, and understood. It is my view that unless education, formal and informal, can produce or allow a general, that is, a related view of man vis-à-vis his fellows and the world of nature, political and economic attempts to achieve such a view are bound to fail. A more clearly perceived and a better world must exist first in one's mental image of it.

The name given to the kind of education I envisage, namely general education, is somewhat misleading. I prefer the term "integrated liberal studies," a pattern for which I shall shortly propose. Education is always specific, as is man's behavior. I do not understand that man is good or bad "in general." Thus, in my view, a body or pattern of integrated liberal studies does not abandon specialisms; it relates them. To miss this truth is to suppose that such a pattern of studies plays with empty abstractions, or worse, with none at all.

The kind and quality of education which I hold to be meet and proper for urban man is one which effects a synthesis of knowledge by bringing separated fields of knowledge together, as coordinated means of discovering the relatedness of things: to draw men and men, things and things, and men and things together. While each exists, it is also true and important to know how they *coexist.* I would call particular attention to my point of departure in such an experience in relatedness. I would start with the great subject-matter *fields,* not with any existing courses which presume to represent their manifold dimensions. I would begin with *generic,* not with particular matters. When the generic centers have been identified it will not be difficult to find suitable particular representatives of them. But to start with particular representatives seems to me to begin an enterprise in integrated liberal studies in ways so diffuse as to occasion bewilderment and confusion. In such an image, an education in integrated liberal studies can be created wherever there is "an explosive mixture of ideas" drawn from any of the substances of the three great fields of knowledge. The pattern of them which I have contrived I shall mention shortly.

My image of an education achieved through an integration of liberal studies comprehends their methods of knowing as well as their substances: thus, as Dewey expresses it, both their *consummatory* and *instrumental* qualities. While I hold it important to know *"that,"* I hold it equally important to know *"how that."* Concurrently, I hold that one's knowledge ought also to comprehend the affective as well as the cognitive aspects of the various fields of knowledge, their value as well as their factual nature.

The universe in which the kind of education I have suggested is located and which fixes its center and its boundaries is the human adventure. Urban man is, willy-nilly, not only caught up in it but deeply involved in it. It ought to be known indigenously, which is to say, that to which man is not only related but in that fullness of relatedness which permits him to know himself. The sense in which an education so located and so conceived will have the attributes of a humane one will, I trust, come clearly to view as I share with you my image of it.

It is the organic nature of the human adventure that I would emphasize. Its unity is revealed, not by one's adding up its parts to achieve some kind of sum of things, but rather by investigating how its parts are interrelated. Several patterns come to mind. I think first of the human adventure as being contained within two realms: the spiritual and the material. But such a two-fold division is both too simple and too vast. I think, next, of the human adventure as consisting of people pursuing moral purposes: all else is apparatus which, at its best, works to unite these two elements and at its worst to separate them. But such a perspective is also both too simple and too vast, although I believe it has the virtue of suggesting relevant modes of analysis.

The pattern on which I have settled has, I believe, the virtue of suggesting the interplay of the constituent realms of the human adventure and thus comes close to what I understand to be its elemental character. It permits me to view people in pursuit of moral purposes in a context which allows also explicit concern with attendant and supporting processes. Its three elements are *religion, social organization,* and *science and technology.*

I choose now to restate and revise this image or pattern in terms of a more abstract trilogy of terms whose alliterative quality may even have some value. This is the trilogy of *poetics, politics,* and *physics.* Let me share with you what these terms symbolize, the meanings they have for me.

First, the realm of *poetics.* Here are to be found the myths by which man in all times and places has lived—which, in being myths, are literally false but poetically true. They are the things on which, through all time, man has set his heart and mind and hand, and whose endless pursuit gives both direction and meaning to his life seen both individually and collectively. This realm, like the other two, is an open one, admitting no closure or finality. The substance of the myths of this realm is a unique kind of facts, a kind not based on empirical proof. John R. Seeley calls them *"fide-facts."* Let him say what he means when he writes that mankind is a *"fide-fact"*:

It is *not* some that that is; it may be some this that might be . . . Mankind is not alone in its nonexistence. Love does not exist; law does not exist; this nation does not exist; the church does not exist; you do not exist and I do not exist—except in so far as we and they are created, constituted, inspirited, cherished, and maintained wholly and solely, in the beginning and continually, altogether and entirely, by the knowledge, faith, and love of men [in Ulich, 1964: 34].

To Seeley's *fide-facts* I would add others which are part of that vision without which "the people perish." They are the Great Oughts of the democratic-humanistic image of the possibilities of man: human dignity, love of beauty, truth, and the good; the religious-poetic sentiment in its million masks, conforming to Santayana's idea of religion as being the poetry in which he believed but the poetry which "intervenes in life" and does not merely "supervene upon it"; political creeds and ideologies among which is democracy itself, which is a *fide-fact* in that it is something men who are good and wise eternally covet but which is always becoming and never complete. Here, also, in this realm are found all the forms of art, the modes of thought, and the rules of construction, criticism, and evaluation proper to its nature.

Here man's mind dwells on the quality of things. Here is the place of the beatific vision. Here are to be found Wordsworth's "awful power of imagination" and Dante's "divine discontent." Here each of these is spent recklessly as the progenitor of man's *fide-facts* which are his goal values. I have called this the poetic realm because, in the

Greek, poetry means that something has "come into being"; the poet makes a world—even that of the "starry heavens above" but perhaps more the quality of "the moral law within." Here, as Shelley understood the matter, the civil order was born and poets became the true but unacknowledged legislators of the world. Likewise its first teachers. Here too, I think must be the place referred to in these lines by a poet whose name I do not know:

> Knowledge, we are not foes,
> I seek thee diligently
> But the world with a great
> Wind blows, shining and
> Not from thee.

Here is the place of those longings in which the wise and good Alexander Meiklejohn taught us we ought to believe, quite as much as in facts. I would name this the realm of *beauty* insofar as the humanities constitute a distinct discipline "concerned with the complex cluster of ideas represented by that vague and ambiguous term." With the logically ideal realm of the good and the true I shall be concerned later (see Peltz, 1968).

Finally, respecting the realm of poetics, now conceived of as containing all humane studies, I am reminded of something which Plato said: "It is impossible to mold men without an ideal of humanity." Moreover, he held the ideal portrait of the "human" or the "human being-like" to be the symbol of the real meaning of civil life. The just state for Plato was the man "writ large," hence his abiding concern with the just man (see Jaeger, 1939: vol. II, 277-278).

In the realm of *physics* are to be found the natural science studies including the so-called life sciences and the natural-science studies proper, especially physics, chemistry, and geology. I trust that you will not quarrel with me over the term physics which I have assigned to this realm of knowledge even though it goes beyond the meaning of that term strictly taken. Here is found the substance of the disciplines named as well as the modes of thought, and the rules of construction, criticism, and evaluation proper to it.

The studies in this realm are, presumably, those into which the purposes of the scholar do not intrude. They are thought by the naive to be free of valuations: thus totally objective. But we know that the standards which discipline those who work in this realm are themselves moral principles. Professor Frank H. Knight, the dean of American economic philosophers, names them as integrity, competence, and humility. Their opposite is fraud. The proposition is undebatable that we can practice science only if we value truth. Furthermore, if we would truly understand how the mind works in this realm we must know as much about the eye that sees as the object seen. Thus the lie is given to the idea of "utter objectivity."

I wish to illustrate something of the kind and degree of involvement of the physical scientist in his work by sharing a paragraph from a review of Michael Polanyi's *Personal Knowledge* (1958). The reviewer wrote as follows:

> How then does the personal factor manifest itself in the very structure of science? Polanyi discovers it wherever there is an act of appraisal, choice, or accreditation. Each science operates within a conceptual framework which it regards as the *"most fruitful,"* for those facts which it *"wishes"* to study because they are *"important,"* and it thereby chooses to ignore other facts which are *"unimportant,"* *"misleading"* and *"of no consequence."*

The italics are mine; they identify some of the major conditions under which a disciplined subjectivity enters. Here the classic epigram of George Geiger is affirmed: "It is not true that science doesn't give a damn."

The realm of physics shares with that of poetics a concern with that most intriguing facet of the mind, namely the imagination. In the realm of poetics it is recklessly spent. There the "if" of imagination is set free with the utmost abandon *except* that the poet's words about what he imagines are not allowed to "ooze off" as Robert Frost puts it: "what he imagines must be dammed back by the wit mill, and not just turned loose in exclamation." In the realm of physics, however, it is demanded that the scholar's "if" be subjected to hard proof so that the reliability of the "then" which follows it may be established. We may then say that in the realms of both poetics and physics (and also in politics) the imagination starts things off. The difference between them lies in where it ends up. In no realm, however, is there any thought except it be enkindled by an emotion.

To the realm of physics I would assign the virtue of truth; *fact* will perhaps do as well. The realm of politics also prizes this virtue, and I shall discuss it shortly. It has been suggested that the only way to get facts of any kind, independent of their value or use, is "merely to look . . . and forthwith remain perpetually dumb, never uttering a word or describing what one sees, after the manner of a calf looking at the moon." But this is a commentary on gaping, not on looking for facts.

Respecting the nature of facts I am disturbed by the conception implied in the language of Thomas Huxley the Elder that it is "necessary to sit down before a fact as a little child, be prepared to give up every preconceived notion, follow wherever nature leads, or you will know nothing." With Huxley's general sentiment I am in accord, but my understanding is that facts are not the result of the dismissal of all preconceptions or that they are simply "accumulated" as if they preexisted someplace "out there." As I understand the matter, facts imply a theoretical, which is to say a symbolic element and most, if not all of them, began as hypotheses. They are the fruit of the perceptive sets of given perceivers and bear the hallmark of their creators. They are, in short, both objective and subjective in their makeup and origin. Furthermore, we know that facts become pertinent to motivation only when their meaning engages our need to decide something.

Here in the realm of physics is potential truth—and power—whose nature the scientist must certify. To what *use* it ought to be put lies with the realm of *politics,* whose concern is with what is good rather than what truth is, as now defined. I think it proper to say that it is the task of physics to discover what is true; it is the task of politics to decide to what social ends it ought to be put. In other words, it is not science that is good or bad; it is men's use of the truths it provides that is good or bad. Legislation may properly be concerned with the use of the findings of science, not with what the scientist studies or the truth at which he arrives by his method of study. It is not given to politics to dictate how "the holy curiosity of inquiry" in the realm of science shall be employed.

The power of the realm of physics lies, then, only partly within it. Science can and does discipline itself internally in order to insure that the laws it states are true, but it cannot discipline the technology by which economic and political forces translate its laws into social power. By this circumstance the realm of physics, however much it be under the aegis of dedicated professionals, is not sole master of its domain. (We also should remember that, just as the the findings of science do not always convert by any direct process into technology, technology also may develop from other sources. Prince Kropotkin in his book, *Fields, Factories and Workshops*

(1901), points out that the main inventions that transformed the society of the eighteenth and nineteenth centuries were the work of artisans and amateurs such as Watt, Stephenson and Fulton.)

But lest it be thought that the findings of science touch and disturb human relations only through technology, let it be remembered how researches in the fields of physiology and biology have contributed to the health and longevity of man. Recall also the impact of the theory of evolution on powerful theological doctrines. So it is not only through the vast technostructures of modern business enterprise that science has changed man's ways. It cannot be doubted that the meaning of scientific knowledge in terms of its emotional significance for living has caused and is increasingly causing an almost total redefintion of ancient views of the "nature of human nature." The scope of the impact of the truths of the natural sciences on the order of human affairs is vast and revolutionary, comprehending as it does the nature of the universe, man's place therein, his relations to the society of his fellows, and human nature and conduct (see Frank, 1948).

The realm of *politics,* although the second named, becomes the third to be discussed because its major function can now be more clearly comprehended. It is the realm of middle principles where the themes and processes of the other realms intersect and interact. It is the realm toward which the methods and substances of the other realms must move if they are to become operational in the human adventure. The goal values of the realm of poetics must find lodgment and expression in a social order if they are to be vital in man's affairs and the instrumental or means values in the realm of physics must, by the same logic, make their contribution to vital human needs. In order to effect the interplay of poetics and physics in the realm of politics, imagination is a prime mover. The "good" which the realm of politics seeks to advance owes its dedication and sense of purpose to the values of "beauty" and the material means for its implementation to the value of "truth."

The meeting of means and ends thus implied comes about within the consensual process which we call politics. Through it, the intellectual, spiritual and material resources of the community interact so as to pattern its human resources and the substance and methods of the realm of physics in such ways as to create a social order which will provide the most hospitable environment for the nurture and growth of the values in the realm of poetics. Thus the "civilization of the dialogue" goes on and is forever renewed. In such an image I have undertaken to amend Sir Charles Snow's concept of the two cultures which, it has always seemed to me, never really met because his analysis provided no place where they could meet in order to effect a working synthesis.

Instead of two cultures I find one with a tripartite pattern whose integrity is contingent on the kinds of interdependencies I have posited. Such a tripartite culture has as its center the political-economic-consensual process. Here, as just noted, the other realms intersect and interact. The task of the consensual process is to advance the good of the community through creating unity out of diversity. It is not gainsaid that something less than a perfect unity is effected. Regardless of the degree of perfection of its operation, such a conception is a humanistic one: man inspired, man oriented, and man inspiring. If Everyman can come into an acquaintance with the human adventure so structured, he may, hopefully, neither be lost in it nor seek only to escape from it. More positively, in the imagery we owe to Robert Redfield, he may experience his education as "a good thing that happens inside people" (Redfield, 1963: 30-73).

In my reference to the almost equally dismal alternatives of Everyman's being lost in the human adventure or seeking only to escape from it, I have in mind his quest for identity. This is, I believe, the consuming fear or the threatening fact which contemporary man almost characteristically experiences. It is my thesis that it is a consequence, in large measure, of the fragmentation which knowledge has undergone and which has, in turn, resulted in a fragmentation of human relations. The wisdom in Louis Wirth's words, cited earlier, comes now into sharper and more illuminating focus. I offer now no easy proposal, no panacea for reconstituting the human community, but make bold to assert that its lack of integrity is the immediate and most powerful causal factor in contemporary man's estrangement from himself because of his estrangement from his fellows.

These estrangements are due to breaks in communication. Perhaps the oldest wisdom in all human lore is implicit in the principle that man is a social being and that whatever mental and spiritual harm befalls him is due to faults in his relations with others. The dependence of man on his fellows is affirmed in these samples of the wisdom of the ages:

First, the Book of Ecclesiastes: "For to him who is joined to all the living, there is hope."

Next, the wisdom of Cicero (*De Officiis*): "Nature urges that a man should wish human society and should wish to enter it."

Then Goethe's understanding that "only in man does man know himself; life alone teaches each one what he is."

The poet Stephen Spender (1962: 26) reflects the profound understanding that the social psychologist George Herbert Mead earlier taught me: "What makes a person real? The sense of existing *in* his own body, *with* his own mind and *within* a community. This is exactly what many persons today lack; they are cut off from half themselves."

There is the reciprocal relation of "I" to "Thou" which Martin Buber explains can exist only in a warm and sympathetic social atmosphere. Jacob Bronowski (1965: 65), as much philosopher as scientist, writes that "by identifying yourself with me you can learn new things about the human self, and about yourself not as one person but as one example. You enter more fully into your own mind by entering through me into the human mind."

And last, in a documentation that could be almost endless, all that has been affirmed is now confirmed in this observation by John Dewey (1927: 219): "We lie, as Emerson said, in the lap of an immense intelligence. But that intelligence is dormant and its communications are broken, inarticulate and faint until it possesses the local community as its medium."

I wish for a moment to return to the realm of politics. This is where man dwells! This is where the integrity of the life of the civil community meets its final test in the quality of assimilation which has been achieved among the three realms of the human adventure. I understand it to be the institutional realm par excellence. However sacred be the community's myths, however grandly humane its sentiments and however advanced its science and elaborate its technology, they all require to be invested in appropriate social institutions. The constituents of these structures are, of course, flesh-and-blood human beings; but they, in turn, must be bound together in manifold ways by joint participation in the poetics, the physics, and the politics of the community.

In this view, the institutional apparatus in the realm of politics provides a constant set of means for realizing the potentials of both poetics and physics. It seems that the world has not suffered from absence of myths and ideals nearly as much as it has suffered from absence of relevant institutional means proper to their realization. The burden of my commentary is this: technique is still something of a novelty in our culture and, as such, is apt to be played with and prized on its own account. Its proper function is, however, to provide the mystique of life with the means relevant to its effective realization. Whatever fault there be in technique comes, I believe, in our unwillingness to use it for ends beyond itself (see Dewey, 1930: 29-30). While man's technical powers are far ahead of the development of his moral faculties, the latter are not capable of self-actualization.

I wish now to illustrate with greater particularity the "altogetherness of things" as it may be realized through the interdependence of the realms of poetics, politics and physics. I shall not, in the following randomly chosen instances attempt to trace in detail the paths which lace together the tripartite divisions of the human adventure. It is insights into a master image of integrated liberal studies rather than blueprints for a curriculum that I mean to suggest. No hierarchy from greatest to least is implied in the order which follows.

If, as Shelley taught, resting on Plato, that "poets are the founders of civil society and the unacknowledged legislators and educators of the world," may it not then be that the social sciences only arrange, within the social order, the elements which poetry has created?

What new light does the foregoing discussion give to Pascal's view of man as a part of nature and though "but a reed, the weakest in nature, is a thinking reed"?

Is it not now clearer that the humane and political realms do not lie merely on top of the physical but incorporate it through its being taken up and transformed by interaction with the economic, the political, the aesthetic, and the religious?

In the second book of the *Republic,* Plato wrote that "the universal voice of mankind is always declaring that justice and virtue are honorable but grievous and toilsome. Is not the difference between myths and *real politics* thereby declared to be universal?

Is it in "the nature of things" that faith must erode as science advances? Are there not perhaps countless instances in which science has served to strengthen our faith in the timeless ideals?

Contemplate the impact of technology on traditional social and moral patterns even to the extent to which the logic and ethos of the natural sciences have been given a place, via the technology they have birthed, in the value systems of all nations and peoples, and have profoundly altered their conceptions of both man and nature.

If a given economic system is defended as insuring the *best* use of a society's resources, does it not thereby become a branch of ethics?

If, as Albert Einstein once observed, "perfection of tools and confusion of aims are characteristic of our time," is a reduced concern with tools sufficient to bring means and ends into a better balance? Does such a quantitative interpretation get at the root of the issue?

Walter Lippmann writes of "the urgency of affairs and the need for detachment" or what he calls the attitude and posture of "disinterestedness." Are urgency and detachment poorly matched in this aphorism, or is he not referring to the delicate balance which the critical mind must maintain between objective and subjective identity with its objects of study?

If, as Kenneth Burke has taught, our everyday vocabularies are not words alone but "the social texture, the local psychoses" and even "the institutional structures and the purposes and practices" which lie behind our words, is not the symbolic nature and function of language made clearer?

> Upon this gifted age. In this dark hour
> Falls from the sky, a meteoric shower of
> Facts; they lie unquestioned, uncommitted
> Wisdom enough to leech us of our ills

wrote Edna St. Vincent Millay. But how do facts become wisdom?

Conceiving man as equipped only with the tools of inquiry, James Branch Cabell responded in the following manner: "And yet more clearly do I perceive that this same man is a maimed God. He is under penalty condemned to compute eternity with false weights and estimate infinity with a yardstick. And he very often does it." Are either eternity or infinity amenable to scientific inquiry?

Respecting the nature of the method of science, might man now better understand that the very idea of science touches his feelings and draws its motive power for thought and the most rigorous and fearless inquiry "from passions which are a compound of reverence, curiosity and limitless hope?"

What light is shed on the relation between reason and passion, the war-and-peace of head and heart, in the face of the pathetic view that the task of reason is to dismiss the passions rather than, as is its proper function, through man's taking thought to choose the best among competing passions?

Might man's understanding of the ways in which the thought processes of poet and scientist bear striking resemblance now be better informed? Scientist and poet alike, "hearken to the illusive," as John Ciardi expresses it. "The poet," he tells us, "does not know what he is going to do until he finds himself knowing it in answer to the demands of his form." Just so, as I understand the matter, does the scientist. "The poem," Ciardi says, "happens to the poet." So too, in my understanding, do the scientist's facts and generalizations "happen to him" according to the discipline of his hypothesis—which also *happens* to him! Truly, man does not see with an empty head and an uncommitted eye.

Is it the mere abundance of material substance that is the moral issue, or is its "divorce from the adventure of living" the issue?

Has not the foregoing discussion of the interrelated nature of the three realms of the human adventure provided a perspective which raises doubt as to their necessary either-or character: in sum, that each must be viewed as standing in a necessarily dualistic (contradictory) relation to the other?

If I now affirm that the community, be it city, commonwealth, nation, or "the great globe itself" may approximate the status of being a democratic one, I would take for granted that its stream of life ran free. I would mean that between its three great realms there was no barrier to their mutually supporting intercourse. If that condition be conceived, for want of a better term, as being one of "balance," it would be one which permitted a moving equilibrium, hence not static or with the valence of any realm fixed and immutable. Nor do I mean to imply anything suggesting a fixed hierarchy of realms. By the demands which an organic life makes, each realm would be sensitive and responsive to the requirements of the others.

A community so conceived would be organized in such a way as to make it possible for every member of it to carry the same response within himself that he calls out in the community of his fellows. This would come about by all possessing a common symbolism which would, ideally, permit each member of the community to put himself in the place of the others' attitude. If my original thesis is a viable one, the preconditions for such a sense of community would be a significant reduction in the degree of fragmentation in the community's division of labor and, concomitantly, a parallel reduction in the degree of fragmentation in social relations. A cyclical process would, thereby, be initiated which could, ideally, make for an ever continuing reduction in the fragmentation of knowledge and human association. Only by such a process of social change and development could a truly democratic community come into being.

CULTURAL CONTINUITY—SCIENTIFIC
AND HUMANISTIC

The foregoing discourse has been set within but one time dimension, the present. But man lives in three time dimensions: past, present, and future. So, if only to locate what Whitehead calls his "insistent present," its connections back with the past and forward to the future must be taken into account. We are, it seems to me, disposed to chop up the continuum of past-present-future into three static parts. I shall undertake to demonstrate that no such cavalier treatment is very enlightening. What I mean to do, above all else, is place contemporary Everyman in the stream of time.

Such a view of contemporary man is, for me, not an option; it is a necessity. I believe that mankind can be understood by knowing contemporary man and quite as truly that contemporary man can be understood by knowing mankind. I speak of the reciprocal relation between knowing an "object" and knowing a "process." In order to know either I must know both. I believe that contemporary man is continuous with the past and that his present is in many ways a portent for the future. I believe, in short, in man's continuity. Unless *it* were a fact I could not imagine mankind to be one.

But now I must make clear why I hold this belief. I accept Edmund Burke's definition of the community as that great partnership which holds among the living, but also among them and the dead and those yet unborn. Any conception of the community of man less inclusive than Burke's would not convey what I understand to be the true, although in part invisible, dimensions of the human adventure whenever and wherever experienced.

Further, I believe in certain common and objective facts which relate to civilized man even though their common quality may be somewhat disguised, even distorted, by the accoutrements of unique times and places. To identify these objective facts I must go behind the exterior facade of past and present cultures and reach down deep into their interiors, into the elemental behaviors of people. In so doing I must divest myself of the shallow and chaotic relativism still in vogue in some precincts of social study.

Finally, I *must* believe these things: for if they are not true, then humanity and its complement mankind is only an accidental convergence of the atoms of man's experience since time began. If I do not believe these things then humanity and mankind are nothing more than ad hoc concatenations of events and experiences with only here and there a quixotic strain of consistency. These things I do not believe and cannot accept. My faith in these matters is sustained by one of Socrates' contemporaries. In *The Gorgias*, Callicles says that "if there is not some community of feeling among mankind however varying in different persons—if every man's feelings were peculiar to himself and were not shared by the rest of the species I do not see how we could communicate our impressions to one another." If perchance his insights were intended to report an understanding only of one's contemporaries, I find great comfort in them because "the variety of men" is, as I understand it, no less characteristic of one's contemporaries than of his "antecedents." I shall shortly adduce proof that Callicles was right, whether what men have in common is limited to but a single age or spans the gap of many ages.

I wish now to treat more fully and critically the continuum of past-present-future. It is the fact of continuity which needs to be established even though the fact of discontinuity must also be acknowledged. What I affirm is the fact of permanence *and* change in human affairs. But change is not total. When Heraclitus taught that "it is never the same river" he did not mean to say that the river ceased to be; only its *sameness* ceased to be. Even so, time flows through the sluice of past-present-future and we would be little short of silly to expect these phases of time to be the same or that as time runs forward it always insures progress.

Whitehead tells us that the present is "holy ground," because it is the "communion of saints. . . . a great and inspiring assemblage." Further he says we ought not to deceive ourselves about how it and the past differ by reason of the "pedantry of dates" for "an age is no less past if it existed two hundred years ago than if it existed two thousand years ago." Furthermore he warns that the main danger in teaching about the past is "the lack of discrimination between the details which are now irrelevant and the main principles which urge forward human existence, ever renewing their vitality by incarnation in novel detail." My later borrowings from the wisdom of the long age past respect these admonitions.

A word of wisdom about the relation of present to past is offered by Dewey (1950: 21): "Piety to the past . . . is for the sake of a present so secure and enriched that it will create a better future." As for his views about the relation of present to future I choose the following two: "thought about future happenings is the only way we can judge the present [for] without such projections there can be . . . no plans for administering present energies, overcoming present obstacles"; and the "future that is foreseen is a future that is sometime to be a present" (Dewey, 1950: 267).

I turn now to the nature of a civilization, for that is what continues in the temporal sequence of past-present-future and continues, as well, in space. It may be seen in sharpest silhouette if contrasted with the state and condition of barbarity which, so Aristotle tells us, is "living as one likes." "Living as one likes" I understand to be living by the undisciplined play of one's impulses—according to what I shall call man's first nature—his undisciplined appetites. Full allegiance to the community or to what Lippmann calls "the demands of civility" requires the development of man's "second nature" by reason of whose discipline he came to prefer the law over the satisfaction of his impulses. The essence of the law is that which man puts upon himself, namely rule by his virtues, not his undisciplined impulses. This is the substance of Aristotle's image of man as "fit to rule and be ruled." It also confirms the quality of Socrates' "second nature" by reason of which inner discipline he chose not to respond to the impulse to run away.

It thus follows that it must be *known* in every civilization what its ideals are both in terms of what claims they make on its citizens now and what the promise of that civilization means and portends for the future. Such knowledge is the principle of all order and certainty in the life of the people whose civilization it is. Its immediate bearing is on the conduct of their life (Lippmann, 1964: 322-323).

Our inheritance from the civilizations of the past comes to us by two processes: one is *cumulative,* the other is *noncumulative.* The first is the scientific, the second, the humanistic heritage. The humanistic heritage is noncumulative by reason of the fact that it carries forward the account of man's responses to the same set of universal circumstances disguised, as I remarked earlier, by the accoutrements of unique times and places. Its themes are universal: doubts and fears, passions, hopes and dreams, victories and defeats, comedy and tragedy, intrigues, loves and hates—in a word, the elemental things which make up the human adventure. They are carried in the discourses of the philosophers, the idylls of the poets, all the art forms that man has made to communicate his visions; all music and all hero and folk tales—every medium through which man has shared with his contemporaries and bequeathed to his successors the meaning of life. The scientific heritage is of a different genre: it is marked by discontinuity in the account of man's ever growing, now slow, now rapid, ever accumulating account of how his knowledge about the natural world has deepened and changed.

The essence of these two heritages and how they differ so profoundly may, I think, be illustrated by two quite different meetings and conversations. Respecting the humanistic heritage, imagine if you will a meeting between Socrates and Santayana as spokesmen for Western philosophy, separated by twenty-three centuries. One may well anticipate their finding no difficulty in plying each other with questions on such themes as the role of knowledge and reason in conduct, and the nature of the universe which makes possible and guarantees conceptions of knowledge and the good that are arrived at. Only the naive would believe that such a meeting would be an exercise in futility after the manner of "but evermore came out by the same door wherein I went." Neither would presume to teach the other; rather they would revel in each others wisdom and criticism. The session would be short but thoroughly delightful.

Imagine now a meeting between Plato and Enrico Fermi, representatives of the field of natural science, also separated by twenty-three centuries. Let them personalize the polarity between science beholden to theology and science totally free from it. If Fermi were to relate to Plato the spiraling cycle of advance in natural science—except for the negative effect on it of Augustine and others because it made no contribution to the salvation of man's soul—one may well suppose that Plato would find the account totally incredible. In such a confrontation Fermi would, indeed, be the teacher and in that role try to trace the changes by which science has freed itself from divine intervention. The meeting would be long, of that you may be sure.

For reasons implicit in the differences in the paths of the humanities and natural sciences as they have led up to our time from the earliest civilizations, I offer now, from a lore much vaster than my knowledge comprehends, some representative expressions of humanistic thought. They confirm what I believe to be the objective truth of the way of mankind. I owe them and the classifications in which they are placed to C. S. Lewis in his engaging and disturbing book, *The Abolition of Man* (1947).

They mark the Way of Mankind, the Tao. They are a lexicon of premises, not a body of conclusions. They are of the same truth as that implied in Jefferson's "We are endowed by the Creator" and the historic Judaic praise of the Law as "true." They postulate that certain attitudes and values are really true and that their opposites are really false.

The Law of Beneficence: "Speak kindness . . . show good will" (Babylonian); "He who is asked for alms should always give" (Hindu); "Thou shalt not bear false witness against thy neighbour" (Jewish, Exodus 20 : 16); and "Never do to others what you would not like them to do to you" (*Analects of Confucius*).

Duty to Parents: "I was a staff at my father's side . . . I went in and out at his command" (Ancient Egyptian); and "Honor thy father and thy mother" (Jewish, Exodus 20 : 12).

The Law of Justice: "Has he drawn false boundaries?" (Babylonian); "Thou shalt not steal" (Jewish, Exodus 20 : 150); "Choose loss rather than shameful gains" (Greek); and "Justice is the settled and permanent intention of rendering to each man his rights" (Justinian, *Institutions*).

The Law of Good Faith and Veracity: I sought no trickery, nor swore false oaths" (Anglo-Saxon, *Beowulf*); "Hateful to me as the gates of Hades, is that man who says one thing, and hides another in his heart" (Homer, *The Iliad*); and "Anything is better than treachery" (Old Norse).

The Law of Mercy: "I have given bread to the hungry, water to the thirsty, clothes to the naked, and a ferry boat to the boatless" (Ancient Egyptian); and, "Nature confesses that she has given to the human race the tenderest hearts, by giving us the power to weep" (Roman, Juvenal 15 : 131).

The Law of Magnanimity: "The Master said, 'Love learning and if attacked be ready to die for the Good Way' " (*Analects of Confucius*); and "Verily, verily I say unto you, unless a grain of wheat falls into the earth and dies, it remains alone, but if it dies it bears much fruit. He who loves his life, loses it" (Christian, John 12 : 24-25).

These are representative indentures in "the book of values" of civilization; they are not merely a set of "interim ethics." Nor are they items in a universal catechism which is only to be recited; they are principles to be lived by. In being principles they are not offered as absolutes, for absolutes by definition are isolated and not amenable to empirical verification. If some can be served as stated, without fail—well and good; if, as is more likely true, they can be served only relatively they are not, by that token, merely expediential devices. We must distinguish between "an idea of perfection and a perfect idea." Charles Frankel (1959: 78) has clarified the difference between them. If there is a "natural Platonism in man, it leads," he says, "to the first; but there is no way in which man can attain the second" (see also Peters, 1966; 1969). My personal subscription is to Frankel's pragmatic position, which Dewey states as follows: "That which guides us *truly,* is true—demonstrated capacity for such guidance is precisely what is meant by truth."

The testaments I have selected are of the essence of humanism, which is the name for those aspirations, activities, and attainments through which man puts on his "second nature" but in so doing does not need to repudiate his natural world. As given, they must be judged, as I have suggested, in the context of their time and place. The appositiveness of some to our time and place may require very little interpretation. For all, however, the degree and conditions under which it were sensible to treat them as absolutes must be soberly considered.

For instance, such an aphorism from the thought of John Locke: "Humanity is to be preserved" is quite as applicable now as in the time he pronounced it. It is, however, very highly generalized. Another noble sentiment which belongs to the days of the glories of the Roman empire, "Dulce et decorum est pro patria mori" (It is sweet and proper to die for one's country) need not be understood as endorsing war but rather as affirming that there are great and noble public causes which may indeed require from modern man supreme sacrifices, even that of life itself. For every generation it must be transposed, transmitted, and transmuted according to the realities of the here-and-now, not that of ancient Rome.

The aphorism given under the rubric of "duty to parents"—"I was a staff at my father's side ... I went in and out at his command"—must not now be taken literally but rather as affirming the imperative of filial respect, if not always for the person of the parent, at least for the concept and institution of parenthood even when those who play that role are not, as individuals, worthy of such respect. It is, furthermore, difficult to conceive of a generation which is not called on to identify its authority figures: those to whom the young look for both guidance and identity. Unless some are found and respected, a harmful and hideous alienation ensues. I mean to suggest that such a stern commitment needs to be recast in terms which make sense in today's world. The need for bridging the gap between generations must be met in terms and with attitudes which are appropriate, not to ancient Egypt, but, let us say, to modern Egypt. Such a requirement means that fixed moral rules must be translated into ethical principles. J. Bronowski's (1965: 97) wisdom on this issue is crisply given: "We can write and rewrite the regulations and the law books, but we cannot define and we cannot impose by edict the values which make up our ethic. They are not written out but acted out."

In the belief that the noncumulative nature of the humanistic heritage has been established I ask what is involved, intellectually, in the mind's moving from static moral rules to dynamic ethical principles. It involves the ability to distinguish between a past good and what would be, if still honored literally, a positive evil; between a wish and a reality system; or between an ancient conviction and the need that it be brought under the most rigorous criticism. I find these opposites suggesting the distinction which Max Weber made between the ethic of conviction and the ethic of responsibility. Under the influence of the first, one's impulse to believe is manifest; under the second, one's ability to question that impulse is tested. The employment of these two ethics would not be necessary were it not for the fact that "time makes ancient good uncouth." For this reason I do not counsel the uncritical passage of humanistic lore from past to present, for there is a profound difference between organic reaction and a critical human response. There is, moreover, a vast and tragic difference between a critical revision and adaptation of the values of past times and their total rejection because they are not given in the idiom of the present. There is, I have insisted, a continuity in man's experience which makes possible the phenomenon we know as civilization. Without these, the concept of mankind would be meaningless, and history but an idiot's dream.

THE EDUCATION OF THE COMMON MAN

Within a common context, namely the human adventure, I have elaborated a concept of integrated liberal education in two time dimensions—the contemporary and the historical. Thus I undertook to set the "relatedness of things" in the present and also within the continuum of past-present-future. In the first perspective the nature of a civilization was taken for granted; in the second perspective it was my focal point.

I now raise two questions. First, for whom is such a liberal integrated education meant? And second, who shall teach it? I ask, as did Crèvecoeur, "Who is this new man?" Or perhaps better, who is this whole generation of men which one hopes might be renewed in mind and spirit through experience in a humanistic education?

I see this generation as one composed of the common man. This is not the common man taken as "the least common denominator" of the mass of men. Nor is he the common man who is said to be "as good as the next man, if not a little bit better." Nor is he the common man taken as a way of referring to a dead level conformity, conceived as a way of resolving the issue of the mass of men living in a world at "sixes and sevens" because they are caught in the "meteoric shower of facts" of Millay's image—caught because they lack the wisdom to use them so as to "leech us of our ills." Nor do I mean the common man as the "average of men."

The generation of the common man of my image is *common* in the sense that all men have a common basic manhood which is a universal fact, the substance of which identifies all men as potential coequals in the brotherhood of mankind. This is at once the Judaic and the Jeffersonian image of man. This is the figure which the present fragmentation of knowledge does not allow contemporary man to know and understand as his birthright.

My image of the common man is related to the kind of education he ought to enjoy. It is not the "socially immanent" education by which, as Emerson expressed it, "we teach boys only to be such men as we are." Such an education can do no better than reproduce "the type" but never provide for growth beyond "the type." The education of the common man ought to aspire to the growth principle. It is well given in the old German aphorism that "Der Mensch ist etwas das Uberwenden werden muss" (We are true men insofar as we become something more).

Finally, respecting the education of the common man there is the issue, also, of the account it ought to take of the so-called liberal-vocational emphases. For an understanding of what these two emphases are and how they might be related I look to fifth-century Greece. Then every citizen belonged to two orders of existence and there was a sharp distinction in his life between what was his own (*idiov*) and what was universal (*xoivov*). As Werner Jaeger (1939: vol. I., 111) explains it, "Man is not only 'idiotic,' he is also 'politic.' As well as his ability in his own profession or trade, he has his share of the universal ability of the citizen, by which he is fitted to cooperate and sympathize with the rest of the citizens in the life of the polis."

In order to comprehend this Greek concept of politics it is necessary to recall that the whole of Greek life and morality was "political" in the sense meant by Socrates and Aristotle, which stands in sharp contrast to the modern technical conception of politics and the state. The Greek conception of politics is revealed in the Platonic dialogues in which such virtues as piety, justice, courage and prudence are discussed. Jaeger's understanding of fifth-century Greek society helps me understand the fairly sharp divisions of concern of contemporary man. A large part of his energy is devoted to money-making activities or cultivating one particular skill. This is for him, as it was for fifth-century Greek man, quite devoid of a "governing spiritual principle" and is hence merely a "tool, a means to an end" (Jaeger, 1939: vol. III, 224).

A quite ideal image of the education of contemporary man would see him as a cultured person in the sense that he would understand his vocation (his role as "idiotic" man) to be a unique expression of his generally humane orientation to the common life. But such an ideally conceived relation between the practical and philosophical phases of Everyman's career would be possible only if the picture which Louis Wirth has given us of the society of our time were untrue: one of minute division of labor, of extreme heterogeneity, and profound conflict of interests. But, regrettably enough, Wirth's analysis is a true one. It follows then that if, for the common man of my image, we covet an education which is able to bridge the gap between the "idiotic" and "politic" phases of his career, his world will have to be materially rebuilt. It seems clear that Plato's conception of the "whole" man, while difficult to come by in his time, is much more difficult to achieve in our time.

In this view the education of the common man may well be taken as the ultimate concern in all of our endeavors better to plan his habitat and institutions. These observations bring us full circle back to the issue with which we began: the fragmentation of knowledge and the consequent fragmentation of human relations—or the other way round if you wish; for which is "cause" and which "effect" is largely a matter of perspective.

To the second question, "Who shall teach the education I have proposed?" there are two answers: the educative school and the educative society. I shall discuss them in that order, finally, however, to learn that no such easy dichotomy of roles agrees with reality.

Under the rubric of the "educative school" I choose to think of the teacher and the institution. I think of the teacher individually as philosopher-teacher, and collectively as a member of the Sacred Remnant not unlike that elite to which Isaiah refers in Chapter 1 : 9. My model of the philosopher-teacher is Socrates, of course. If I now affirm that he failed in that role it was because the City, or the "educative society," was the more powerful political force and hence the more powerful educative agent, as it still is. The charge against Socrates was that he had corrupted the youth of Athens. His defense was that it was not he who was their corrupter but the very establishment which made the charge, brought him to trial and condemned him to death. Who had the last word I do not remember but I remember the indictment which Socrates made against his accusers, which was that they "had no care for the soul but only about wealth, power, honor, or some other less important matter."

There is something unmistakably contemporary in the confrontation between Socrates and the Athenian civic establishment. In our society, preoccupied as it is with getting rich and powerful, an education conceived of as a grand adventure in improvement in the making of judgments about humane values I find either grossly neglected or seldom even contemplated. I shall shortly adduce evidence in support of this view.

The rank and file of American teachers in our public schools (kindergarten through the twelfth year) are more representative of the rank and file of the whole population in their talent and dedication than any other profession. Because this is true I believe that improvement in the quality of America's teacher corps will depend, in the long run, on improvement in the moral and intellectual quality of the great mass of the American people. This, it seems to me, is the way it ought to be in a democracy.

But the Sacred Remnant of which I have spoken is not made up of the rank and file although it is recruited from it. Its numbers are made up of those who have been willing to give up intellectual and moral ordinariness. They excel their fellows in being more discriminating in mind, more dedicated to their craft, more self-critical and self-reliant, less frightened by those in authority, more gifted in the talents of leadership, and more fully understanding of the role which public education ought to play in a democracy and must play in America if our urban order is to be reconstructed. They are largely hidden from public view and seek little or no acclaim, but, in sum, perform the function of a sacred elite.

My conception and assessment of the "educative school" is hardly to be distinguished from my conception and assessment of the "educative society" for, in a democracy, the school as institution is and must be according to the image which the "educative society" has of it. My commentary on the "educative school" will be entirely from a critical perspective, for I am now more concerned to offer sample evidence of its shortcomings than its excellencies. I mean to identify important aspects of it which ought to be changed if it is to be the means for transforming our culture; it is not my intent to condemn it out of hand. Nevertheless my view is an ex parte one, for the prosecution.

It is understaffed, underfinanced, its teachers grossly underpaid; in these respects and others it stands well below the level of excellence which it is touted to have attained. Far too frequently—especially in the high school—its curriculum in not unlike the one which Alfred N. Whitehead found to be dominant in the British secondary schools of the late 1920's, i.e., "a rapid table of contents which a deity might run over in his mind while he was thinking of creating a universe and had not yet decided how to put it together." In its excessive concern to certify or license its students to meet the requirements for various status and role "places" in our culture, it finds very little time to acquaint them with their humanistic heritage, as it may be shown to be relevant to their time and circumstances, and with helping them master the arts of self-discovery and self-discipline. Remembering often sterile and unrelated facts seems typically to take precedence over a concern with the growth of the mind. The major emphasis is on making a living and getting ahead—a narrowly conceived vocationalism run amuck (Miller, 1967; Friedenberg, 1968).

The social organization of the American public school perhaps best instanced by the high school is, characteristically, a kind of educational colonial system ruled over by principals or superintendents who treat the teachers as their subjects who, in turn, treat the students as their subjects enjoying still less personal dignity and freedom than they themselves do. With only a slight shift in ways of looking at the American secondary school as a social system, it fits any one or all of the following patterns of human relations: a factory, a department store, or a bureaucratized office. An excerpt from Warner Bloomberg's paper, "The School as Factory," is illustrative:

Its classes begin and end on a shift-change schedule; with a few special exceptions its rooms are completely standardized; its work is so arranged that teachers and students may be treated as easily replaceable and frequently inter-changeable parts. Order within this system rests more upon authority and potential coercion than upon commitment to common values, a sense of mutuality, and social responsibility.

To such shortcomings of the "educative school" I wish to add two more which seem to me to confirm even more positively my view that the faults of the "educative school" are those of the "educative society" or the City. The dialogue which goes on between the two is a very uneven one due, of course, to the fact that the school is the society's school and not its own.

I offer now two illustrations of the prostitution of the "educative school" by the "educative society." Whenever the elementary "neighborhood school" in the white community is defended, it is the exclusion of nonwhite students which is the underlying reason for its defense. Such a posture has the effect of continuing the segregation of school children on extrinsic grounds, namely their skin color. But the rationale for defending the "neighborhood school," as such, contains a hidden economic factor: the school is seen as the first line of defense against putting "the neighborhood" on an open real estate market, and this policy of exclusion is justified by the definition of property rights as a moral absolute. Thus the perpetuation of segregated education and the absolutism of property rights are the basic issues. They are sicklied over by the sentimental and nostalgic but quite effective image of "the neighborhood school." Both the moral imperatives of democratic philosophy and the facts produced by research are ignored in an alleged "defense of education." (Almost innumerable statistical studies prove that change in the "color" of a residential community does not necessarily result in falling real estate values. It is, in most instances, due to "panic selling" induced by unscrupulous real estate agencies.) If the schools were not utilized as a means to create the new citizenry for a democratic metropolis, the students would at least learn how the school system itself was being used to deny part of what is best in our heritage.

My second illustration involves the Head Start program. It is designed to help the children of the City's pariah people catch up with the children of their socioeconomic "betters," who were born with a "head start." While the attempt to equalize the education of these two social segments of the City is a noble one which promises some measure of success, it gives the impression, intended or unintended as the case may be, that the roots of social inequality run no deeper than a "good education." They run, as we well know, to such roots as the crass and evil prejudices of the "whites" who have "got it made" and to such attendant stubborn factors as lack of skills, parental illiteracy, insufficient child care, weak family structure, poor diet, and a sense of defeat and deep despair. It is such factors as these which we now know weight heavily among the determinants of the achievement scores of elementary school children. If political and educational leaders permit the lay public to believe that gross social inequities may be removed through the mere instrumentation of a Head Start program in the elementary school, they are, in substance, parties to the perpetuation of "quick and easy" solutions to what is democracy's most hideous enemy, social inequality.

In concluding this chapter I offer no apocalyptic wisdom as to what the leaders in public education, whether they be spokesmen for the School or the City, may do to improve the education of Everyman. I suggest that the issue is not whether the School shall or shall not influence the course of future social life but in what direction. This view offers the following alternatives which John Dewey set forth over thirty years ago:

(1) The leaders may act so as to perpetuate the present confusion about the relation between humane and vocational education. This would be to choose not to choose, the result of which is to drift.

(2) They may espouse and practice a philosophy of conservatism and thus attempt to make the School a force in maintaining the old order, the status quo.

(3) They may seek to make the newer scientific, technological, and spiritual forces the School's ally, for it is these forces which must be understood and directed if the School is to be an effective agent in social change.

If the task of education be seen from the single perspective of what the City can do, it is sufficient to insist that the climate which its whole complement of institutions establishes must be one in which the claims of humanism are made prior to all other claims. Only then may Everyman be secure in anticipating that, in truth, "die Stadt luft macht frei."

REFERENCES

BRONOWSKI, JACOB (1965) The Identity of Man. Garden City, N.Y.: Natural History Press.

DEWEY, JOHN (1950) Human Nature and Conduct. New York: The Modern Library.

———(1930) Individualism: Old and New. New York: Minton, Balch.

———(1927) The Public and Its Problems. New York: H. Holt.

FRANK, LAWRENCE K. (1948) Society as the Patient. New Brunswick: Rutgers Univ. Press.

FRANKEL, CHARLES (1956) The Case For Modern Man. New York: Harper.

FRIEDENBERG, EDGAR (1968) "Status and role in education." The Humanist 29 (September-October): 13.

HOLTON, GERALD (1963) "Modern science and the intellectual tradition." In Paul C. Obler and Herman A. Estrin [eds.] The New Scientist: Essays on the Methods and Values of Modern Science. Garden City, N.Y.: Doubleday.

JAEGER, WERNER (1939) Paideia: The Ideals of Greek Culture. New York: Oxford Univ. Press.

KROPOTKIN, PETER ALEKSEEVICH (1901) Fields, Factories, and Workshops. New York: G. Putnam.

LEWIS, C. S. (1947) The Abolition of Man. New York: Macmillan.

LIPPMANN, WALTER (1964) A Preface to Morals. New York: Lime.

MILLER, S. M. (1967) "Breaking the credential barriers." New York: Ford Foundation, Office of Reports.

PELTZ, RICHARD (1968) "The True, the Good, and the humanities." Journal of Aesthetic Education 2 (January).

PETERS, RICHARD (1969) "Rules with reasons: the bases of moral education." The Nation (January 13): 49-52.

———(1966) Ethics and Education. Atlanta: Scott, Foresman.

POLANYI, MICHAEL (1958) Personal Knowledge. Chicago: Chicago Univ. Press.

REDFIELD, MARGARET PARK [ed.] (1963) The Social Uses of Social Science: The Papers of Robert Redfield, Vol. II. Chicago: Chicago Univ. Press.

SPENDER, STEPHEN (1962) "The connecting imagination." Adventures of the Mind. New York: Vintage Books.

ULICH, ROBERT [ed.] (1964) Education and the Idea of Mankind. New York: Harcourt, Brace & World.

WALLAS, GRAHAM (1914) The Great Society. New York: Macmillan.

WHITEHEAD, ALFRED NORTH (1925) Science and the Modern World. New York: Macmillan.

BIBLIOGRAPHY

Bibliography

A

ADRIAN, CHARLES R. (1968) "The autonomy of local political leadership."
Pp. 1-9 in Richard W. Taylor (ed.) Political Leaders in Action. Iowa City:
Sernoll.

ABRAMS, CHARLES (1965) The City is the Frontier. New York: Harper &
Row.

———(1964) Man's Struggle for Shelter in an Urbanizing World. Cambridge,
Mass.: MIT Press.

ADLER, MORTIMER J. (1967) The Difference of Man and the Difference It
Makes. New York: Holt, Rinehart & Winston.

ALONSO, WILLIAM (1964) Location and Land Use. Cambridge, Mass.: Harvard
Univ. Press.

ALTSCHULER, ALAN (1965) "The City Planning Process: A Political Analysis.
Ithaca: Cornell Univ. Press.

American Academy of Arts and Sciences (1968) The Conscience of the City.
Daedalus (Fall), issued as Vol. 97, No. 4 of Proceedings of the American
Academy of Arts and Sciences.

The American Behavioral Scientist (1966) Special issue on "Environment and
Behavior." Vol. 10 (September).

ANDERSON, MARTIN (1964) The Federal Bulldozer. Cambridge, Mass.: MIT
Press.

ANDERSON, STANFORD [ed.] (1968) Planning for Diversity and Choice:
Possible Futures and Their Relations to the Man-Controlled Environment.
Cambridge, Mass.: MIT Press.

ANDERSON, STANLEY (1962) Britain in the Sixties: Housing. London: Penguin Books.
ASHWORTH, WILLIAM (1954) The Genesis of Modern British Town Planning. London: Routledge and Kegan Paul.
ASTIN, ALEXANDER W. (1962) "Productivity of undergraduate institutions." Science 136 (April): 129-135.

B

BABCOCK, RICHARD F. (1966) The Zoning Game: Municipal Practices and Policies. Madison, Wisc.: Univ. of Wisconsin Press.
——— and F. BOSSELMAN (1967) "Citizen participation: a suggestion for the central city." Law and Contemporary Problems 32 (Spring): 220-228.
BACON, EDMUND N. (1967) Design of Cities. New York: Viking Press.
———(1957) "Design and Changing Values." Pp. 1-5 in Seventh Aspen International Design Conference: Conference Papers, June, 1957. Aspen, Colo.: Aspen International Design Conference.
BANFIELD, EDWARD C. (1961) Political Influence. New York: Free Press.
——— and JAMES Q. WILSON (1963) City Politics. Cambridge, Mass.: Harvard Univ. Press.
BANKS, J. A. (1968) "Population change and the Victorian city." Victorian Studies 11 (March).
———(1954) Prosperity and Parenthood: A Study of Family Planning Among the Victorian Middle Classes. London: Routledge and Kegan Paul.
BARNARD, CHESTER I. (1947) The Functions of the Executive. Cambridge, Mass.: Harvard Univ. Press.
Barron's National Business and Financial Weekly (1968) "Little red school house? Community control of education invites mob rule." (Vol. 48, September 2): 1.
BAUER, RAYMOND A. [ed.] (1966) Social Indicators. Cambridge, Mass.: MIT Press.
BEARDWOOD, ROGER (1968) "The new Negro mood." Fortune 77 (January): 146-151.
BECKINSALE, R. P. and J. M. HOUSTON [eds.] (1968) Urbanization and its Problems: Essays Presented to E. W. Gilbert. New York: Barnes & Noble.
BELL, DANIEL (1968) "The adequacy of our concepts." Pp. 127-161 in Bertram M. Gross (ed.) A Great Society? New York: Basic Books.
———(1964) "The post industrial society." In Eli Ginzburg (ed.) Technology and Social Change. New York: Columbia Univ. Press.
BENNIS, WARREN (1966) Changing Organizations—Essays on the Development and Evolution of Human Organizations. New York: McGraw-Hill.
BERRY, BRIAN J. L. and JACK MELTZER [eds.] (1967) Goals for Urban America. Englewood Cliffs: Prentice-Hall.
BERTALANFFY, LUDWIG VON (1967) Robots, Men and Minds. New York: Braziller.

BEYER, GLENN H. [ed.] (1967) The Urban Explosion in Latin America: A Continent in Process of Modernization. Ithaca: Cornell Univ. Press.
———(1965) Housing and Society. New York: Macmillan.
BIDERMAN, ALBERT (1966) "Social indicators and goals." Pp. 111-129 in Raymond A. Bauer (ed.) Social Indicators. Cambridge, Mass.: MIT Press.
BLOOM, BENJAMIN S. (1964) Stability and Change in Human Characteristics. New York: John Wiley.
BLOOMBERG, WARNER, JR. and HENRY J. SCHMANDT [eds.] (1968) Power, Poverty, and Urban Policy. Beverly Hills: Sage Publications.
BLUMENFELD, HANS (1967) The Modern Metropolis: Its Origins, Growth, Characteristics, and Planning (Selected essays edited by Paul D. Spreiregen). Cambridge, Mass.: MIT Press.
BOLLENS, JOHN C. and HENRY J. SCHMANDT (1965) The Metropolis: Its People, Politics, and Economic Life. New York: Harper & Row.
BOWLES, SAMUEL and HENRY M. LEVIN (1968) "The determinants of scholastic achievement—an appraisal of some recent evidence." The Journal of Human Resources 3 (Winter): 3-24.
BRACEY, HOWARD E. (1964) Neighbours: Subdivision Life in England and the United States. Baton Rouge: Louisiana State Univ. Press.
BRAGER, GEORGE A. and FRANCIS P. PURCELL [eds.] (1967) Community Action Against Poverty: Readings from the Mobilization Experience. New Haven, Conn.: College & University Press.
BRANCH, MELVILLE C. (1966) Planning: Aspects and Applications. New York: John Wiley.
BRANCH, MELVILLE C., JR. (1942) Urban Planning and Public Opinion. Princeton: Princeton Univ. Press.
BREDEMEIER, H. C. (1968) "The politics of the poverty cold war." Urban Affairs Quarterly 3 (June): 3-35.
BREESE, GERALD et al. (1969) The Impact of Large-Scale Installations on Nearby Areas: Accelerated Urban Growth. Beverly Hills: Sage Publications.
BRENNAN, T. (1948) Midland City. London: Dennis Dobson.
BRIGGS, ASA (1968) "The Victorian city: quantity and quality." Victorian Studies 11 (Summer Supplement).
———(1968) Victorian Cities. London: Pelican Books.
BRITTON, LEONARD M. (1964) Report of Visitation and Sources Investigated. Cleveland: Martha Holden Jennings Foundation.
BRONOWSKI, JACOB (1965) The Identity of Man. Garden City, N.Y.: Natural History Press.
BRUNER, JEROME S. et al. (1956) A Study of Thinking. New York: John Wiley.
Buchanan Report (1963) Traffic in Towns. London: H.M. Stationery Office.
BURD, GENE (1968) Voters, the Press and Urban Renewal. Milwaukee: Center for the Study of the American Press, Marquette Univ.
———(1968) "Magazines and the Metropolis." Address delivered to the Annual Convention of the Association for Education in Journalism, Univ. of Kansas, August 27-28.
———(1968) "Urban crises and criticisms of the mass media." Paper presented to Mass Media and Education Institute, Marquette Univ., July 24.

———(1968) "Magazines and public images of cities." Address delivered to the Annual Convention of the American Association of Commerce Publications, July 15, Milwaukee, Wisconsin.

———(1968) "The metropolitan press and urban problems." Paper presented to the American Political Science Association's Public Affairs Reporting Awards Seminar, June 26, Sun Valley, Idaho.

———(1968) "Urban renewal in the city room." Quill 56 (May): 12-13.

———(1968) "Urban alienation and the underground press." Lecture to Alpha Kappa Delta, April 16, Milwaukee, Wisconsin.

———(1968) "Media in metropolis." National Civic Review 58 (January): 138-143.

———(1968) "Media view the ghetto." New City 6 (January): 8-11.

———(1967) "Pickets, political power and the urban press." Address delivered to Kiwanis Club, National Newspaper Week, October 10, Menominee Falls, Wisconsin.

———(1967) "Cities and the press." Paper presented to the National Convention of the Association for Education in Journalism, Univ. of Colorado, August 30.

———(1966) "The suburban community press." Catholic School Editor 35 (June): 3-5.

BURKHEAD, JESSE et al. (1967) Input and Output in Large City High Schools. Syracuse: Syracuse Univ. Press.

BUSHNELL, DON and DWIGHT W. ALLEN [eds.] (1966) The Computer in American Education. New York: John Wiley.

C

CALHOUN, JOHN B. (1962) "Population density and social pathology." Scientific American 206: 139-148.

CAMPBELL, ALAN K. and SEYMOUR SACKS (1967) Metropolitan America: Fiscal Patterns and Governmental Systems. New York: Free Press.

CARR, STEPHEN (1967) "The city of the mind." Chap. 8 in W. R. Ewald, Jr. (ed.) Environment for Man: The Next Fifty Years. Bloomington: Indiana Univ. Press.

CARVER, HUMPHREY (1962) Cities in the Suburbs. Toronto: Univ of Toronto Press.

CHADWICK, GEORGE F. (1966) The Park and the Town. Public Landscape in the 19th and 20th Centuries. London: Architectural Press.

CHAPMAN, S. D. (1963) "Working-class housing in Nottingham during the Industrial Revolution." Transactions of the Thoroton Society 67.

City (1968) "Analysis: Should the schools decentralize?" (Vol. 2, March): 30-32.

CLARK, JOSEPH (1967) "The new urbanism." Pp. 38-51 in B. Berry and J. Meltzer (eds.) Goals for Urban America. Englewood Cliffs, N.J.: Prentice-Hall.

CLARK, KENNETH (1968) "Alternative public school systems." Harvard Educational Review 38 (Winter): 100-113.

CLARK, PETER B. (1959) The Chicago Big Business Man as a Civic Leader. Ph.D. dissertation, Univ. of Chicago (unpub.).

CLARK, S. D. (1966) The Suburban Society. Toronto: Univ. of Toronto Press.

CLARK, TERRY N. [ed.] (1968) Community Structure and Decision-Making: Comparative Analyses. San Francisco: Chandler Publishing.

CLAY, GRADY (1960) "Planning design and public opinion." Journal of the American Society of Planning Officials (Proceedings of the Conference at Bal Harbour, Florida, May 22-26): 129-139.

CLINARD, MARSHALL B. (1966) Slums and Community Development. London: Collier-Macmillan.

COHEN-PORTHEIM, P. (1930) England, The Unknown Isle (Alan Harris, translator). London: Duckworth.

COLEMAN, JAMES S. (1957) Community Conflict. Glencoe, Ill.: Free Press.

——— et al. (1966) Equality of Educational Opportunity. Washington, D.C.: Government Printing Office.

CONANT, JAMES B. (1961) Slums and Suburbs. New York: McGraw-Hill.

———(1959) The American High School Today. New York: McGraw-Hill.

CONANT, RALPH W. [ed.] (1965) The Public Library and the City. Cambridge, Mass.: MIT Press.

CONNERY, ROBERT H. [ed.] (1968) Urban Riots: Violence and Social Change. (Proceedings of the Academy of Political Science, Vol. XXIX, No. 1). New York: Academy of Political Science.

COOK, DONALD (1963) "Cultural innovation and disaster in the American city." Pp. 87-93 in L. Duhl (ed.) The Urban Condition. New York: Basic Books.

COTTRELL, FRED (1955) Energy and Society. New York: McGraw-Hill.

The Cox Commission (1968) Crisis at Columbia. New York: Random House.

COX, HARVEY (1965) The Secular City. New York: Macmillan.

COX, JAMES L. (1967) Metropolitan Water Supply: The Denver Experience. Boulder, Colo.: Bureau of Governmental Research and Service.

D

DAVIES, J. CLARENCE (1966) Neighborhood Groups and Urban Renewal. New York: Columbia Univ. Press.

DAVIS, KINGSLEY (1965) "Some demographic aspects of poverty in the United States." Pp. 299-319 in Margaret S. Gordon (ed.) Poverty in America. San Francisco: Chandler Publishing.

DENTLER, ROBERT A. (1968) American Community Problems. New York: McGraw-Hill.

——— et al. [eds.] (1967) The Urban R's. New York: Frederick A. Praeger.

DEWEY, JOHN (1950) Human Nature and Conduct. New York: The Modern Library.

———(1930) Individualism: Old and New. New York: Minton, Balch.

———(1927) The Public and Its Problems. New York: H. Holt.

DIETZE, GOTTFRIED (1963) In Defense of Property. Chicago: Henry Regnery.

DOBRINER, WILLIAM [ed.] (1958) The Suburban Community. New York: G. P. Putnam.

DOLL, RUSSELL C. (1968) Types of Elementary Schools in a Big City. Ph.D. dissertation, Univ. of Chicago (unpub.).

DOREMUS, J. C. (1968) "Upward bound in transition." Educational Opportunity Forum 1 (No. 2): 51-61.

DOWNES, BRYAN T. (1968) "The social characteristics of riot cities." Social Science Quarterly 49 (December): 504-520.

DOXIADIS, C. A. (1968) Ekistics: An Introduction to the Science of Human Settlements. New York: Oxford Univ. Press.

———(1966) Urban Renewal and Future of the American City. Chicago: Public Administration Service.

———(1963) Architecture in Transition. London: Hutchison.

DUHL, LEONARD [ed.] (1963) The Urban Condition: People and Policy in the Metropolis. New York: Basic Books.

DURKHEIM, EMILE (1933) The Division of Labour in Society (George Simpson, translator). New York: Macmillan.

DYOS, H. J. (1968) "The slum attacked." New Society 280 (February 8).

———(1968) "The slum observed." New Society 279 (February 1).

———[ed.] (1968) The Study of Urban History. London: Edward Arnold.

———(1961) Victorian Suburb: A Study of the Growth of Camberwell. Leicester: Leicester Univ. Press.

E

EDELMAN, MURRAY (1964) The Symbolic Uses of Politics. Urbana: Univ. of Illinois Press.

The Editors of Fortune (1968) The Exploding Metropolis. Garden City, N.Y.: Doubleday.

EICHLER, EDWARD P. and MARSHALL KAPLAN (1967) The Community Builders. Berkeley: Univ. of California Press.

ELAZAR, DANIEL J. (1968) "Are we a nation of cities?" Pp. 89-97 in Robert A. Goldwin (ed.) A Nation of Cities. Chicago: Rand McNally.

ELDREDGE, H. WENTWORTH (1967) Taming Megalopolis. Vol. I: What Is and What Could Be; Vol. II: How to Manage an Urbanized World. Garden City, N.Y.: Doubleday.

ELIAS, C. E. JR. et al. [eds.] (1965) Metropolis: Values in Conflict. Belmont, Calif.: Wadsworth Publishing.

ELIOT, THOMAS H. (1959) "Toward an understanding of public school politics." American Political Science Review 53 (December): 1032-1051.

ELULL, JACQUES (1967) The Technological Society. New York: Vintage Books.

ENDLEMAN, SHALOM [ed.] (1968] Violence in the Streets. Chicago: Quadrangle Books.

EWALD, WILLIAM R., JR. [ed.] (1967) Environment for Man: The Next Fifty Years. Bloomington: Indiana Univ. Press.

F

FAGIN, HENRY and ROBERT C. WEINBERG (1958) Planning and Community Appearance. New York: Regional Plan Association.

FALTERMAYER, EDMUND K. (1968) Redoing America: A Nationwide Report on How to Make Our Cities and Suburbs Livable. New York: Harper & Row.

FESLER, JAMES (1949) Area and Administration. University, Ala.: Univ. of Alabama Press.

FINLAY, DAVID J. et al. (1967) Enemies in Politics. Chicago: Rand McNally.

FISCHER, JOHN H. (1968) "Fischer on decentralization." Education News 3 (August 5): 16.

FISHER, ROBERT M. [ed.] (1955) The Metropolis in Modern Life. Garden City, N.Y.: Doubleday.

FLEISHER, AARON (1966) "Technology and urban form: the influence of changing communications, transportation, and occupational structures." Pp. 37-52 in Marcus Whiffen (ed.) The Architect and the City. Cambridge, Mass.: MIT Press.

FLINN, M. W. (1965) Report on the Sanitary Condition of the Labouring Population of Great Britain by Edward Chadwick, 1842. Edinburgh: Edinburgh Univ. Press.

FOOTE, NELSON N. (1960) Housing Choices and Constraints. New York: McGraw-Hill.

FOX, DAVID J. (1967) Expansion of the More Effective School Program. New York: Center for Urban Education.

FRANK, LAWRENCE K. (1948) Society as the Patient. New Brunswick: Rutgers Univ. Press.

FRANKEL, CHARLES (1956) The Case for Modern Man. New York: Harper.

FRAZIER, E. FRANKLIN (1931) The Negro Family in the United States. Chicago: Univ. of Chicago Press.

FRIED, MARC (1963) "Grieving for a lost home." Chap. 12 in L. Duhl (ed.) The Urban Condition. New York: Basic Books.

FRIEDEN, BERNARD J. (1964) The Future of Old Neighborhoods: Rebuilding for a Changing Population. Cambridge, Mass.: MIT Press.

——— and ROBERT MORRIS [eds.] (1968) Urban Planning and Social Policy. New York: Basic Books.

FRIEDENBERG, EDGAR (1968) "Status and role in education." The Humanist 29 (September-October): 13.

FRIEDMAN, LAWRENCE M. (1968) Government and Slum Housing: A Century of Frustration. Chicago: Rand McNally.

FRIEDMANN, JOHN (1966) Regional Development Policy: A Case Study of Venezuela. Cambridge, Mass.: MIT Press.
———(1961-1962) "Cities in social transformation." Comparative Studies in Society and History 4.
——— and WILLIAM ALONSO [eds.] (1964) Regional Development and Planning. Cambridge, Mass.: MIT Press.
FUTRELL, ASHLEY B. (1960) "What the press expects of city hall." Southern City 12 (December): 1.

G

GANS, HERBERT J. (1968) People and Plans: Essays on Urban Problems and Solutions. New York: Basic Books.
———(1967) The Levittowners. New York: Pantheon Books.
———(1962) The Urban Villagers. New York: Free Press.
GARDNER, JOHN W. (1968) "City Hall can't go it alone." Public Management 50 (November): 266-267.
GEORGE, M. DOROTHY (1968) London Life in the 18th Century. New York: Harper & Row.
GESCHWENDER, JAMES (1968) "Civil rights protests and riots." Social Science Quarterly 49 (December): 474-484.
GIBBS, JACK P. and LEO F. SCHNORE (1960) "Metropolitan growth: an international study." American Journal of Sociology (September): 160-170.
GIEBER, WALTER and WALTER JOHNSON (1961) "The City hall 'beat': a study of reporter and source roles." Journalism Quarterly 38 (Summer): 296.
GINZBERG, ELI [ed.] (1964) Seminar on Technology and Social Change. New York: Columbia Univ. Press.
——— et al. (1968) Manpower Strategy for the Metropolis. New York: Columbia Univ. Press.
GITTELL, MARILYN [ed.] (1967) Educating an Urban Population. Beverly Hills: Sage Publications.
——— and T. EDWARD HOLLANDER (1967) Six Urban School Districts. New York: Frederick A. Praeger.
GIURGOLA, ROMALDO (1966) "Architecture in change: European experiments pointing a new direction for city design." Pp. 103-119 in Marcus Whiffen (ed.) The Architect and the City. Cambridge, Mass.: MIT Press.
GLAAB, CHARLES N. and THEODORE A. BROWN (1967) A History of Urban America. New York: Macmillan.
GLASS, DAVID (1935) The Town and a Changing Civilization. London: John Lane, Bodley Head.
GLADWIN, THOMAS (1967) Poverty U.S.A. Boston: Little, Brown.
GLAZER, NATHAN and DANIEL P. MOYNIHAN (1963) Beyond the Melting Pot: The Negroes, Puerto Ricans, Jews, Italians, and Irish of New York City. Cambridge, Mass.: MIT Press.

GLOAG, JOHN (1949) The Englishman's Castle. London: Eyre and Spottiswoode.

GOLDBERG, JERRY et al. [ed.] (1962) Education for Urban Design: Proceedings of a Conference, January 8-10. St. Louis, Mo.: Washington Univ. School of Architecture (mimeo).

GOODMAN, PERCIVAL (1962) "The population explosion as it affects teaching of urban design and architecture." Pp. 14-24 in Jerry Goldberg et al. (eds.) Education for Urban Design: Proceedings of a Conference, January 8-10. St. Louis, Mo.: Washington Univ. School of Architecture (mimeo.).

GORDON, KERMIT [ed.] (1968) Agenda for the Nation. Washington, D.C.: Brookings Institution.

GORDON, MARGARET S. [ed.] (1965) Poverty in America. San Francisco: Chandler Publishing.

GORDON, MITCHELL (1965) Sick Cities. Baltimore: Penguin Books.

GOTTMANN, JEAN (1966) "Why the skyscraper?" Geographical Review 56 (April): 190-212.

————(1966) "The ethics of living at high densities." Ekistics 21 (February): 141-145.

————(1961) Megalopolis: The Urbanized Northeastern Seaboard of the United States. New York: Twentieth Century Fund.

———— and ROBERT A. HARPER [eds.] (1967) Metropolis on the Move: Geographers Look at Urban Sprawl. New York: John Wiley.

Greater London Committee on Housing (1965) [Milner-Holland Report, Cmd. 2605]. London: H. M. Stationery Office.

GREER, SCOTT (1965) Urban Renewal and American Cities: A Sociological Critique. Indianapolis: Bobbs-Merrill.

————(1963) Metropolitics: A Study of Political Culture. New York: John Wiley.

————(1962) Governing the Metropolis. New York: John Wiley.

————(1962) The Emerging City, Myth and Reality. New York: Free Press.

————(1955) Social Organization. New York: Random House.

GRIMSHAW, ALLEN (1959) "Lawlessness and violence in American and their special manifestations in changing Negro-white relationships." Journal of Negro History 44 (January): 52-72.

GRODZINS, MORTON (1959) The Metropolitan Area as a Racial Problem. Pittsburgh: Pittsburgh Univ. Press.

GROSS, BERTRAM [ed.] (1968) A Great Society? New York: Basic Books.

GRUBB, JEANETTE (1940) "A study of the editorial policies of the Indianapolis News and its relationship to the growth of the city of Indianapolis." Master's thesis, Northwestern Univ. (unpub.).

GRUEN, VICTOR D. (1966) "Environmental architecture: a team approach to design on a city scale." Pp. 67-87 in Marcus Whiffen (ed.) The Architect and the City. Cambridge, Mass.: MIT Press.

————(1964) Heart of Our Cities. New York: Simon & Schuster.

GRUNDSTEIN, NATHAN D. (1966) "Urban information systems and urban management decisions and control." Urban Affairs Quarterly 1 (June): 21-32.

GURR, T. (1968) "Urban disorder; perspectives from the comparative study of civil strife." American Behavioral Scientist 11 (March-April): 51-54.

GUTHEIM, FREDERICK (1962) "The next 50 million Americans—where will they live?" Progressive Architecture (August): 98-99.
GUTKIND, E. A. (1962) The Twilight of Cities. New York: Free Press.

H

HADDEN, JEFFREY K. and EDGAR F. BORGATTA (1965) American Cities: Their Social Characteristics. Chicago: Rand McNally.
HADDEN, JEFFREY K. et al. (1968) "The making of Negro mayors, 1967." Trans-Action 5 (January): 29-31.
—— et al. [eds.] (1967) Metropolis in Crisis: Social and Political Perspectives. Itasca, Ill.: F. E. Peacock Publishers.
HAIG, R. M. (1926) "Toward an understanding of the metropolis." Quarterly Journal of Economics 11.
HANDLIN, OSCAR (1966) "Negro American." Daedalus 95 (Winter): 268-283.
—— and JOHN BURCHARD [eds.] (1963) The Historian and the City. Cambridge, Mass.: MIT and Harvard Univ. Press.
HANSEN, WALTER G. (1957) "Traffic approaching cities." Public Roads 31 (April): 155-158.
HARRIS, BRITTON (1966) Comprehensive Transportation Planning: Report to the California State Office of Planning. Berkeley: Center for Planning and Development Research.
———(1966) Urban Transportation Planning, Philosophy of Approach. Philadelphia: Univ. of Pennsylvania Institute for Environmental Studies.
—— [ed.] (1965) Urban Development Models: New Tools for Planning. Special issue of the Journal of the American Institute of Planners 31 (May).
HATT, PAUL K. and ALBERT J. REISS, JR. [eds.] (1957) Cities and Society. Glencoe, Ill.: Free Press.
HAUSER, PHILIP M. (1960) Population Perspectives. New Brunswick: Rutgers Univ. Press.
—— and LEO F. SCHNORE [eds.] (1965) The Study of Urbanization. New York: John Wiley.
HAVIGHURST, ROBERT J. (1966) Education in Metropolitan Areas. Boston: Allyn & Bacon.
———(1964) The Public Schools of Chicago. Chicago: Board of Education.
HAWLEY, WILLIS D. and FREDERICK M. WIRT [eds.] (1968) The Search for Community Power. Englewood Cliffs: Prentice-Hall.
HAWORTH, LAWRENCE (1963) The Good City. Bloomington: Indiana Univ. Press.
HECKSCHER, AUGUST (1962) The Public Happiness. New York: Atheneum.
HERRING, FRANCES W. [ed.] (1965) Open Space and the Law. Berkeley: Univ. of California Institute of Governmental Studies.
HILL, ROSCOE and MALCOLM FEELEY [eds.] (1969) Affirmative School Integration: Efforts to Overcome De Facto Segregation in Urban Schools. Beverly Hills: Sage Publications.

HILLERY, GEORGE A., JR. (1968) Communal Organizations: A Study of Local Societies. Chicago: Univ. of Chicago Press.

HIRSCH, WERNER Z. [ed.] (1963) Urban Life and Form. New York: Holt, Rinehart & Winston.

—— et al. (1964) Spillover of Public Education Costs and Benefits. Los Angeles: Univ. of California Institute of Government and Public Affairs.

HODGE, PATRICIA LEAVEY and PHILIP M. HAUSER (1968) The Challenge of America's Metropolitan Population Outlook, 1960 to 1985. New York: Frederick A. Praeger.

HOLDEN, MATTHEW, JR. (forthcoming) "Politics, public order, and pluralism." In James R. Klonoski and Robert I. Mendelsohn (eds.) The Allocation of Justice. Boston: Little, Brown.

——(1969) The Republic in Crisis. San Francisco: Chandler Publishing (forthcoming).

HOLLAND, LAURENCE B. [ed.] (1966) Who Designs America? Garden City, N.Y.: Doubleday.

HOLTON, GERALD (1963) "Modern science and the intellectual tradition." In Paul C. Obler and Herman A. Estrin (eds.) The New Scientist: Essays on the Methods and Values of Modern Science. Garden City, N.Y.: Doubleday.

HOOVER, EDGAR M. and RAYMOND VERNON (1959) Anatomy of a Metropolis: The Changing Distribution of People and Jobs Within the New York Metropolitan region. Cambridge, Mass.: Harvard Univ. Press.

HOWARD, EBENEZER (1946) Garden Cities of Tomorrow. London: Faber and Faber.

HUGHES, EVERETT (1943) French Canada in Transition. Chicago: Univ. of Chicago Press.

HUXLEY, ALDOUS (1958) Brave New World Revisited. New York: Harper & Row.

I

International Council on Social Welfare (1967) Urban Development: Its Implications for Social Welfare (Proceedings of the XIIIth International Conference of Social Work). New York: Columbia Univ. Press.

J

JABLONSKY, ADELAIDE (1968) "Some trends in education for the disadvantaged." IRCD Bulletin 4 (March): 1-7.

JACKSON, JOHN B. (1966) "The purpose of the city: changing city landscapes as manifestations of cultural values." Pp. 13-36 in Marcus Whiffen (ed.) The Architect and the City. Cambridge, Mass.: MIT Press.

JACOBS, JANE (1961) The Death and Life of Great American Cities. New York: Random House.

JAEGER, WERNER (1939) Paideia: The Ideals of Greek Culture. New York: Oxford Univ. Press.

JAMES, H. THOMAS (1963) Wealth, Expenditures and Decision-Making for Education. Stanford: Stanford Univ. School of Education.

––– et al. (1966) The Determinants of Educational Expenditures in Large Cities of the U.S. Stanford: Stanford Univ. School of Education.

JENCKS, CHRISTOPHER (1968) "Private schools for black children." New York Times Magazine (November).

JENNINGS, HILDA (1962) Societies in the Making: A Study of Development and Redevelopment Within a Country Borough. New York: Humanities Press.

JENSEN, ROLF (1966) High Density Living. New York: Frederick A. Praeger.

JEPHSON, HENRY (1907) The Sanitary Evolution of London. London: Fisher Unwin.

JOHNS, EWART (1965) British Townscapes. London: Edward Arnold.

JOHNSON, EARL S. (1957) "The function of the central business district in the metropolitan community." Pp. 248-259 in Paul K. Hatt and Albert J. Reiss, Jr. (eds.) Cities and Society. Glencoe, Ill.: Free Press.

Joint Economic Committee, Subcommittee on Urban Affairs, U.S. Congress, House and Senate (1967) Hearings, Urban America: Goals and Problems. Washington, D.C.

JOSEPHSON, MATTHEW (1934) The Robber Barons. The Great American Capitalists, 1861-1901. New York: Harcourt, Brace & World.

K

KAHN, HERMAN and ANTHONY J. WIENER (1967) The Year 2,000. A Framework for Speculation on the Next 33 Years. New York: Macmillan.

KAIN, JOHN F. (1966) "The big cities' big problem." Challenge (September-October): 5-8.

––– and JOSEPH J. PERSKY (1967) "The North's stake in southern rural poverty." Harvard Program in Regional and Urban Economics, Discussion Paper No. 18.

Kansas City Star Staff (1915) William Rockhill Nelson: The Story of a Man, a Newspaper and a City. Cambridge, Mass.: Riverside Press.

KANTOR, N. B. [ed.] (1965) Mobility and Mental Health. Springfield, Ill.: C. C. Thomas.

KATZ, IRWIN (1967) "The socialization of academic motivation in minority group children." Pp. 133-190 in Nebraska Symposium on Motivation. Lincoln: Univ. of Nebraska Press.

KAUFMAN, HERBERT and WALLACE S. SAYRE (1960) Governing New York City. New York: Russell Sage Foundation.

KEY, V. O. (1949) Southern Politics in State and Nation. New York: Alfred A. Knopf.

KLAPPER, JOSEPH T. (1960) The Effects of Mass Communications. Glencoe, Illinois.: Free Press.

KOZOL, JONATHAN (1967) Death at an Early Age. Boston: Houghton Mifflin.

KRISTENSSON, FOLKE (1967) "People, firms, and regions." Pp. 1-10 in a publication of The Economic Research Institute. Stockholm: Stockholm School of Economics.

KRISTOL, IRVING (1968) "Decentralization, for what?" Public Interest 2 (Spring): 17-25.

KROPOTKIN, PETER ALEKSELVICH (1901) Fields, Factories, and Workshops. New York: G. Putnam.

KRUYTBOSCH, CARLOS E. and SHELDON L. MESSINGER [eds.] (1968) The state of the university: authority and change. The American Behavioral Scientist 11 (May-June).

L

LAMPARD, ERIC E. (1968) "The evolving system of cities in the United States: urbanization and economic development." Pp. 80-141 in Harvey S. Perloff and Lowdon Wingo, Jr. (eds.) Issues in Urban Economics. Baltimore: Johns Hopkins Press.

LANSING, JOHN BAND et al. (1964) Residential Location and Urban Mobility. Ann Arbor, Mich.: Univ. of Michigan Survey Research Center.

LAPIN, HOWARD S. (1964) Structuring the Journey to Work. Philadelphia: Univ. of Pennsylvania Institute of Urban Studies.

LAPP, RALPH E. (1965) The New Priesthood. New York: Harper & Row.

LAZARSFELD, PAUL F. and ROBERT MERTON (1948) "Mass communications, popular taste and organized social action." Pp. 147-153 in Lyman Bryson (ed.) The Communication of Ideas. New York: Harper & Row.

LESHER, RICHARD L. and GEORGE J. HOWICH (1966) Assessing Technology Transfer. Washington, D.C.: Government Printing Office.

LEWIS, C. S. (1947) The Abolition of Man. New York: Macmillan.

LESSINGER, JACK (1962) "The case for scatteration." Journal of the American Institute of Planners 28 (August): 159-169.

LIEBERSON, STANLEY (1963) Ethnic Patterns in American Cities. New York: Free Press.

LINDSTROM, CARL E. (1960) The Fading American Newspaper. New York: Doubleday.

LIPPMANN, WALTER (1964) A Preface to Morals. New York: Lime.

LOHMAN, JOSEPH D. (1967) Cultural Patterns in Urban Schools: A Manual for Teachers, Counselors, and Administrators. Berkeley: Univ. of California Press.

LONG, NORTON E. (1962) The Polity (A collection of papers edited by Charles Press). Chicago: Rand McNally.

LOWRY, IRA S. (1966) Migration and Metropolitan Growth: Two Analytical Models. San Francisco: Chandler Publishing.

LYNCH, KEVIN (1960) The Image of the City. Cambridge, Mass.: MIT and Harvard Univ. Press.

LYND, ROBERT S. and HELEN M. LYND (1929) Middletown. New York: Harcourt, Brace & World.

M

McCLOSKY, HERBERT et al. (1960) "Issue conflict and consensus among party leaders and followers." American Political Science Review 54 (June): 406-427.

McDONALD, NEIL A. (1955) Study of Political Parties. Garden City, N.Y.: Doubleday.

McFEE, JUNE KING (1965) "Poverty and urban aesthetics." Paper prepared for a conference on "The Education of Children and Adults in Aesthetic Awareness of the Environment of Man," held by the Department of Art and Art Education, Univ. of Wisconsin, Madison, October 23-30 (mimeo.).

McHARG, IAN (1966) "The ecology of the city: a plea for environmental consciousness of the city's physiological and psychological impacts." Pp. 53-65 in Marcus Whiffen (ed.) The Architect and the City. Cambridge, Mass.: MIT Press.

MACK, RAYMOND W. [ed.] (1968) Our Children's Burden: Studies of Desegregation in Nine American Communities. New York: Random House.

———(1955-56) "Do we really believe in the Bill of Rights?" Social Problems 3 (Winter): 264-267.

McKISSICK, FLOYD (1966) Hearings Before Subcommittee on Executive Reorganization of Committee on Government Operations, 90th Congress. Pp. 2289-2331 in Federal Role in Urban Affairs, Part 2. Washington, D.C.: Government Printing Office.

McLUHAN, MARSHALL (1966) Understanding Media: The Extensions of Man. New York: McGraw-Hill.

———(1951) The Mechanical Bride. New York: Vanguard Press.

MAKIELSKI, STANISLAW J., JR. (1966) The Politics of Zoning: The New York Experience. New York: Columbia Univ. Press.

MANN, PETER H. (1965) An Approach to Urban Sociology. London: Routledge and Kegan Paul.

MANSTEIN, BODO (1968) "Shaping the future in a rational manner." Perspectives 6 (June): 26-27.

MARRIS, PETER and MARTIN REIN (1967) Dilemmas of Social Reform. New York: Atherton Press.

MARSHALL, LEON S. (1940) "The emergence of the first industrial city: Manchester, 1780-1850." In Caroline F. Ware (ed.) The Cultural Approach to History. New York: Columbia Univ. Press.

MARTIN, JOHN S. and E. CURTIS HENSON (1968) "Atlanta takes the initiative." NEA Journal 57 (May): 28-29.

MARTIN, ROSCOE C. (1965) Cities and the Federal System. New York: Atherton Press.

MARTINDALE, DON (1958) "Prefatory remarks: the theory of the city." Pp. 9-62 in Don Max Weber, The City (Don Martindale and Gertrud Neuwirth, translators). Glencoe, Ill.: Free Press.

MASOTTI, LOUIS H. (1967) Education and Politics in Suburbia. Cleveland: Case Western Reserve Univ. Press.

——— and DON R. BOWEN [eds.] (1968) Riots and Rebellion: Civil Violence in the Urban Community. Beverly Hills: Sage Publications.

MASSERMAN, J. H. (1943) Behavior and Neurosis. Chicago: Univ. of Chicago Press.

MAYER, ALBERT (1967) The Urgent Future: People, Housing, City, Region. New York: McGraw-Hill.

MAYER, MARTIN (1969) "The full and sometimes very surprising story of Ocean Hill, the teachers' union and the teachers' strikes of 1968." New York Times Magazine (February 2): 18.

Mayor's Advisory Panel of Decentralization of the New York City Schools (1967) Reconnection for Learning: A Community School System for New York City. New York: Ford Foundation.

MERTON, THOMAS (1956) Thoughts in Solitude. New York: Dell Publishing.

MERVIN, JACK C. and RALPH W. TYLER (1966) "What the assessment of education will ask." Nation's Schools 78 (November): 77-79.

MESTHENE, EMMANUEL G. (1968) "How technology will shape the future." Science (July): 161.

MEYER, J. R. et al. (1965) The Urban Transportation Problem. Cambridge, Mass.: Harvard Univ. Press.

MICHAEL, DONALD N. (1966) "Some speculations on the social impact of technology." In Dean Morse and Aaron W. Warner (eds.) Technological Innovation and Society. New York: Columbia Univ. Press.

MILLER, S. M. (1967) "Breaking the credential barriers." New York: Ford Foundation, Office of Reports.

MILLER, WALTER B. (1967) "Violent crimes in city gangs." Pp. 127-141 in Thomas R. Dye and Brett W. Hawkins (eds.) Politics in the Metropolis. Columbus: Charles E. Merrill.

MILLS, C. WRIGHT (1956) The Power Elite. New York: Oxford Univ. Press.

MINAR, DAVID W. (1967) "The politics of education in large cities." Pp. 308-320 in Marilyn Gittell (ed.) Educating an Urban Population. Beverly Hills: Sage Publications.

MINER, JERRY (1963) Social and Economic Factors in Spending for Public Education. Syracuse: Syracuse Univ. Press.

Ministry of Housing and Local Government (1964) The South-East Study: 1961-1981. London: H. M. Stationery Office.

Ministry of Town and Country Planning of Great Britain (1946) Final Report of the New Town Committee. London: H. M. Stationery Office.

MINUCHIN, SALVADOR et al. (1967) Families of the Slums: An Exploration of Their Structure and Treatment. New York: Basic Books.

MITCHELL, R. J. and M. D. R. Leys (1958) A History of London Life. London: Penguin Books.

MITCHELL, ROBERT and CHESTER RAPKIN (1954) Urban Traffic: A Function of Land Use. New York: Columbia Univ. Press.

MOHRING, HERBERT (1961) "Land values and the measurement of highway benefits." Journal of Political Economy 69 (June): 236-249.

MOLEY, RAYMOND (1964) The Republican Opportunity. New York: Duell, Sloan & Pearce.

MOORE, WILBUR and ARNOLD FELDMAN (1962) Labor Commitment and Economic Development. Princeton: Princeton Univ. Press.

MORRIS, R. N. (1968) Urban Sociology. London: Allen and Unwin.

MORRIS, RICHARD B. (1965) Government and Labor in Early America. New York: Octagon Books.

MORT, PAUL R. (1952) Educational Adaptability. New York: Metropolitan School Study Council.

――― and ORLANDO F. FURNO (1960) Theory and Synthesis of a Sequential Simplex. New York: Columbia Teachers College.

MOSES, LEON N. (1962) "Income, leisure, and wage pressure." The Economic Journal 72 (June): 320-334.

MOYNIHAN, DANIEL P. (1965) "Employment, income, and the ordeal of the Negro family." Daedalus 94 (Fall): 745-770.

MUMFORD, LEWIS (1968) The Urban Prospect. New York: Harcourt, Brace & World.

―――(1966) The Myth of the Machine. New York: Harcourt, Brace & World.

―――(1961) The City in History: Its Origins, Its Transformations and Its Prospects. New York: Harcourt, Brace & World.

Municipal Manpower Commission (1962) Governmental Manpower for Tomorrow's Cities. New York: McGraw-Hill.

N

National Commission on Community Health Services (1966) Health is a Community Affairs. Cambridge, Mass.: Harvard Univ. Press.

National Commission on Urban Problems (1969) Building the American City. Washington, D.C.: Government Printing Office.

National Council of Social Service (1943) The Size and Social Structure of a Town. London: Allen and Unwin.

National School Public Relations Association (1968) Computers: New Era for Education. Washington, D.C.

Nation's Cities (1967) "What kind of city do we want?" (April): 17-47.

New York City Planning Commission (1964) The Future by Design: A Symposium on the Considerations Underlying the Development of a Comprehensive Plan for the City of New York.

New York Regional Plan Association (1967) The Region's Growth.

——(1967) Regional Plan News, No. 86 (October).

——(1967) Public Participation in Regional Planning (October).

The New York Times (1968) "Special training program at City University's center in Harlem sharpens children's perceptions." (September 1) 46, cols. 4-5.

NEWSON, JOHN and ELIZABETH NEWSON (1968) Four Years Old in an Urban Community. Chicago: Aldine Publishing.

NEWTON, V. M. (1961) Crusade for Democracy. Ames: Iowa State Univ. Press.

NICHOLSON, MEREDITH (1940) "Indianapolis: a city of homes." Atlantic Monthly 93 (June): 836-845.

NIEBANCK, PAUL L. with MARK R. YESSIAN (1968) Relocation in Urban Planning: From Obstacle to Opportunity. Philadelphia: Univ. of Pennsylvania Press.

NIEBURG, H. L. (1968) "Violence, law and the social process." American Behavioral Scientist 2 (March-April): 17-19.

NIEDERHOFFER, ARTHUR (1967) Behind the Shield. Garden City, N.Y.: Doubleday.

NISBET, ROBERT (1953) The Quest for Community. New York: Oxford Univ. Press.

NORTON, JOHN D. (1966) Dimensions in School Finance. Washington, D.C.: National Education Association.

O

Ohio State University Advisory Commission on Problems Facing the Columbus Public Schools (1968) Recommendations to the Columbus Board of Education.

OSBORN, FREDERICK J. (1946) Green Belt Cities. London: Faber.

—— and ARNOLD WHITBECK (1963) The New Towns, the Answer to Megalopolis. London: Leonard Hill.

OWEN, WILFRED (1966) The Metropolitan Transportation Problem. Washington, D.C.: Brookings Institution.

P

PARK, ROBERT E. (1952) Human Communities: The City and Human Ecology. New York: Free Press.

PASSOW, A. HARRY (n.d.) Toward Creating a Model School System. A Study of the Washington, D.C. Public Schools. New York: Teachers College, Columbia Univ.

PEETS, ELBERT (1968) On the Art of Designing Cities (Selected essays edited by Paul D. Spreiregen). Cambridge, Mass.: MIT Press.

PELL, CLAIBORNE (1966) Megalopolis Unbound: The Supercity and the Transportation of Tomorrow. New York: Frederick A. Praeger.

PELTZ, RICHARD (1968) "The True, the Good, and the humanities." Journal of Aesthetic Education 2 (January).

PERLMAN, SELIG and PHILIP TAFT (1935) Labor Movements: History of Labor in the United States, 1896-1932. New York: Macmillan.

PERLOFF, HARVEY S. (forthcoming) "Frameworks for thinking about U.S. government in the year 2000." In Harvey S. Perloff (ed.) volume on problems of American government, under the auspices of the Commission on the Year 2000.

———(1963) "Social Planning in the metropolis." Pp. 331-347 in Leonard J. Duhl (ed.) The Urban Condition: People and Plan Policy in the Metropolis. New York: Basic Books.

——— and LOWDON WINGO, JR. [eds.] (1968) Issues in Urban Economics. Baltimore: Johns Hopkins Press.

PERRUCCI, ROBERT and MARC PILISUK [compilers] (1968) The Triple Revolution: Social Problems in Depth. Boston: Little, Brown.

PETERS, RICHARD (1969) "Rules with reasons: the bases of moral education." The Nation (January 13): 49-52.

———(1966) Ethics and Education. Atlanta: Scott, Foresman.

PFEIFFER, JOHN (1968) New Look at Education. New York: Odyssey Press.

PIMLOTT, J. A. R. (1947) The Englishman's Holiday. London: Faber.

POLANYI, MICHAEL (1958) Personal Knowledge. Chicago: Chicago Univ. Press.

POOL, ITHIEL DE SOLA (1968) "Behavioral technology." Foreign Policy Association, Toward the Year 2018. New York: Cowles Education.

POUNDS, RALPH L. (1968) The Development of Education in Western Culture. New York: Appleton-Century-Crofts.

Premier Ministre (1963) Avant-Projet de Programme Duodécennal pour la Région de Paris. Paris.

President's Commission on Law Enforcement and the Administration of Justice (1967) The Challenge of Crime in a Free Society. Washington, D.C.: Government Printing Office.

PRESTHUS, ROBERT (1964) Men at the Top. New York: Oxford Univ. Press.

Project TALENT (1962) Studies of the American High School. Pittsburgh: Univ. of Pittsburgh.

PURDOM, C. B. [ed.] (1921) Town Theory and Practice. London: Benn Brothers.

R

RAINWATER, LEE and WILLIAM L. YANCEY (1967) The Moynihan Report and the Politics of Controversy. Cambridge, Mass.: MIT Press.

RAND, CHRISTOPHER (1967) Los Angeles: The Ultimate City. New York: Oxford Univ. Press.

RASMUSSEN, STEEN EILER (1967) London: The Unique City. Cambridge, Mass.: MIT Press.

———(1962) Experiencing Architecture. Cambridge, Mass.: MIT Press.

REDFIELD, MARGARET PARK [ed.] (1963) The Social Uses of Social Science: The Papers of Robert Redfield, Vol. II. Chicago: Chicago Univ. Press.

REHFUSS, JOHN A. (1968) "Metropolitan government: four views." Urban Affairs Quarterly 3 (June): 91-111.

REIN, MARTIN and S. M. MILLER (1966) "Poverty, policy, and purpose: the dilemmas of choice." Pp. 20-64 in L. H. Goodman (ed.) Economic Progress and Social Welfare. New York: Columbia Univ. Press.

REINER, THOMAS A. (1963) The Place of the Ideal Community in Urban Planning. Philadelphia: Univ. of Pennsylvania Press.

REISS, ALBERT J., JR. (1957) "The sociology of urban life: 1946-56." Pp. 3-15 in Paul K. Hatt and Albert J. Reiss, Jr. (eds.) Cities and Society: The Revised Reader in Urban Sociology. New York: Free Press.

———(1955) "An analysis of urban phenomena." Pp. 41-49 in Robert Fisher (ed.) The Metropolis in Modern Life. Garden City, N.Y.: Doubleday.

REISSMANN, LEONARD (1964) The Urban Process: Cities in Industrial Societies. New York: Free Press.

Resources for the Future (1967) Environmental Quality in a Growing Economy. Washington, D.C.: Resources for the Future.

Revue Française (1966) Special issue on urbanism. Vol. 56, No. 2.

RICHARDS, J. M. (1946) The Castles on the Ground. London: Architectural Press.

RIDGEWAY, JAMES (1968) The Closed Corporation. New York: Random House.

RIDKER, RONALD G. (1967) Economic Costs of Air Pollution: Studies in Measurement. New York: Frederick A. Praeger.

RIESMAN, DAVID et al. (1950) The Lonely Crowd. New Haven: Yale Univ. Press.

RIKER, WILLIAM (1964) Federalism: Origins, Operations, Significance. Boston: Little, Brown.

RIMMER, W. G. (1961) "Working men's cottages in Leeds, 1770-1840." Thoresby Society Publications 46.

ROBERTS, DAVID (1960) Victorian Origins of the British Welfare State. New Haven: Yale Univ. Press.

ROBSON, WILLIAM A. (1955) Great Cities of the World: Their Government, Politics and Planning. New York: Macmillan.

ROETHLISBERGER, F. S. and W. J. DICKSON (1939) Management and the Worker. Cambridge, Mass.: Harvard Univ. Press.

ROGERS, DAVID (1968) 110 Livingston Street. New York: Random House.

———(1968) "New York City schools: a sick bureaucracy." Saturday Review (July): 47.

ROSENTHAL, ROBERT and LENORE JACOBSON (1968) Pygmalion in the Classroom. New York: Holt, Rinehart & Winston.

ROSS, DONALD H. [ed.] (1958) Administration for Adaptability. New York: Metropolitan School Study Council.

ROWLAND, HOWARD S. and RICHARD L. WING (1967) Federal Aid for Schools. New York: Macmillan.

Royal Australian Institute of Architects (1967) Australian Outrage: The Decay of a Visual Environment. New York: Taplinger Publishing.

Royal Commission (1940) Report on the Distribution of the Industrial Population. London: H. M. Stationery Office.

RUBLOWSKY, JOHN (1967) Nature in the City. New York: Basic Books.

RUDE, GEORGE (1964) The Crowd in History. New York: John Wiley.

S

SAYRE, WALLACE S. and HERBERT KAUFMAN (1965) Governing New York City. New York: W. W. Norton.

SCHLIVEK, LOUIS B. (1965) Man in Metropolis: A Book About the People and Prospects of a Metropolitan Region. Garden City, N.Y.: Doubleday.

SCHMOOKLER, JACOB (1966) Invention and Economic Growth. Cambridge, Mass.: Harvard Univ. Press.

SCHNORE, LEO F. (1965) The Urban Scene: Human Ecology and Demography. New York: Free Press.

———(1958) "Social morphology and human ecology." American Journal of Sociology 63: 620-634.

———and HENRY FAGIN [eds.] (1967) Urban Research and Policy Planning. Beverly Hills: Sage Publications.

School Management (1966) "The national cost of education index—1965-66." (January): 115-120.

SCHORR, ALVIN L. (1968) Explorations in Social Policy. New York: Basic Books.

SCHRAG, PETER (1968) "The end of the common school." Saturday Review 51 (April 20): 68.

———(1967) The Village School Downtown. Boston: Beacon Press.

SCHRIEVER, BERNARD A. and WILLIAM W. WEIFERT [eds.] (1968) Air Transportation 1975 and Beyond A Systems Approach: Report of the Transportation Workshop, 1967. Cambridge, Mass.: MIT Press.

SCHULTZ, D. P. (1965) Sensory Restriction: Effects on Behavior. New York: New York City Academic Press.

SCHULZE, ROBERT O. (1961) "The bifurcation of power in a satellite community." Pp. 19-80 in Morris Janowitz (ed.) Community Political Systems. New York: Free Press.

SCHUR, EDWIN M. (1965) Crimes Without Victims. Englewood Cliffs, N.J.: Prentice-Hall.

Scientific American (1965) Cities. New York: Alfred A. Knopf.

SELF, PETER (1961) Cities in Flood: The Problems of Urban Growth. London: Faber.

SELIGMAN, BEN B. (1968) Permanent Poverty: An American Syndrome. Chicago: Quadrangle Books.

———[ed.] (1965) Poverty as a Public Issue. New York: Free Press.

SENIOR, DEREK [ed.] (1966) The Regional City: An Anglo-American Discussion of Metropolitan Planning. London: Longmans.

SHANNON, LYLE W. (1963) "The public's perception of social welfare agencies and organizations in an industrial community." Journal of Negro Education (Summer): 276-285.

———and ELAINE M. KRASS (1964) "The economic absorption of immigrant laborers in a northern industrial community." American Journal of Economics and Sociology 23 (January): 65-84.

———and ELAINE M. KRASS (1963) "The urban adjustment of immigrants: the relationship of education to occupation and total family income." Pacific Sociological Review 6 (Spring): 37-42.

SHARP, HARRY and LEO F. SCHNORE (1962) "The changing color composition of metropolitan areas." Land Economics 38 (May): 169-185.

SHARP, THOMAS (1968) Town and Townscape. London: John Murray.

SHERRARD, THOMAS D. [ed.] (1968) Social Welfare and Urban Problems. New York: Columbia Univ. Press.

SHOVER, JAMES L. (1966) Cornbelt Rebellion. Urbana: Univ. of Illinois Press.

SILVER, ALLAN (1967) "The demand for order in civil society." Pp. 1-24 in David J. Bordua (ed.) The Police: Six Sociological Essays. New York: John Wiley.

SIMMEL, GEORG (1957) "The metropolis and mental life." Pp. 635-647 in P. K. Hatt and A. J. Reiss (eds.) Cities and Society. Glencoe: Free Press.

SITTE, CAMILLO (1965) City Planning According to Artistic Principles. (George R. Collins and Christiane Crasemann Collines, translators). New York: Random House.

SIZER, THEODORE R. and PHILIP WHITTEN (1968) "A proposal for a poor children's Bill of Rights." Psychology Today 2 (August): 59-63.

SJOBERG, GIDEON (1960) The Preindustrial City. Glencoe: Free Press.

SKINNER, B. F. (1968) The Technology of Teaching. New York: Appleton-Century-Crofts.

SKOLNICK, JEROME (1966) Justice Without Trial. New York: John Wiley.

SLOANE, JOSEPH C. (1966) "Beauty and anti-beauty in the American city." Pp. 35-41 in Robert E. Stipe (ed.) Perception and Environment: Foundations of Urban Design (Proceedings of a 1962 Seminar on Urban Design). Chapel Hill: Univ. of North Carolina Institute of Government.

SMIGIELSKI, W. K. (1968) Leicester Today and Tomorrow. London: Pyramid Press.

SNOW, C. P. (1962) Science and Government. New York: New American Library.

SOMMERS, WILLIAM (1958) "Council-manager government, a review." Western Political Quarterly 11 (March): 145-155.

SORENSON, ROY (1961) "A citizen looks at metropolitan problems." Pp. 167-174 in First Conference on Environmental Engineering and Metropolitan Planning. Evanston: Northwestern Univ. Press.

SPECTORSKY, A. C. (1955) The Exurbanites. New York: Berkeley Publishing.

SPIEGEL, HANS B. C. [ed.] (1968) Citizen Participation in Urban Development, Vol. 1: Concepts and Issues. Washington, D.C.: NTL Institute for Applied Behavioral Science.

SPILHAUS, ATHELSTAN (1968) "The experimental city." Science (February): 159.

SPREIREGEN, PAUL D. (1963) "The practice of urban design: some basic principles." Journal of the American Institute of Architects 39 (June): 59-74.

STARR, ROGER (1966) The Living End: The City and Its Critics. New York: Coward-McCann.

STEIN, CLARENCE S. (1966) Toward New Towns for America. Cambridge, Mass.: MIT Press.

STEWART, CECIL (1952) A Prospect of Cities, Being Studies towards a History of Town Planning. London: Longsman, Green.

STIPE, ROBERT E. [ed.] (1966) Perception and Environment: Foundations of Urban Design (Proceedings of a 1962 seminar on urban design). Chapel Hill: Univ. of North Carolina Institute of Government.

STRAUSS, ANSELM L. [ed.] (1968) The American City: A Sourcebook of Urban Imagery. Chicago: Aldine Publishing.

———(1961) Images of the American City. New York: Free Press.

———and RICHARD WOHL (1958) "Symbolic representation and the urban milieu." American Journal of Sociology 63 (March): 523-532.

STROLE, LEO et al. (1962) Mental Health in the Metropolis: The Midtown Manhattan Study. New York: McGraw-Hill.

Subcommittee on Urban Affairs of the Joint Economic Committee, Congress of the U.S. (1967) Urban America: Goals and Problems. Washington, D.C.: Government Printing Office.

SULLIVAN, NEIL V. (1968) "Discussion." Harvard Educational Review 38 (Winter): 148-155.

SUPPES, PATRICK (1966) "The uses of computers in education." Information edited by Scientific American. San Francisco: W. H. Freeman.

T

Task Force on Economic Growth and Opportunity (1967) America's Cities: Current Problems and Trends. Washington, D.C.: U.S. Chamber of Commerce.

TAYLOR, A. J. P. (1957) "The world's cities (1): Manchester." Encounter 8 (March).

THELEN, HERBERT A. (1960) Education and the Human Quest. New York: Harper & Row.

THOLFSEN, T. R. (1956) "The artisan and the culture of early Victorian Birmingham." Univ. of Birmingham Historical Journal 4.

THOMAS, J. ALAN (1968) "Modernizing state finance programs." Paper presented at the NEA School Finance Conference held in Dallas, March 31-April 2.

———(1964) Administrative Rationality and the Productivity of School Systems. Chicago: Univ. of Chicago Midwest Administration Center.

THOMPSON, WILBUR (1968) "The city as distorted price system." Psychology Today 23 (August): 28-33.

———(1966) "Urban economic development." Pp. 81-121 in Werner Z. Hirsch (ed.) Regional Accounts for Policy Decisions. Baltimore: Johns Hopkins Press for Resources for the Future.

———(1965) A Preface to Urban Economics. Baltimore: Johns Hopkins Press for Resources for the Future.

———(1965) "Urban economic growth and development in a national system of cities." Pp. 431-491 in Philip M. Hauser and Leo F. Schnore (eds.) The Study of Urbanization. New York: John Wiley.

TOCQUEVILLE, ALEXIS DE (1948) The Recollections of Alexis de Tocqueville (J. P. Mayer, ed.). London: Harvill Press.

TOFFLER, ALVIN [ed.] (1968) The Schoolhouse in the City. New York: Frederick A. Praeger.

TOMLINSON, T. M. (1968) "Riot ideology in Los Angeles: a study of Negro attitudes." Social Science Quarterly 49 (December): 485-503.

Town and Country Planning Association (1968) New Towns Come of Age. London: Town and Country Planning Association.

TUNNARD, CHRISTOPHER (1968) The Modern American City. Princeton: D. Van Nostrand.

———(1953) The City of Man. New York: Charles Scribner.

———and BORIS PUSHKAREV (1963) Man-Made America: Chaos or Control. New Haven: Yale Univ. Press.

TURNER, RALPH (1941) The Great Cultural Traditions. New York: McGraw-Hill.

U

ULICH, ROBERT [ed.] (1964) Education and the Idea of Mankind. New York: Harcourt, Brace & World.

ULLMAN, EDWARD L. (1954) "Amenities as a factor in regional growth." Geographical Review 44 (January): 119-132.

U.S. Commission on Civil Rights (1967) Racial Isolation in the Public Schools. Washington, D.C.: Government Printing Office.

U.S. Commission on Law Enforcement and Justice (1967) Task Force Report: Assessment of Crime. Washington, D.C.: Government Printing Office.

U.S. Committee on Government Operations, Subcommittee on Executive Reorganization (1966) Hearings, Federal Role in Urban Affairs. Washington, D.C.

U.S. Committee on Science and Astronautics (1966) Proceedings, Sixth Panel on Science and Technology. Washington, D.C.

U.S. House of Representatives, 89th Congress, 1st Session (1965) "Message from the President of the United States relative to the problems and future of the central city and its suburbs." Washington, D.C.: Government Printing Office.
U.S. Office of Education (1967) Title I/Year II: The Second Annual Report of Title I of the Elementary and Secondary Education Act of 1965. Washington, D.C.: Government Printing Office.
———(1966) Digest of Educational Statistics: 1966. Washington, D.C.: National Center for Educational Statistics.
U.S. Office of Planning and Research, Department of Labor (1965) The Negro Family. Washington, D.C.: Department of Labor.
U.S. White House Conference (1965) Beauty for America: Proceedings of the White House Conference on Natural Beauty, May 24-25. Washington, D.C.: Government Printing Office.
University of Pennsylvania Graduate School of Fine Arts (1963) Civic Design Symposium I: Emerging Forces and Forms in the City Today. Philadelphia: Univ. of Pennsylvania Graduate School of Fine Arts.

V

VERNON, RAYMOND (1962) The Myth and Reality of Our Urban Problems. Cambridge, Mass.: Joint Center for Urban Studies of MIT and Harvard Univ.
———(1960) Metropolis 1985: An Interpretation of the Results of the New York Metropolitan Region Study. Cambridge, Mass.: Harvard Univ. Press.
———(1959) The Changing Function of the Central City. New York: Committee for Economic Development.
VICKERS, GEOFFREY (1965) The Art of Judgment. London: Chapman and Hall.
VINCENT, HAROLD S. (1965) Quality Education. Milwaukee: Milwaukee Public Schools.

W

WALL, NED L. (1960) "The press, the public and planning." P. 5 of Information Report No. 134 for American Society of Planning Officials. May.
WALLAS, GRAHAM (1914) The Great Society. New York: Macmillan.
WARNER, SAM BASS, JR. [ed.] (1966) Planning for a Nation of Cities. Cambridge, Mass.: MIT Press.
———(1962) Streetcar Suburbs: The Process of Growth in Boston, 1870-1900. Cambridge, Mass.: MIT and Harvard Univ. Press.

WARREN, ROLAND L. [ed.] (1966) Perspectives on the American Community. Chicago: Rand McNally.

WASKOW, ARTHUR (1966) From Race Riot to Sit-In. Garden City, N.Y.: Doubleday.

WEBBER, MELVIN M. (1964) "The urban place and the nonplace urban realm." Pp. 79-153 in Melvin M. Webber et al. Explorations Into Urban Structure. Philadelphia: Univ. of Pennsylvania Press.

——— et al. (1967) Explorations Into Urban Structure. Philadelphia: Univ. of Pennsylvania Press.

WEBER, ADNA FERRIN (1963) The Growth of Cities in the Nineteenth Century. A Study of Statistics. Ithaca: Cornell Univ. Press.

WEBSTER, DONALD (1958) Urban Planning and Municipal Public Policy. New York: Harper & Row.

WEINBERG, ALVIN N. (1967) Reflections on Big Science. Cambridge, Mass.: MIT Press.

WEIZSACKER, C. F. VON (1964) The Relevance of Science. New York: Harper & Row.

WENTWORTH, ELDREDGE H. [ed.] (1967) Taming Megalopolis. New York: Frederick A. Praeger.

WESTIN, ALAN F. (1967) Privacy and Freedom. New York: Atheneum.

WHEATON, WILLIAM (1964) "Public and private agents of change in urban expansion." Pp. 154-191 in Melvin M. Webber et al. Explorations Into Urban Structure. Philadelphia: Univ. of Pennsylvania Press.

——— et al. [eds.] (1966) Urban Housing. New York: Free Press.

WHIFFEN, MARCUS [ed.] (1969) The Architect and the City. Cambridge, Mass.: MIT Press.

WHITE, MORTON and LUCIA WHITE (1962) The Intellectual Versus the City. Cambridge, Mass.: MIT and Harvard Univ. Press.

WHITEHEAD, ALFRED NORTH (1925) Science and the Modern World. New York: Macmillan.

WHYTE, WILLIAM H., JR. (1956) The Organization Man. Garden City, N.Y.: Doubleday.

WILHELM, SIDNEY M. (1962) Urban Zoning and Land Use Theory. New York: Free Press.

WILKES, L. and G. DODDS (1964) Tyneside Classical: The Newcastle of Grainger, Dobson and Clayton. London: John Murray.

WILLBERN, YORK (1964) The Withering Away of the City. University, Ala.: Univ. of Alabama Press.

WILLIAMS, OLIVER P. and CHARLES R. ADRIAN (1963) Four Cities: Philadelphia: Univ. of Pennsylvania Press.

WILLIAMS, OLIVER P. et al. (1965) Suburban Differences and Metropolitan Policies. Philadelphia: Univ. of Pennsylvania Press.

WILLIAMS, RAYMOND (1960) Culture and Society, 1780-1950. Garden City, N.Y.: Doubleday.

WILLIAMS, ROBIN, JR. et al. (1964) Strangers Next Door. Englewood Cliffs, N.J.: Prentice-Hall.

WILSON, GODFREY and MONICA WILSON (1945) The Analysis of Social Change. London: Cambridge Univ. Press.

WILSON, JAMES Q. (1968) Varieties of Police Behavior. Cambridge, Mass.: Harvard Univ. Press.
———(1966) "The Negro in American politics." Pp. 431-457 in John P. Davis (ed.) The American Negro Reference Book. Englewood Cliffs, N.J.: Prentice-Hall.
———(1966) "The war on cities." The Public Interest. No. 3 (Spring): 27-44.
——— [ed.] (1967) The Metropolitan Enigma: Inquiries into the Nature and Dimensions of America's "Urban Crisis." Washington, D.C.: Chamber of Commerce of the United States.
WOOD, ROBERT C. (1963) "The contributions of political science to urban form." Pp. 99-128 in Werner Z. Hirsch (ed.) Urban Life and Form. New York: Holt, Rinehart & Winston.
———(1961) 1400 Governments. Cambridge, Mass.: Harvard Univ. Press.
———(1959) Metropolis Against Itself. New York: Committee for Economic Development.
———(1958) Suburbia: Its People and Their Politics. Boston: Houghton Mifflin.
WRIGHT, LAWRENCE (1964) Home Fires Burning: The History of Domestic Heating and Cooking. London: Routledge and Kegan Paul.
———(1960) Clean and Decent: The Fascinating History of the Bathroom and the Water Closet and of Sundry Habits, Fashions and Accessories of the Toilet, Principally in Great Britain, France and America. London: Routledge and Kegan Paul.

Y

YLVISAKER, PAUL N. (1961) "Diversity and the public interest." Journal of the American Institute of Planners 27 (May): 107-117.
YOUNG, M. and P. WILMOTT (1957) Family and Kinship in East London. London: Routledge and Kegan Paul.
YOUNG, WHITNEY (1964) To Be Equal. New York: McGraw-Hill.

Z

ZALD, MAYER N. [ed.] (1967) Organizing for Community Welfare. Chicago: Quadrangle Books.
ZIMMER, BASIL G. and AMOS H. HAWLEY (1968) Metropolitan Area Schools: Resistance to District Reorganization. Beverly Hills: Sage Publications.

THE AUTHORS

The Authors...

CHARLES R. ADRIAN is Professor of Political Science and Chairman of that department of the University of California at Riverside. He received his Ph.D. from the University of Minnesota, and has taught at Wayne State University and Michigan State University. He served as Research Consultant to the Michigan Constitutional Convention and as Administrative Assistant in the Office of the Governor of Michigan. Among his many publications are *Four Cities: A Study in Comparative Politics* (with Oliver Williams) and *Governing Urban America.* He has contributed extensively to such journals as the *American Political Science Review* and the *Midwest Journal of Political Science.*

WARNER BLOOMBERG, JR. is Professor of Urban Affairs and Chairman of that department of the University of Wisconsin at Milwaukee. He received his Ph.D. in sociology from the University of Chicago. He has taught at Syracuse University and was Co-director there of the summer workshop on Human Relations and Social Conflict. He has also served as consultant and visiting faculty for a number of institutes dealing with education of the disadvantaged.

Among Dr. Bloomberg's publications are *Local Community Leadership*, coauthor (University College of Syracuse University, 1960); *Suburban Power Structures and Public Education*, coauthor (Syracuse University Press, 1963); and "Community Organization," in Howard S. Becker, editor, *Social Problems: A Modern Approach* (Wiley, 1966). He has also contributed to such journals as the *American Sociological Review, Young Children,* and *Graduate Comment.*

HANS BLUMENFELD is Consultant to the Metropolitan Toronto Planning Board and to the Service d'Urbanisme of the City of Montreal. He is also a Senior Lecturer in the Division of Town and Regional Planning, University of Toronto. A graduate in architecture from the Polytechnic Institute of Darmstadt, he has lectured at Columbia University, the University of Pennsylvania, and many other institutions in Canada and abroad. He has served as Chief of the Division of Planning Analysis for the Philadelphia Planning Commission and as Assistant Director of the Metropolitan Toronto Planning Board. His writings in the planning field are numerous and have appeared in such journals as *Ekistics, The Town Planning Review, Canadian Architect, Annals of the American Academy of Political and Social Science,* and the *Journal of the American Institute of Planners.* Many of his articles have been assembled in a volume edited by Paul D. Spreiregen entitled *The Modern Metropolis: Its Origin, Growth, Characteristics, and Planning.*

GENE BURD is Assistant Professor of Journalism at Marquette University and a former newsman who covered urban problems for the Kansas City, Missouri, *Star;* Houston, Texas, *Chronicle;* Albuquerque, New Mexico, *Journal;* and Three Rivers, Michigan, *Commercial.* He also was editor of several Northwest suburban Chicago weeklies and has written articles on his specialty of cities and the press in *New City, National Civic Review, Quill,* and *Progressive* magazines. He is the author of *Voters, the Press and Urban Renewal,* a monograph published by the Center for the Study of the American Press, where he is a Research Associate. He earned his Ph.D. in urban affairs and journalism at Northwestern University, worked for the Chicago Department of City Planning and Encyclopedia Britannica, and was the last Resident-in-Research at Chicago's Hull House before it was demolished for urban renewal. He is conducting research on "magazines and the metropolis" for the

Magazine Publishers Association of New York, and from 1953 to 1954 studied the Japanese press in Los Angeles while on a fellowship from the *Los Angeles Times*.

MICHAEL DECKER is Assistant Professor of Political Science and Public Administration, at the Institute of Public Administration of Pennsylvania State University. Currently completing his doctoral thesis concerning the use of the planning heuristic by public schools, his principal interests are the study of human problem-solving and the relationship between contemporary management technology and decision-making in the public sector.

H. J. DYOS is Reader in Economic History in the University of Leicester, England. He received his Ph.D. in modern economic history from the University of London in 1952. He served as Dean of the Faculty of the Social Sciences in the University of Leicester from 1965 to 1968. He is Convenor of the Urban History Group in Britain and editor of its *Urban History Newsletter*. His main field of research has been the history of London in the nineteenth century. He has written *Victorian Suburb: A Study of the Growth of Camberwell* and is coauthor (with Derek H. Aldcroft) of the forthcoming book, *British Transport: An Economic Survey from the Seventeenth Century to the Twentieth*. He has contributed numerous articles to such journals as *Victorian Studies* and the *Journal of Transport History*.

JOHN C. GOLDBACH is Assistant Professor of Political Science at the University of Wisconsin at Milwaukee. He received his Ph.D. from Claremont Graduate School and has taught at Los Angeles City College and San Fernando Valley State College. His principal field of interest has been urban and metropolitan politics. He is the author of *Boundary Change in California* and has also contributed articles to the *Public Administration Review* and the *Western Political Quarterly*.

JEAN GOTTMANN is Professor of Geography at the University of Oxford, England. He was previously Professor at the École des Hautes Études and the School of Political Science of the University of Paris. He has taught at many American universities including Johns Hopkins, Columbia, Pittsburgh, California at Berkeley, and Wisconsin at Milwaukee, and has several times served as a member of the Institute of Advanced Studies at Princeton. His period of service as Director of Research and Studies for the United Nations Department of Social Affairs and as Chairman for the study of Regional Planning of the International Geographical Union have provided him with unique opportunities for studying the problems of metropolitan growth. Among the books he has authored are the *Geography of Europe* (1950), *Virginia at Midcentury* (1955), and *Megalopolis: The Urbanized Northeastern Seaboard of the United States* (1961). He has also contributed extensively to professional journals in the fields of geography and planning.

SCOTT GREER is Director of the Center for Metropolitan Studies and Professor of Political Science and Sociology at Northwestern University. He received his Ph.D. from the University of California at Los Angeles in 1952. He has an extensive background in the problems of the metropolis. He served as Research Director of the Laboratory in Urban Culture at Occidental College from 1952 to 1955 and as Chief Sociologist for the Metropolitan St. Louis Survey from 1956 to 1957. Prior to his present appointment, he was Research Professor at the Transportation Center at Northwestern University. He is the author of numerous books, including *Governing the Metropolis* (1962), *Metropolitics* (1963), *The Emerging City* (1962), and *Urban Renewal and American Cities* (1965). He has also contributed extensively to professional journals, including the *American Sociological Review, American Journal of Sociology,* and *Social Forces.*

NATHAN D. GRUNDSTEIN is Professor of Political Science and Director of the Graduate Program in Public Management Science at Case Western Reserve University. He received his Ph.D. from Syracuse University and an LL.B. from George Washington University. He has served in various administrative positions in the federal government and has taught at Wayne State University and the University of Pittsburgh. His publications range over the fields of public management and public law, with particular emphasis on executive problems of management organization design, managerial development, and regulatory agencies. He has contributed to such journals as the *Public Administration Review* and the *Journal of Public Law.*

WALTER D. HARRIS, JR. is Associate Professor of City Planning in the School of Art and Architecture at Yale University. He holds degrees in architecture and city planning, and in 1960 was awarded an honorary Ph.D. by the National Engineering University in Lima, Peru. He has taught at Carnegie Institute of Technology and at various other universities in the United States and abroad. From 1959 to 1961 he was Director of the Inter-American Housing and Planning Center of the Organization of American States in Bogotá, and from 1961 to 1966 was Principal Advisor for Housing and Planning to the OAS Secretariat in Washington. He is the author of three books and numerous articles on Latin American and United States housing and planning subjects.

ROBERT J. HAVIGHURST is Professor of Education and Human Development at the University of Chicago. He received his Ph.D. from Ohio State University and has taught at Miami University (Ohio), the University of Wisconsin, and Ohio State University. He served as Director for General Education of the Rockefeller Foundation and as Co-director of the Brazilian Government Center of Education Research. He is the author of numerous books in the field of education including *Human Development and Education* (1953), *American Higher Education in the 1960's* (1960), *The Public Schools of Chicago* (1964), *The Educational Mission of the Church* (1965), and *Education in in Metropolitan Areas* (1966). He has also contributed extensively to such professional journals as the *International Education Review, Comparative Education Review, American Journal of Sociology,* and the *American Sociological Review.*

MATTHEW HOLDEN, JR. is Professor of Political Science at Wayne State University. He received his Ph.D. from Northwestern University in 1961. His research and writing deals primarily with decision-making behavior, with special interest in urbanization and the politics of urbanization, and the processes of political integration and disintegration. He is the author of the forthcoming book, *The Republic in Crisis,* and has contributed to numerous journals, including the *American Political Science Review, Western Political Science Quarterly,* and the *Urban Affairs Quarterly.*

EARL S. JOHNSON is Professor Emeritus of the Social Sciences at the University of Chicago and since 1959 has been a Visiting

Professor at the School of Education of the University of Wisconsin at Milwaukee. He received his Ph.D. in sociology from the University of Chicago. He is the author of *Theory and Practice of the Social Studies* (Macmillan, 1956) and has contributed many essays and articles to such diverse publications as the *American Behavioral Scientist, American Sociological Review, Christian Century, Encyclopedia Britannica, Journal of Social Forces,* and *Social Education,* dealing with the professions, the urban community, general education, and the teaching of the social studies. He is a member of the international Board of Advisors of the Council for the Study of Mankind and an advisor to the World Law Fund.

DANIEL U. LEVINE is Associate Director of the Center for the Study of Metropolitan Problems in Education at the University of Missouri, Kansas City. He is the author of numerous works on the problem of urban education, including "Black Power: Implications for the Urban Educator" and (with Robert J. Havighurst) "Negro Population Growth and Enrollment in the Public Schools: A Case Study and Its Implications" (both in *Education and Urban Society*).

LEO LEVY is Director of the Division of Planning and Evaluation Services of the Illinois Department of Mental Health. He received his Ph.D. in clinical psychology from the University of Washington, Seattle. He served as Chief Psychologist-Administrator at the Pueblo, Colorado, Guidance Center, and has taught at the University of Michigan and the University of Illinois Graduate College at the Medical Center in Chicago. He served as consulting editor to the *Community Mental Health Journal,* and has contributed to various journals, including the *Journal of Psychology, Journal of Health and Human Behavior, American Journal of Public Health,* and *Mental Hospitals.*

LOUIS H. MASOTTI is Associate Professor of Political Science and Director of the Civil Violence Research Center at Case Western Reserve University. Among his publications are *Education and Politics in Suburbia* and a chapter in *Educating an Urban Population.* He is Editor-in-Chief of a new journal devoted to the study of urban education (*Education and Urban Society*) and sits as an elected member of the Cleveland Heights-University Heights Board of Education. He has contributed to various journals including the *Urban Affairs Quarterly* and the *Journal of Urban Law.*

H. L. NIEBURG is Professor of Political Science at the University of Wisconsin at Milwaukee. He received his Ph.D. from the University of Chicago, and has taught at the University of Chicago and Case Western Reserve. He was a newspaper journalist prior to his academic career. He is the author of *Nuclear Secrecy and Foreign Policy* (1964), *In the Name of Science* (1966), and *Political Violence: The Behavioral Process* (1969). He has also contributed to various journals, including the *American Political Science Review, World Politics,* and the *Journal of Conflict Resolution.*

LESLIE PAFFRATH has been President and a Member of the Board of Trustees of the Johnson Foundation at Racine, Wisconsin, since 1959. For eight years prior to that he was an officer of the Carnegie Endowment for International Peace, New York City. He holds a B.A. from Union College, Schenectady, New York, and pursued his interest in international relations through graduate study at Columbia University and in Geneva, Switzerland. In 1964 and 1965 he served as Secretary General for the International Convocation on the Requirements for Peace and he was Co-chairman of the World Assembly for Human Rights in 1968. His leadership of the Johnson Foundation has reflected, not only his evident concern for international conciliation and peace, but also compelling commitments to the achievement of justice and equity for the deprived within the United States.

HENRY J. SCHMANDT is Professor of Urban Affairs and Political Science at the University of Wisconsin at Milwaukee. He served as Assistant Director of the Missouri State Reorganization Commission, Associate Director of the Metropolitan St. Louis Survey, and Associate Director of Metropolitan Community Studies, Dayton. He was the first Chairman of the Southeastern Wisconsin Regional Planning Commission and is presently on the Advisory Council to the Wisconsin Department of Local Affairs and Development. Among the books he has coauthored are *Metropolitan Reform in St. Louis* (1961), *Exploring the Metropolitan Community* (1961), and *The Metropolis: Its People, Politics, and Economic Life* (1965). He is also author of *The Milwaukee Metropolitan Study Commission* (1965) and *Courts in the American Political System* (1968).

PETER J. O. SELF is Professor of Public Administration at the London School of Economics. He is a member of the South East

Regional Economic Planning Council and Chairman of Executive, Town and Country Planning Association. He served on the editorial staff of the *Economist* from 1944-1962. Professor Self is author of *Regionalism* (1949) and *Cities in Flood: The Problems of Urban Growth* (Second edition, 1961). He has also contributed numerous articles to various periodicals on public administration, politics, and planning.

JOHN G. SUESS is presently Professor of Music and Chairman of that department at Case Western Reserve University in Cleveland, Ohio. He received his Bachelor of Science degree from Northwestern University in political science and international relations and his Ph.D. from Yale University in music history. His broad range of interest in the social sciences and humanities in an urban framework are manifested in his multi-faceted involvement in aiding the development of interdisciplinary fine arts programs related to an urban environment. Dr. Suess is the author of several books and articles on the application of social phenomena to musical phenomena.

HAROLD M. VISOTSKY is Director of the Illinois Department of Mental Health. He received his Doctor of Medicine degree from the University of Illinois College of Medicine and from 1952 to 1955 engaged in graduate training in psychiatry at the University of Illinois Neuropsychiatric Institute. He has served as director of psychiatry residency education and training at the University of Illinois and as Director of Mental Health for the Chicago Board of Health. He is a diplomate of the American Board of Psychiatry and Neurology and a member of the editorial boards of the *American Journal of Social Psychiatry* and the *Psychiatry Digest*. He has published a number of research papers in the field of mental health.

JAMES Q. WILSON is Professor of Government at Harvard University, where he has taught since 1961. From 1963 to 1966 he was Director of the Joint Center for Urban Studies at MIT and Harvard. He received his Ph.D. degree from the University of Chicago in 1959. He has written *Negro Politics, The Amateur Democrat, City Politics* (with Edward C. Banfield), and, most recently, *Varieties of Police Behavior*. Among the volumes he has edited are *City Politics and Public Policy* and *The Metropolitan Enigma: Inquiries Into the Nature and Dimensions of America's "Urban Crisis."*